Reagents for Organic Synthesis

Fiesers'

Reagents for Organic Synthesis

VOLUME TWENTY FIVE

Tse-Lok Ho

A JOHN WILEY & SONS, INC., PUBLICATION

For general information on our other products and services or for technical support, please contact
our Customer Care Department within the United States at (800) 762-2974, outside the United States
at (317) 572-3993 or fax (317) 572-4002.

Wiley also publishes its books in a variety of electronic formats. Some content that appears in print
may not be available in electronic format. For more information about Wiley products, visit our web
site at www.wiley.com.

ISBN 978-0-470-43375-1
ISSN 0271-616X

Printed in the United States of America

10 9 8 7 6 5 4 3 2 1

CONTENTS

PREFACE

In the Preface of ROS-24 I mentioned Ji Hsiao-Lan with the profoundest of admiration because of his role in editing the encyclopedic "Four Libraries of Books". During preparation of the present volume I happened to be reading "The Meaning of Everything. The Story of the Oxford English Dictionary" by Simon Winchester. The heart-wrenching journey that lasted 71 years for the completion of the first edition of the chef-d'oeuvre strikes a resonance in my heart.

This volume covers chemical literature from the beginning of 2007 to the end of June, 2008. From this period the most glaring mosaic of chemical vision scintillates with an aura of aurum.

GENERAL ABBREVIATIONS

Ac	acetyl
acac	acetylacetonate
ADDP	1,1′-(azodicarbonyl)dipiperidine
AIBN	2,2′-azobisisobutyronitrile
An	*p*-anisyl
aq	aqueous
Ar	aryl
ATPH	aluminum tris(2,6-diphenylphenoxide)
9-BBN	9-borabicyclo[3.3.1]nonane
BINOL	1,1′-binaphthalene-2,2′-diol
Bn	benzyl
Boc	*t*-butoxycarbonyl
bpy	2,2′-bipyridyl
BSA	*N,O*-bis(trimethylsilyl)acetamide
Bt	benzotriazol-1-yl
Bu	*n*-butyl
Bz	benzoyl
18-c-6	18-crown-6
c-	cyclo
CAN	cerium(IV)ammonium nitrate
Cap	caprolactamate
cat	catalytic
Cbz	benzyloxycarbonyl
Chx	cyclohexyl
cod	1,5-cyclooctadiene
cot	1,3,5-cyclooctatriene
Cp	cyclopentadienyl
Cp*	1,2,3,4,5-pentamethylcyclopentadienyl
CSA	10-camphorsulfonic acid
Cy	cyclohexyl
cyclam	1,4,8,11-tetraazacyclotetradecane
DABCO	1,4-diazobicyclo[2.2.2]octane
DAST	(diethylamino)sulfur trifluoride
dba	dibenzylideneacetone
DBN	1,5-diazobicyclo[4.3.0]non-5-ene
DBU	1,8-diazobicyclo[5.4.0]undec-7-ene

DCC	*N,N'*-dicyclohexylcarbodiimide
DDQ	2,3-dichloro-5,6-dicyano-1,4-benzoquinone
de	diastereomer excess
DEAD	diethyl azodicarboxylate
DIAD	diisopropyl azodicarboxylate
Dibal-H	diisobutylaluminum hydride
DMA	*N,N*-dimethylacetamide
DMAD	dimethyl acetylenedicarboxylate
DMAP	4-dimethylaminopyridine
DMD	dimethyldioxirane
DME	1,2-dimethoxyethane
DMF	*N,N*-dimethylformamide
DMPU	*N,N'*-dimethylpropyleneurea
DMSO	dimethyl sulfoxide
dpm	dipivaloylmethane
dppb	1,4-bis(diphenylphosphino)butane
dppe	1,2-bis(diphenylphosphino)ethane
dppf	1,2-bis(diphenylphosphino)ferrocene
dppp	1,3-bis(diphenylphosphino)propane
dr	diastereomer ratio
DTTB	4,4'-di-*t*-butylbiphenyl
E	COOMe
ee	enantiomer excess
en	ethylenediamine
er	enantiomer ratio
Et	ethyl
EVE	ethyl vinyl ether
Fc	ferrocenyl
Fmoc	9-fluorenylmethoxycarbonyl
Fu	furanyl
HMDS	hexamethyldisilazane
HMPA	hexamethylphosphoric amide
hv	light
Hx	*n*-hexyl
i	iso
Ipc	isopinocampheyl
kbar	kilobar
L	ligand
LAH	lithium aluminum hydride
LDA	lithium diisopropylamide
LHMDS	lithium hexamethyldisilazide

LTMP	lithium 2,2,6,6-tetramethylpiperidide
LN	lithium naphthalenide
lut	2,6-lutidine
M	metal
MAD	methylaluminum bis(2,6-di-*t*-butyl-4-methylphenoxide)
MCPBA	*m*-chloroperoxybenzoic acid
Me	methyl
MEM	methoxyethoxymethyl
Men	menthyl
Mes	mesityl
Mexyl	3,5-dimethylphenyl
MOM	methoxymethyl
Ms	methanesulfonyl (mesyl)
MS	molecular sieves
MTO	methyltrioxorhodium
MVK	methyl vinyl ketone
nbd	norbornadiene
NBS	*N*-bromosuccinimide
NCS	*N*-chlorosuccinimide
NIS	*N*-iodosuccinimide
NMO	*N*-methylmorpholine *N*-oxide
NMP	*N*-methylpyrrolidone
Np	naphthyl
Ns	*p*-nitrobenzenesulfonyl
Nu	nucleophile
Oc	octyl
PCC	pyridinium chlorochromate
PDC	pyridinium dichromate
PEG	poly(ethylene glycol)
Ph	phenyl
phen	1,10-phenenthroline
Pht	phthaloyl
Piv	pivaloyl
PMB	*p*-methoxybenzyloxymethyl
PMHS	poly(methylhydrosiloxane)
PMP	*p*-methoxyphenyl
Pr	*n*-propyl
py	pyridine
Q^+	quaternary onium ion
RAMP	(*R*)-1-amino-2-methoxymethylpyrrolidine
RaNi	Raney nickel

RCM	ring closure metathesis
R^f	perfluoroalkyl
ROMP	ring opening metathesis polymerization
s-	secondary
(s)	solid
salen	N,N'-ethylenebis(salicylideneiminato)
SAMP	(S)-1-amino-2-methoxymethylpyrrolidine
sc	supercritical
SDS	sodium dodecyl sulfate
sens.	sensitizer
SEM	2-(trimethylsilyl)ethoxymethyl
SES	2-[(trimethylsilyl)ethyl]sulfonyl
TASF	tris(dimethylamino)sulfur(trimethylsilyl)difluoride
TBAF	tetrabutylammonium fluoride
TBDPS	t-butyldiphenylsilyl
TBDMS	t-butyldimethylsilyl
TBS	t-butyldimethylsilyl
TEMPO	2,2,6,6-tetramethylpiperidinooxy
Tf	trifluoromethanesulfonyl
THF	tetrahydrofuran
THP	tetrahydropyranyl
Thx	t-hexyl
TIPS	triisopropylsilyl
TMEDA	N,N,N',N'-tetramethylethylenediamine
TMS	trimethylsilyl
Tol	p-tolyl
TON	turn over numbers
Tp	tris(1-pyrazolyl)borato
tpp	tetraphenylporphyrin
Ts	tosyl (p-toluenesulfonyl)
TSE	2-(trimethylsilyl)ethyl
TTN	thallium trinitrate
Z	benzyloxycarbonyl
Δ	heat
))))	microwave

REFERENCE ABBREVIATIONS

A

Acetic anhydride.

   Dehydration.[1]   Ketoximes of alkyl aryl ketones afford pyrrolines on heating with
Ac$_2$O in dimethylacetamide. Cyclization probably proceeds via H-abstraction after the
nitrenium ions are formed.

68%

[1]Savarin, C.G., Grise, C., Murry, J.A., Reamer, R.A., Hughes, D.L. *OL* **9**, 981 (2007).

Acetylacetonato(1,5-cyclooctadiene)rhodium(I).

   Aryltrialkoxysilanes.   Preparation of ArSi(OR)$_3$ from ArX and HSi(OR)$_3$ is readily
accomplished with the aid of (acac)Rh(cod) in DMF.[1]

[1]Murata, M., Yamasaki, H., Ueta, T., Nagata, M., Ishikura, M., Watanabe, S., Masuda, Y. *T* **63**, 4087
(2007).

Acetylacetonato(dicarbonyl)rhodium(I).

   Alkynylation.   Addition of 1-alkynes to α-keto esters is catalyzed by (acac)Rh(CO)$_2$
in the presence of a hindered phosphine ligand [e.g., 2-(di-*t*-butylphosphino)biphenyl].[1]
Complexes containing more electron-rich analogues of the acetylaetonato ligand favor
the reaction.

   Coupling.   Allylic carbonylation and coupling with boronic acids transform 2,3-diaza-
bicyclo[2.2.1]hept-5-enes into 5-hydrazinyl-2-cyclopentenyl ketones.[2]

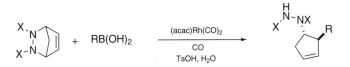

Fiesers' Reagents for Organic Synthesis, Volume 25. By Tse-Lok Ho
Copyright © 2010 John Wiley & Sons, Inc.

Addition to α-dicarbonyl compounds.[3]   α-Diketones and α-keto esters react in aqueous DME with ArB(OH)₂ to produce the monadducts.

Reduction.   Conjugated acids are converted to saturated aldehydes by syngas at room temperature, using (acac)Rh(CO)₂ in conjunction with a special guanidine as catalyst.[4] Only CO is liberated as stoichiometric side product. Furthermore, conditions for this highly selective reaction do not disturb acetals, esters, carbamates, ethers, silyl ethers, sulfides and many other functional groups.

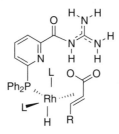

Hydroformylation.   With the Rh complex as catalyst (and a phosphite ligand) enamides and *N*-vinylimides are converted under syngas to α-amidoacetaldehydes.[5]

[1]Dhondi, P.K., Carberry, P., Choi, L.B., Chisholm, J.D. *JOC* **72**, 9590 (2007).
[2]Menard, F., Weise, C.F., Lautens, M. *OL* **9**, 5365 (2007).
[3]Ganci, G.R., Chisholm, J.D. *TL* **48**, 8266 (2007).
[4]Smejkal, T., Breit, B. *ACIE* **47**, 3946 (2008).
[5]Saidi, O., Ruan, J., Vinci, D., Wu, X., Xiao, J. *TL* **49**, 3516 (2008).

Acetyl chloride.

Nitration of arylamines.   Nitration is performed by treatment of the [ArNHR₂]NO₂ salts with two equivalents of AcCl.[1] Apparently, the active nitrating agent, AcONO₂, is formed.

[1]Zhang, P., Cedilote, M., Cleary, T.P., Pierce, M.E. *TL* **48**, 8659 (2007).

N-Alkoxycarbonylazoles.

Allyl carbonates.   1-Allyloxycarbonylimidazole is an allyloxycarbonylating agent for enolate ions (e.g., generated from ketones and NaHMDS in DME, −78°).[1] *O*-Allylation occurs under the influence of BF₃·OEt₂. Substituted allyl groups are similarly transferred from homologous reagents.

Carbamates, carbonates, and thiocarbonates are also readily prepared from the highly stable, nonhygroscopic, and usually crystalline mixed carbamates **1** of 3-nitro-1,2,4-triazole.[2]

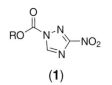

(1)

[1]Trost, B.M., Xu, J. *JOC* **72**, 9372 (2007).
[2]Shimizu, M., Sodeoka, M. *OL* **9**, 5231 (2007).

Alkylaluminum chlorides.

Rearrangement. α-Siloxyarylacetaldehydes give aryl ketones on treatment with Me$_2$AlCl. On the other hand, chloroaluminum biphenyl-2,2′-bis(triflylamide) catalyzes an alternative rearrangement pathway.[1]

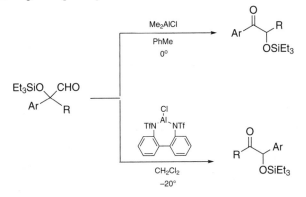

[1]Ohmatsu, K., Tanaka, T., Ooi, T., Maruoka, K. *ACIE* **47**, 5203 (2008).

S-Alkylisothiouronium salts.

Thiol surrogates. These readily available compounds (RX + thiourea) release RSH in the presence NaOH for conjugate addition. Essentially they are odorless thiolating agents.[1]

[1]Zhao, Y., Ge, Z.-M., Cheng, T.-M., Li, R.-T. *SL* 1529 (2007).

η³-Allyl(1,5-cyclooctadiene)palladium tetrafluoroborate.

Allylation.[1] The Pd salt in the presence of 6-diphenylphosphino-2-pyridone catalyzes *C*-allylation of indoles (at C-3) and pyrroles (at C-2) with allyl alcohol in toluene at 50°, generating water as the only byproduct. The key to activation of the allylating agent is by H-bonding.

Nucleophilic substitution.[2]   Benzylic acetates react with nucleophiles such as amines, sodium arenesulfonates, and malonic esters under the influence of the title reagent together with DPPF and a mild base [Et_3N in EtOH or K_2CO_3 in *t*-AmOH].

[1]Usui, I., Schmidt, S., Keller, M., Breit, B. *OL* **10**, 1207 (2008).
[2]Yokogi, M., Kuwano, R. *TL* **48**, 6109 (2007).

η³-Allyl(cyclopentadienyl)palladium.

Cycloaddition.   The Pd complex is useful for generating internal salts containing a π-allylpalladium complex from (ω-1)-methylene lactones. Trapping of the intermediates by other 1,3-dipoles such as nitrones results in the products of different types of heterocycles (with larger ring size).[1]

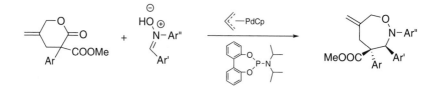

The subtle ligand effects are manifested in the reaction of dipolar species with acrylic esters, apparently due to different number of P-ligands on the π-allylpalladium complex. With two additional ligands (phosphites) on Pd the π-allyl segment suffers attack at the central carbon to eventually generate spiro[2.4]heptanes, whereas only one additional ligand (phosphine) engenders an electronic bias toward bond formation at the terminus.[2]

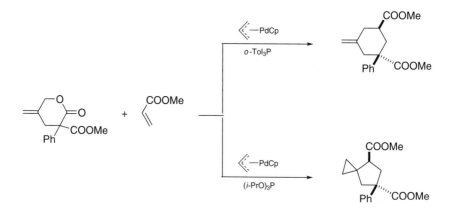

Carboboration.[3]   An alkyl group is delivered from (alkyl)zirconocene chlorides to a triple bond accompanied by the formation of an oxaborolidine unit. Remarkably, Me_3P (vs. other phosphine ligands) has a unique stereochemical influence.

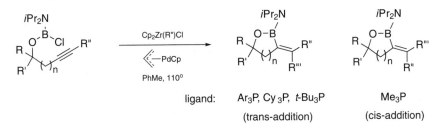

ligand: Ar₃P, Cy ₃P, *t*-Bu₃P   Me₃P

(trans-addition)   (cis-addition)

Elimination.[4]   *o*-Quinodimethane is generated from (*o*-trimethylsilylmethyl)benzyl methyl carbonate on heating with the Pd complex and DPPE in DMSO at 120°.

[1]Shintani, R., Murakami, M., Hayashi, T. *JACS* **129**, 12356 (2007).
[2]Shintani, R., Park, S., Hayashi, T. *JACS* **129**, 14866 (2007).
[3]Daini, M., Yamamoto, A., Suginome, M. *JACS* **130**, 2918 (2008).
[4]Giudici, R.E., Hoveyda, A.H. *JACS* **129**, 3824 (2007).

η³-Allyldichloro(triphenylphosphine)palladium.

Borylsilylation.[1]   (Chlorodimethylsilyl)pinacolatoborane adds to 1-alkynes to give 1-pinacolatoboryl-2-silylalkenes. The relative amount of the addends is the determinant factor in the stereochemical outcome of the reaction

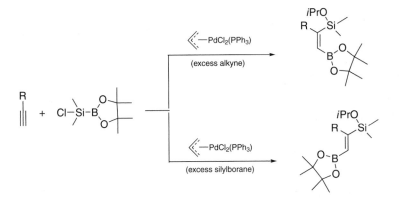

[1]Ohmura, T., Oshima, K., Suginome, M. *CC* 1416 (2008).

Allylstannanes.

Allyl addition.[1]   Diastereoselectivity for the addition of an allyl group to hexacarbonyl-dicobalt complexes of 4-hydroxy-2-alkynals is much higher using allyltriphenylstannane instead of the tributyl congener.

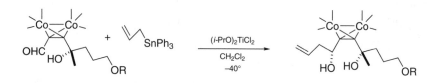

[1]Hayashi, Y., Yamaguchi, H., Toyoshima, M., Okado, K., Toyo, T., Shoji, M. *OL* **10**, 1405 (2008).

Aluminum bromide.

Reductive phenylation.[1]   Naphthalenediols and benzene combine to afford hydroxy-tetralones. The transformation occurs when the mixtures of the aromatic compounds are treated with an excess of AlBr$_3$.

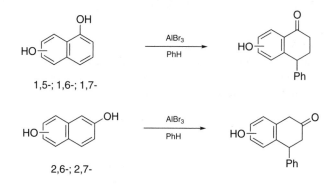

[1]Koltunov, K.Yu. *TL* **49**, 3891 (2008).

Aluminum chloride.

Friedel–Crafts acylation.   A synthesis of chilenine is completed by a two-fold Friedel–Crafts acylation of an *N*-(arylethyl)amide with oxalyl chloride.[1]

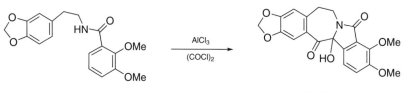

chilenine

Acylation of arylidenecyclobutanes is accompanied by ring expansion.[2] A route to norbornen-7-ones entails an intramolecular desilylative Friedel–Crafts acylation.[3] Such compounds are not directly accessible by a Diels–Alder reaction.

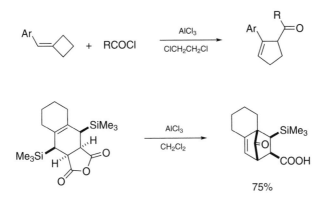

75%

Carbimination.   Thiophene and *N*-substituted pyrroles and indoles undergo electrophilic substitution with ArNC at room temperature. The reaction gives imines as products.[4]

Aromatization.   Treatment of 6-hydroxy-1,2,3,6-tetrahydro-*N*-tosyl-3-pyridones with AlCl$_3$ in MeNO$_2$ at $-78°$ brings about dehydration and *O*-tosylation to give 3-tosyloxypyridines.[5]

[1]Kim, G., Jung, P., Tuan, L.A. *TL* **49**, 2391 (2008).
[2]Jiang, M., Shi, M. *OL* **10**, 2239 (2008).
[3]Li, D., Liu, G., Hu, Q., Wang, C., Xi, Z. *OL* **9**, 5433 (2007).
[4]Tobisu, M., Yamaguchi, S., Chatani, N. *OL* **9**, 3351 (2007).
[5]Hodgson, R., Kennedy, A., Nelson, A., Perry, A. *SL* 1043 (2007).

Aluminum dimethylamide.

Transamination.[1]   Tertiary amides are converted to secondary amides on reaction with secondary amines in the presence of Al$_2$(NMe$_2$)$_6$.

[1]Hoerter, J.M., Otte, K.M., Gellman, S.H., Cui, Q., Stahl, S.S. *JACS* **130**, 647 (2008).

Aluminum iodide.

Baylis–Hillman reaction.   Ethyl propynoate apparently undergoes iodoalumination to generate a nucleophilic species that adds onto carbonyl compounds. (*Z*)-β-Iodoacrylic esters are produced.[1]

[1]Lee, S.I., Hwang, G.-S., Ryu, D.H. *SL* 59 (2007).

Aluminum tris(2,6-diphenylphenoxide), ATPH.

Macrolide synthesis.[1]   By way of an intramolecular aldol reaction using ATPH and LiTMP, macrocyclic (10-, 12-, and 14-membered) lactones are formed.

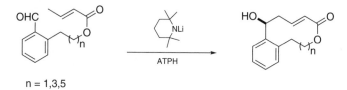

n = 1,3,5

[1]Abramite, J.A., Sammakia, T. *OL* **9**, 2103 (2007).

Aluminum triflate.

Cycloisomerization.[1]   An oxime function is liable to add to a double bond at an appropriate distance and the reaction is realized by heating unsaturated oximes with Al(OTf)$_3$ in MeNO$_2$.

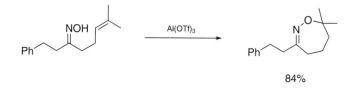

84%

[1]Cheminade, X., Chiba, S., Narasaka, K., Dunach, E. *TL* **49**, 2384 (2008).

Aminocarbenes.

Reviews.[1,2]   Applications of heterocyclic carbenes in organic synthesis have been reviewed.

Aldol reactions.   Enolization of ketones at room temperature (and ensuing silylation) is readily effected by 1,3-bis(1-adamantyl)imidazol-2-ylidene.[3] Accordingly, Mukaiyama aldol reaction is accomplished under the appropriate conditions.[4]

Baylis–Hillman reaction products are obtained in an unconventional manner from α-silylpropargyl alcohols and aldehydes, using 1,3-bis(2,6-diisopropylphenyl)imidazol-2-ylidene as catalyst.[5]

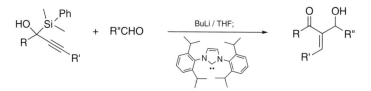

Acyloin condensation.   Carbene species (for promoting intramolecular acyloin condensation) are more readily generated from 1,2,4-triazolium salts when one of the N-substituents is highly electron-deficient (e.g., **1**).[6] The bicyclic triazolium salt **2** derived from pyroglutamic acid catalyzes benzoin condensation in modest yields, in which electron-rich ArCHO is less reactive but better asymmetric induction is observed.[7]

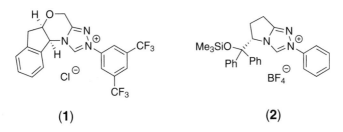

(1)   **(2)**

Analogous condensation of ArCHO and aldimines gives α-amino ketones.[8]

Carboxylic derivatives. A mixture of an aldehyde and a nitrosoarene is converted into an *N*-arylhydroxamic acid on treatment with **3** and DBU,[9] whereas α,α-dichloro aldehydes gives α-chloro carboxamides in the presence of amines under similar conditions.[10] A mild organic base is needed to generate the carbene (and a slight variation of the catalyst system for the same reaction comprises the *N*-mesityltriazolium chloride and imidazole base.[11])

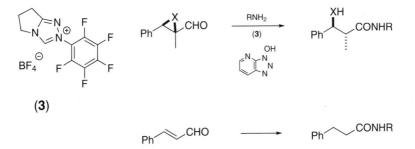

(3)

There is a significant difference in reaction profile for the reaction of enals with nitroso-arenes. Isoxazolidin-5-ones are formed and alcoholysis of which leads to β-arylamino esters.[12] With the nitroarenes replaced by arylazo carbonyl compounds to perform the reaction 3-oxopyrazoldinones result.[13]

As a redox process, the ring expansion of β-formyl-β-lactams to furnish succinimides[14] and the ring scission of 2-nitrocyclopropanecarbaldehydes[15] are also mediated by an azocarbene.

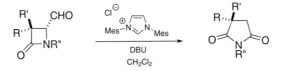

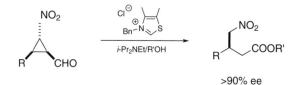

>90% ee

Enals generated by oxidation of allylic alcohols with MnO_2 in the presence of azolium ylides are trapped to form secondary allylic alcohols. These are subject to further oxidation and the resulting ketones undergo alcoholysis in situ.[16]

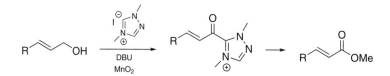

[1]Hahn, F.E., Jahnke, M.C. *ACIE* **47**, 3122 (2008).
[2]Marion, N., Diez-Gonzalez, S., Nolan, S.P. *ACIE* **46**, 2988 (2007).
[3]Song, J.J., Tan, Z., Reeves, J.T., Fandrick, D.R., Yee, N.K., Senanayake, C.H. *OL* **10**, 877 (2008).
[4]Song, J.J., Tan, Z., Reeves, J.T., Yee, N.K., Senanayake, C.H. *OL* **9**, 1013 (2007).
[5]Reynolds, T.E., Stern, C.A., Scheidt, K.A. *OL* **9**, 2581 (2007).
[6]Takikawa, H., Suzuki, K. *OL* **9**, 2713 (2007).
[7]Enders, D., Han, J. *TA* **19**, 1367 (2008).
[8]Li, G.-Q., Dai, L.-X., You, S.-L. *CC* 852 (2007).
[9]Wong, F.T., Patra, P.K., Seayad, J., Zhang, Y., Ying, J.Y. *OL* **10**, 2333 (2008).
[10]Vora, H.U., Rovis, T. *JACS* **129**, 13796 (2007).
[11]Bode, J.W., Sohn, S.S. *JACS* **129**, 13798 (2007).
[12]Seayad, J., Patra, P.K., Zhang, Y., Ying, J.Y. *OL* **10**, 953 (2008).
[13]Chan, A., Scheidt, K.A. *JACS* **130**, 2740 (2008).
[14]Li, G.-Q., Li, Y., Dai, L.-X., You, S.-L. *OL* **9**, 3519 (2007).
[15]Vesely, J., Zhao, G.-L., Bartoszewicz, A., Cordova, A. *TL* **49**, 4209 (2008).
[16]Maki, B.E., Chan, A., Phillips, E.M., Scheidt, K.A. *OL* **9**, 371 (2007).

Antimony(V) chloride.

Indanones.[1]   *trans*-2,3-Disubstituted indanones are produced in reasonably good yields from a mixture of arylalkynes and aldehydes with EtOH (1 equiv.) as additive, by treatment with $SbCl_5$.

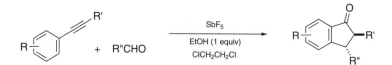

[1]Saito, A., Umakoshi, M., Yagyu, N., Hanzawa, Y. *OL* **10**, 1783 (2008).

Arylboronic acids.

Amide formation.   *o*-Halophenylboronic acids catalyze the Diels–Alder reaction of acrylic acid as well as condensation of carboxylic acids with amines at room temperature (in the presence of 4A-molecular sieves).[1]

A thorough study indicates that (1-methyl-4-pyridinio)boronic acid iodide is a superior catalyst for amidation under azeotropic conditions, and esterification of 2-hydroxyalkanoic acids.[2]

[1]Al-Zoubi, R.M., Marion, O., Hall, D.G. *ACIE* **47**, 2876 (2008).
[2]Maki, T., Ishihara, K., Yamamoto, H. *T* **63**, 8645 (2007).

N-**Arylsulfinylimines.**

Imido transfer.[1]   Aldehydes are converted into RCH=NAr on reaction with ArN=S=O, using catalysts such as $VOCl_3$, $MoOCl_3$, and MoO_2Cl_2.

[1]Zhizhin, A.A., Zarubin, D.N., Ustynyuk, N.A. *TL* **49**, 699 (2008).

Azobisisobutyronitrile.

Deallylation.   Allyl carboxylates are hydrolyzed under neutral conditions on treatment with AIBN (10 mol%) and water. This radical deallylation generally proceeds in high yields.[1]

Oxidative cyclization.   Alkynyllactams cyclize by reaction with PhSH and AIBN, involving carbon radical shuffle.[2]

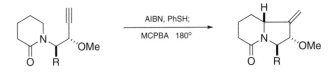

[1]Perchyonok, V.T., Ryan, S.J., Langford, S.J., Hearn, M.T., Tuck, K.L. *SL* 1233 (2008).
[2]Denes, F., Beaufis, F., Renaud, P. *OL* **9**, 4375 (2007).

B

Barium alkoxides.

Aminoalkylation.[1]   The use of (ArO)₂Ba in THF to deprotonate 3-butenoic esters for reaction with *N*-phosphinylaldimines gives α-substituted crotonates.

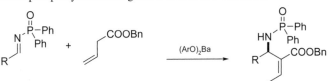

Aldol + Michael reactions.[2]   A 2 : 1 condensation between ArCOMe and Ar'CHO is observed when the mixtures are treated with (*i*-PrO)₂Ba.

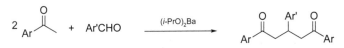

[1]Yamaguchi, A., Aoyama, N., Matsunaga, S., Shibasaki, M. *OL* **9**, 3387 (2007).
[2]Yanagisawa, A., Takahashi, H., Arai, T. *T* **63**, 8581 (2007).

Barium hydride.

Michael reaction.[1]   2-Cycloalkenones dimerize in the presence of BaH₂. However, 2-cyclopentenone condenses with chalcone to form a bicyclo[2.2.1]heptanone.

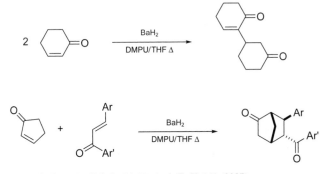

[1]Yanagisawa, A., Shinohara, A., Takahashi, H., Arai, T. *SL* 141 (2007).

Fiesers' Reagents for Organic Synthesis, Volume 25. By Tse-Lok Ho
Copyright © 2010 John Wiley & Sons, Inc.

Benzenesulfonic anhydride.

Amide formation.[1]   Activation of carboxylic acids by $(PhSO_2)_2O$ (with catalytic DMAP) as mixed anhydrides for acylation of R_2NH is a very simple operation.

[1]Funasaka, S., Kato, K., Mukaiyama, T. *CL* **36**, 1456 (2007).

Benzyl *N*-phenyl-2,2,2-trifluoroacetimidate.

O-Benzylation.[1]   Benzyl ethers of base-sensitive hydroxy esters and hindered alcohols are formed by reaction with the title reagent (Me_3SiOTf as catalyst). The reagent is more stable than the trichloro analogue and it can be prepared from $CF_3C(=NPh)Cl$ and BnOH.

[1]Okada, Y., Ohtsu, M., Bando, M., Yamada, H. *CL* **36**, 992 (2007).

1,1′-Binaphthalene-2-amine-2′-phosphines.

Substitution reactions.   An S_N2 reaction between 2-trimethylsiloxyfuran and acetylated Baylis–Hillman adducts is induced by the amine/phosphine **1**.[1]

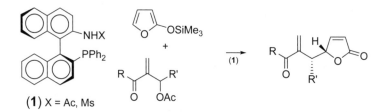

Actually the *N*-acetyl derivative catalyzes the aza-Baylis–Hillman reaction.[2]

[1]Jiang, Y.-Q., Shi, Y.-L., Shi, M. *JACS* **130**, 7202 (2008).
[2]Qi, M.-J., Ai, T., Shi, M., Li, G. *T* **64**, 1181 (2008).

1,1′-Binaphthalene-2,2′-bis(*p*-toluene sulfoxide).

Michael reaction.   The title compound is a bidentate S,S-ligand for Rh. Complexes of the sort are used in mediating aryl transfer from $ArB(OH)_2$ to 2-cycloalkenones and conjugated lactones under basic conditions.[1]

[1]Mariz, R., Luan, X., Gatti, M., Linden, A., Dorta, R. *JACS* **130**, 2172 (2008).

1,1′-Binaphthalene-2,2′-diamine derivatives.

Aldol reaction.   Asymmetric aldol reaction of chloroacetone with electron-deficient ArCHO gives mainly the *anti*-3-chloro-4-hydroxy-2-butanones, in the presence of **1**.[1] The protocol is valid for other ketones,[2] and aldol reactions catalyzed by **2** have also been reported.[3]

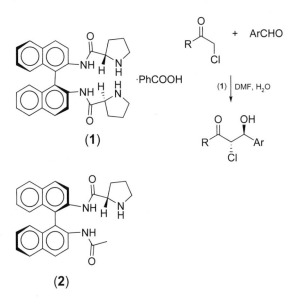

(1)

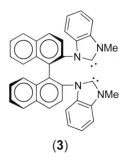

(2)

Using bisthiourea derived from a chiral octahydro-BINAMINE as catalyst (with DABCO base) Baylis–Hillman reaction proceeds in 60–88% ee.[4]

Conjugate addition. The parent chiral BINAMINE is an excellent ligand for $CuCl_2$ to promote the conjugate addition of diorganozinc reagents.[5] When complexed to carbene **3** palladium dicarboxylates exhibit catalytic activity in the aryl transfer from $ArB(OH)_2$ to 2-cycloalkenones.[6]

(3)

Addition to multiple bonds. The semilabile acyloxy ligands in Pd complexes of **3** are exchangeable. In the reaction of allyltributylstannane with RCHO, π-allylpalladium species are formed via such an exchange.[7]

The aluminum complex **4** of the salen prepared from BINAMINE catalyzes a very interesting and useful reaction. It turns propargylsilanes into α-silylallylidenation agents for aldehydes such as glyoxamides.[8]

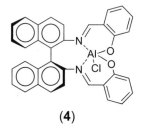

(4)

Exquisite diastereochemical control is attained by tuning the relative bulk of the two alkoxy groups in ketene silyl acetals derived from α-alkoxyacetic esters, during aldol reaction with aldehydes. A chiral version is promoted SiCl$_4$ in the presence of the phosphotriamide **5**.[9] The same set of reaction conditions is also applicable to create asymmetric quaternary carbon centers, for example, in the reaction of *N*-silyl ketenimines with ArCHO.[10]

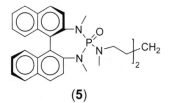

(5)

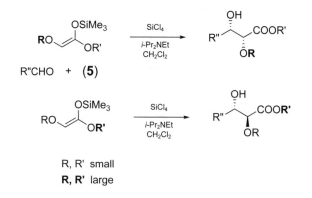

With complex **6** asymmetric addition of α-nitroalkanoic esters to imines is achieved.[11]

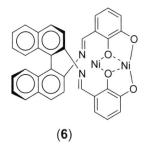

(6)

The rather unusual tetramidoytterbate anion **7** is responsible for asymmetric induction in an intramolecular hydroamination.[12]

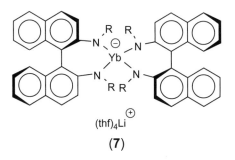

$(thf)_4Li^{\oplus}$

(7)

Redox reactions. With PdI_2 complexed to **3** kinetic resolution of secondary benzylic and allylic alcohols can be carried out via enantioselective oxidation (O_2, Cs_2CO_3, PhMe, 80°).[13] The acetoxydiiodo-Rh carbene complex of **3** is a catalyst for asymmetric reduction of aroylacetic esters with Ph_2SiH_2.[14]

Substitution. Ullmann diaryl ether synthesis catalyzed by $Cu(OTf)_2$–BINAMINE occurs at a relatively low temperature (in dioxane, 110°, base: Cs_2CO_3).[15]

[1]Guillena, G., del Carmen Hita, M., Najera, C. *TA* **18**, 1272 (2007).
[2]Guillena, G., del Carmen Hita, M., Najera, C., Viozquez, S.F. *TA* **18**, 2300 (2007).
[3]Guizzetti, S., Benaglia, M., Raimondi, L., Celentano, G. *OL* **9**, 1247 (2007).
[4]Shi, M., Liu, X.-G. *OL* **10**, 1043 (2008).
[5]Hatano, M., Asai, T., Ishihara, K. *TL* **48**, 8590 (2007).
[6]Zhang, T., Shi, M. *CEJ* **14**, 3759 (2008).
[7]Zhang, T., Shi, M., Zhao, M. *T* **64**, 2412 (2008).
[8]Evans, D.A., Aye, Y. *JACS* **129**, 9606 (2007).
[9]Denmark, S.E., Chung, W.-j. *ACIE* **47**, 1890 (2008).
[10]Denmark, S.E., Wilson, T.W., Burk, M.T., Heemstra Jr, J.R. *JACS* **129**, 14864 (2007).
[11]Chen, Z., Morimoto, H., Matsunaga, S., Shibasaki, M. *JACS* **130**, 2170 (2008).
[12]Aillaud, I., Collin, J., Duhayon, C., Guillot, R., Lyubov, D., Schulz, E., Trifonov, A. *CEJ* **14**, 2189 (2008).
[13]Chen, T., Jiang, J.-J., Xu, Q., Shi, M. *OL* **9**, 865 (2007).

[14]Xu, Q., Gu, X., Liu, S., Dou, Q., Shi, M. *JOC* **72**, 2240 (2007).
[15]Naidu, A.B., Raghanath, O.R., Prasad, D.J.C., Sekar, G. *TL* **49**, 1057 (2008).

1,1′-Binaphthalene-2,2′-dicarboxylic acids.

Addition to imines. Functionalized secondary amines are formed by addition of hydrazones[1] and diazo compounds[2] to aldimines, and these reactions are subject to asymmetric induction by **1**.

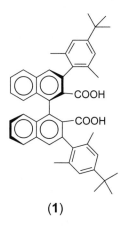

(1)

[1]Hashimoto, T., Hirose, M., Maruoka, K. *JACS* **130**, 7556 (2008).
[2]Hashimoto, T., Maruoka, K. *JACS* **129**, 10054 (2007).

1,1′-Binaphthalene-2,2′-diol and analogues.

Strecker synthesis. The 3,3′-disubstituted BINOL **1** is used in promoting addition of Me$_3$SiCN to *N*-tosylketimines. Adding one equivalent of 1-adamantanol enhances reaction rates and enantioselectivity.[1]

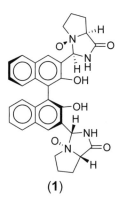

(1)

Addition reactions.   Asymmetric allyl transfer from allyl boronates to *N*-acyl imines is assisted by (*S*)-3,3'-diphenyl-BINOL.[2] Alkenyldimethoxyboranes react with conjugated carbonyl compounds with excellent enantioselectivity in the presence of a chiral 3,3'-diiodo-BINOL.[3]

Allyl addition to hydrazones in the presence of 3,3'-bissulfonyl-BINOLs gives products in low to moderate ee (10–68%), but much improvement(95–98% ee) is observed for using fluorinated organosulfonyl analogues.[4]

For regioselective introduction of a chiral sidechain to C-2 of the indole nucleus the higher nucleophilicity of C-3 must be overcome. Employing the 4,7-dihydro derivatives the preferred reaction site is moved (to the active α-position of 4,5-disubstituted pyrroles), and asymmetric Michael reaction has been demonstrated with a chiral 3,3'-dibromo-BINOL as catalyst.[5]

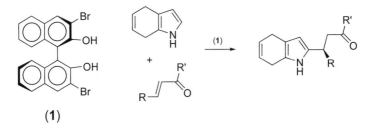

(1)

Substitution reactions.[6]   3,3'-Bis(2-hydroxy-3-isopropylbenzyl)-BINOL causes opening of *meso*-epoxides by ArNH$_2$ asymmetrically.

[1]Hou, Z., Wang, J., Liu, X., Feng, X. *CEJ* **14**, 4484 (2008).
[2]Lou, S., Moquist, P.N., Schaus, S.E. *JACS* **129**, 15398 (2007).
[3]Wu, T.R., Chong, J.M. *JACS* **129**, 4908 (2007).
[4]Kargbo, R., Takahashi, Y., Bhor, S., Cook, G.R., Lloyd-Jones, G.C., Shepperson, I.R. *JACS* **129**, 3846 (2007).
[5]Blay, G., Fernandez, I., Pedro, J.R., Vila, C. *TL* **48**, 6731 (2007).
[6]Arai, K., Salter, M.M., Yamashita, Y., Kobayashi, S. *ACIE* **46**, 955 (2007).

1,1'-Binaphthalene-2,2'-diol – copper complexes.

N-Arylation.   Reaction of R$_2$NH with ArI is completed at room temperature using the BINOL-CuBr complex as catalyst.[1]

[1]Jiang, D., Fu, H., Jiang, Y., Zhao, Y. *JOC* **72**, 672 (2007).

1,1'-Binaphthalene-2,2'-diol (modified) – hafnium complexes.

Mannich reaction.   A complex derived from (*t*-BuO)$_4$Hf, imidazole and 6,6'-dibromo-BINOL is air-stable. It is capable of asymmetric induction in catalyzing the Mannich reaction (80–90% ee).[1]

[1]Kobayashi, S., Yazaki, R., Seki, K., Ueno, M. *T* **63**, 8425 (2007).

1,1'-Binaphthalene-2,2'-diol – iridium complexes.

Allylation.   A highly selective monoallylation of ketone enamines with allylic carbonates by the S_N2' pathway is observed with a complex of BINOL of iridium(I).[1] In the reaction $ZnCl_2$ is also present.

[1]Weix, D.J., Hartwig, J.F. *JACS* **129**, 7720 (2007).

1,1'-Binaphthalene-2,2'-diol – magnesium complexes.

Hetero-Diels–Alder reaction.   The complex formed on treatment of BINOL with *i*-Bu₂Mg shows excellent performance in catalyzing enantioselective cycloaddition of Danishefsky's diene with aldehydes to give 2,3-dihydro-4*H*-pyran-4-ones.[1]

[1]Du, H., Zhang, X., Wang, Z., Bao, H., You, T., Ding, K. *EJOC* 2248 (2008).

1,1'-Binaphthalene-2,2'-diol – niobium complexes.

Aminolysis.[1]   The 3,3'-disubstituted BINOL **1** forms a complex with $Nb(OMe)_5$ that has found use in catalyzing the opening of epoxides and aziridines with $ArNH_2$.

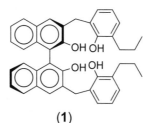

(1)

[1]Arai, K., Lucarini, S., Salter, M.W., Ohta, K., Yamashita, Y., Kobayashi, S. *JACS* **129**, 8103 (2007).

1,1'-Binaphthalene-2,2'-diol – titanium complexes.

Addition to C=O.   Asymmetric addition reactions involving tetraallylstannane,[1] 2-furyldiethylalane[2] and (thf)AlAr₃[3] to ketones in the presence of a titanium complex of BINOL has been studied. The unsymmetrical BINOL **1** and its octahydro derivative form Ti complexes that have been used in reactions with Grignard reagents[4] and organozincs,[5] respectively.

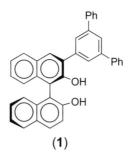

(1)

A chiral catalyst system for aldol reaction of conjugated thioketene silyl acetals (i.e., from thio esters) consists of Ti-BINOL and $(MeO)_3B$.[6] The teranuclear Ti complex is air-stable and its use in aldol reactions requires low loading.[7]

A polymer with repeating 6,6′-dibutyl-BINOL units that are linked to each other at C-5 and C-5′ has been synthesized. The Ti-complex of the polymer catalyzes the addition of alkynylzinc species to aldehydes.[8]

α-Cyanohydrin derivatives. BINOL **2**, (1*R*,2*S*)-2-acetamino-1,2-diphenylethanol, and (*i*-PrO)$_4$Ti self-assemble on admixture. The ensuing complex is a good catalyst for cyanoethoxycarbonylation of aldehydes.[9] A simpler Ti catalyst is that obtained from 3-(1-imidazolyl)-BINOL **3**, which serves in derivatization of ArCHO with Me$_3$SiCN.[10]

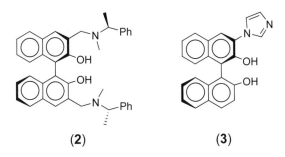

(2)   **(3)**

Cycloaddition. Through empirical screening the dinuclear Ti complex **4** of 6,6′-diiodo-BINOL and the complex prepared from **5** have been chosen to promote 1,3-dipolar cycloaddition (nitrone + enal)[11] and hetero-Diels–Alder reaction (Danishefsky's diene + RCHO),[12] respectively.

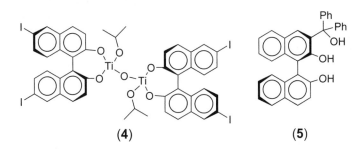

(4)   **(5)**

[1]Wooten, A.J., Kim, J.G., Walsh, P.J. *OL* **9**, 381 (2007).
[2]Wu, K.-H., Chuang, D.-W., Chen, C.-A., Gau, H.-M. *CC* 2343 (2008).
[3]Chen, C.-A., Wu, K.-H., Gou, H.-M. *ACIE* **46**, 5373 (2007).
[4]Muramatsu, Y., Harada, T. *ACIE* **47**, 1088 (2008).
[5]Harada, T., Ukon, T. *TA* **18**, 2499 (2007).
[6]Heumann, L.V., Keck, G.E. *OL* **9**, 4275 (2007).
[7]Schetter, B., Ziemer, B., Schnakenburg, G., Mahrwald, R. *JOC* **73**, 813 (2008).
[8]Wu, L., Zheng, L., Zong, L., Xu, J., Cheng, Y. *T* **64**, 2651 (2008).

[9]Gou, S., Liu, X., Zhou, X., Feng, X. *T* **63**, 7935 (2007).
[10]Yang, F., Wei, S., Chen, C.-A., Xi, P., Yang, L., Lan, J., Gau, H.-M., You, J. *CEJ* **14**, 2223 (2008).
[11]Hashimoto, T., Omote, M., Kano, T., Maruoka, K. *OL* **9**, 4805 (2007).
[12]Yang, X.-B., Feng, J., Wang, N., Wang, L., Liu, J.-L., Yu, X.-Q. *OL* **10**, 1299 (2008).

1,1′-Binaphthalene-2,2′-diol – vanadium complexes.

Oxidative coupling. Vanadium complex **1A**[1] or **1B**[2] can be used in converting 2-naphthols to (*R*)-BINOLs and (*S*)-BINOLs, respectively, in air.

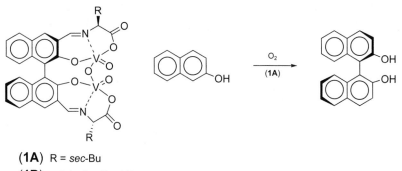

(1A) R = *sec*-Bu
(1B) octahydro, R = *t*-Bu

[1]Guo, Q.-X., Wu, Z.-J., Luo, Z.-B., Liu, Q.-Z., Ye, J.-L., Luo, S.-W., Cun, L.-F., Gong, L.-Z. *JACS* **129**, 13927 (2007).
[2]Mikami, M., Yamataka, H., Jayaprakash, D., Sasai, H. *T* **64**, 3361 (2008).

1,1′-Binaphthalene-2,2′-diol (modified) – zinc complexes.

Addition to aldehydes. Organozinc addition to aldehydes with BINOLs as catalysts likely involves precoordination. 3,3′-Disubstituted BINOLs, especially with substituents providing additional ligating groups, are found to be highly effective, as exemplified by the use of **1** and **2** in reaction of Et₂Zn and alkynylzincs (in situ), respectively.[1,2]

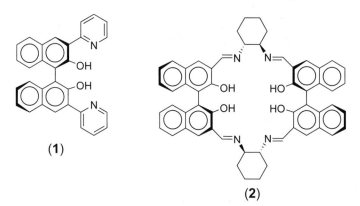

(1)

(2)

Alkynylzinc addition to ArCHO can also be carried out in the presence of **3**,[3] whereas the disilyl derivative **4** catalyzes enantioselective Reformatsky reaction on ketones (with ee up to 90%)[4] which operates by a free radical mechanism (requiring air to initiate the reaction).

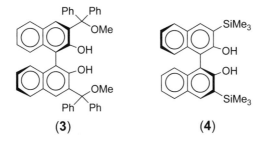

(3)   **(4)**

(*S*)-BINOL complexed to Et_2Zn shows catalytic activity in the hetero-Diels–Alder reaction of Danishefsky's diene and imine derived from ethyl glyoxylate.[5]

[1]Milburn, R.M., Hussain, S.M.S., Prien, O., Ahmed, Z., Snieckus, V. *OL* **9**, 4403 (2007).
[2]Li, Z.-B., Liu, T.-D., Pu, L. *JOC* **72**, 4340 (2007).
[3]Wang, Q., Chen, S.-Y., Yu, X.-Q., Pu, L. *T* **63**, 4422 (2007).
[4]Fernandez-Ibanez, M.A., Macia, B., Minnaard, A.J., Feringa, B.L. *CC* 2571 (2008).
[5]Di Bari, L., Guillarme, S., Hanan, J., Henderson, A.P., Howard, J.A.K., Pescitelli, G., Probert, M.R., Salvadori, P., Whiting, A. *EJOC* 5771 (2007).

1,1'-Binaphthalene-2,2'-diol (modified) – zirconium complexes.

Michael reaction.   A complex derived from (*t*-BuO)$_4$Zr and chiral 3,3'-dibromo-BINOL induces the enantioselective conjugate addition of indole (at C-3) to enones.[1]

[1]Blay, G., Fernandez, I., Pedro, J.R., Vila, C. *OL* **9**, 2601 (2007).

1,1'-Binaphthalene-2,2'-diol ethers.

Epimerization.   The ether **1** forms imines with α-amino acids. Imines of L-amino acids suffer from $A^{1,3}$-strain when maintaining a hydrogen-bonded conformation with the urea unit, therefore they are prone to undergo epimerization.[1]

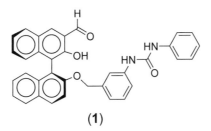

(1)

[1]Park, H., Kim, K.M., Lee, A., Ham, S., Nam, W., Chin, J. *JACS* **129**, 1518 (2007).

1,1'-Binaphthalene-2,2'-diyl *N*-alkylaminophosphites.

Cycloaddition. The *N*-triflyl derivative **1** that has very bulky substituents at C-3 and C-3' is air-stable. It is an effective Bronsted acid for catalyzing 1,3-dipolar cycloaddition (of nitrones and vinyl ethers).[1] An *endo* transition state is adopted for the reaction in which the proton simultaneously coordinates with oxygen atoms of both addends. In contrast, Lewis acids tend to favor the *exo* transition state.

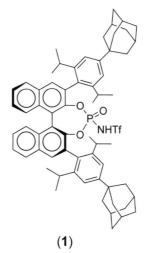

(1)

[1]Jiao, P., Nakashima, D., Yamamoto, H. *ACIE* **47**, 2411 (2008).

Copper(I) complexes.

Substitution reactions. Preparation of 2-branched 3-buten-1-yl bromides in the chiral form is conveniently accomplished by a Cu-catalyzed Grignard reaction in the presence of **2B** or *ent*-**2B**.[1] The valuable α-substituted allyl boronates are similarly accessed, although a report describes the use of the octahydro derivative of **2B**.[2]

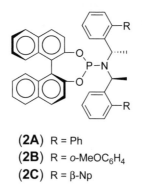

(2A) R = Ph
(2B) R = o-MeOC$_6$H$_4$
(2C) R = β-Np

Addition reactions. Chiral *N*-formylbenzylamines are formed by reaction of a Cu-catalyzed (ligand: **2A** or *ent*-**2A**) organozinc reaction. It involves generation of *N*-formylaldimines from the α-sulfonylamine derivatives.[3] Interestingly, the same system is applicable to imine trapping following conjugate addition to enones.[4]

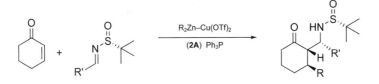

[1]Falciola, C.A., Alexakis, A. *ACIE* **46**, 2619 (2007).
[2]Carosi, L., Hall, D.G. *ACIE* **46**, 5913 (2007).
[3]Pizzuti, M.G., Minnaard, A.J., Feringa, B.L. *JOC* **73**, 940 (2008).
[4]Gonzales-Gomez, J.C., Foubelo, F., Yus, M. *TL* **49**, 2343 (2008).

Iridium complexes.

Allylic substitution. The iridium complex of **2A** is effective for catalyzing allylic substitution reactions, for example, in reaction of enamines with allylic carbonates to yield branched products.[1] Chiral allylic ethers are similarly prepared.[2]

The reaction of allyl carbonates with arylzinc reagents also pursues an S_N2' pathway preferentially, with Ir-complex of **1b** as promoter, but asymmetric induction is only moderate.[3] Using indole as nucleophile, substitution also proceeds.[4]

Primary allylic alcohols activated by $(EtO)_5Nb$ in situ are converted into branched allylic amines using iridium complex derived from $[(cod)IrCl]_2$ and the (R,R,S_a) isomer of **2A**.[5]

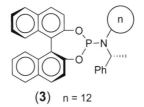

(3)   n = 12

The enantiomer of **2A** catalyzes allylation of ammonia to provide branched diallylamines.[6] If the nucleophile is changed to CF_3CONHK or Boc_2NLi a better ligand is **4**.

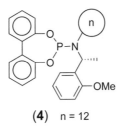

(4)   n = 12

The double inversion mechanism that operates in the Ir-catalyzed decarboxylative decomposition of secondary allylic carbamates effectively converts allylic alcohols into the corresponding amine derivatives with complete retention of configuration.[7]

Rearrangement. Transformation of 2-alkenols to 3-amino-1-alkenes can be performed via decarboxylative rearrangement of the derived carbamates, the iridium complex of **2B** possesses activity for endowing chirality to the amines.[8] When crotyl β-ketoalkanoate and homologues are exposed to the iridium complex of **2A** in the presence of DBU, rearrangement and decarboxylation occur, forming optically active 1-alken-5-ones.[9]

[1]Weix, D.J., Hartwig, J.F. *JACS* **129**, 7720 (2007).
[2]Ueno, S., Hartwig, J.F. *ACIE* **47**, 1928 (2008).
[3]Alexakis, A., El Hajjaji, S., Polet, D., Rathgeb, X. *OL* **9**, 3393 (2007).
[4]Liu, W.-B., He, H., Dai, L.-X., You, S.-L. *OL* **10**, 1815 (2008).
[5]Yamashita, Y., Gopalarathnam, A., Hartwig, J.F. *JACS* **129**, 7508 (2007).
[6]Pouy, M.J., Leitner, A., Weix, D.J., Ueno, S., Hartwig, J.F. *OL* **9**, 3949 (2007).
[7]Singh, O.V., Han, H. *OL* **9**, 4801 (2007).
[8]Singh, O.V., Han, H. *JACS* **129**, 774 (2007).
[9]He, H., Zheng, X.-J., Li, Y., Dai, L.-X., You, S.-L. *OL* **9**, 4339 (2007).

Nickel complexes.

Hydrovinylation.[1]   Addition of ethylene to styrenes occurs in the presence of the Ni-complex of phosphoramidites. Tuning of the catalysts indicates the unsymmetrical aminophosphite **5** is a good performer for asymmetric induction.

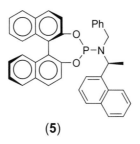

(5)

[1]Smith, C.R., RajanBabu, T.V. *OL* **10**, 1657 (2008).

Palladium complexes.

Hydrosilylation.   Chiral α-arylethanol can be synthesized from styrenes via hydrotrichlorosilylation and oxidative desilylation. The first step is accomplished with a Pd catalyst containing ligand **6**.[1]

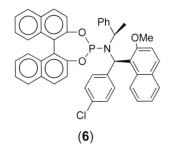

(6)

Cycloaddition reactions. The aminophosphite **H₈-2C** derived from octahydro-BINOL is found to promote the [3+3]cycloaddition of nitrones and trimethylenemethane derivatives to furnish 1,2-oxazines.[2] Remarkable ligand effects have been observed in the spiroannulation of oxindoles: products possessing opposite configuration at the spirocyclic center arise by changing the naphthyl substituents on the pyrrolidine ring (**7A** [α-Np] vs. **7B** [β-Np]).[3]

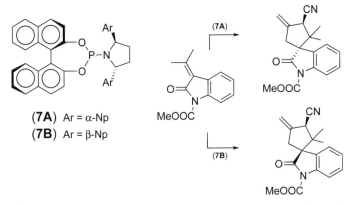

(7A) Ar = α-Np
(7B) Ar = β-Np

The bis-(β-naphthyl)pyrrolidinyl-containing ligand also finds use to induce chirality in the trimethylenemethane cycloaddition to imines, which leads to 2-substituted 4-methylenepyrrolidines.[4]

Another Pd-catalyzed reaction involves 1,2-di-*t*-butyldiaziridinone with dienes and it employs ligand **8**.[5]

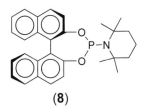

(8)

Allylation.   The multidentate ligand **9** has been developed for regioselective and diastereoselective allylation of ketones.[6]

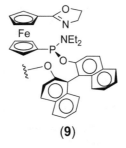

(9)

[1]Li, X., Song, J., Xu, D., Kong, L. *S* 925 (2008).
[2]Shintani, R., Park, S., Duan, W.-L., Hayashi, T. *ACIE* **46**, 5901 (2007).
[3]Trost, B.M., Cramer, N., Silverman, S.M. *JACS* **129**, 12396 (2007).
[4]Trost, B.M., Silverman, S.M., Stambuli, J.P. *JACS* **129**, 12398 (2007).
[5]Du, H., Yuan, W., Zhao, B., Shi, Y. *JACS* **129**, 11688 (2007).
[6]Zheng, W.-H., Zheng, B.-H., Zheng, Y., Hou, X.-L. *JACS* **129**, 7718 (2007).

Rhodium complexes.

Hydrogenation.   Phosphoramidite ligands to make up a Rh catalyst for enantioselective hydrogenation of dehydroamino acid derivatives include **10**, which is derived from 3,3′-bis(diphenylphosphino)-BINOL,[1] and **11** that contains two binaphthyl groups.[2]

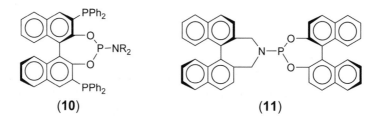

(10)   **(11)**

Hydrogenation of alkenes by Rh catalysis is said to benefit from multidentate sulfonamide-based flexible phosphorus ligands such as **12** that are adaptive to hydrogen bondings.[3]

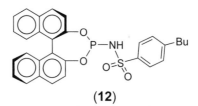

(12)

Hydroboration.   Regioselective and enantioselective hydroboration of 3-alkenamides is accomplished with a Rh-catalyzed process. Formation of C—B bond at C-3 is due to amide group direction, and asymmetric induction originates from ligand **13**.[4]

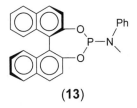

(13)

[1]Zhang, W., Zhang, X. *JOC* **72**, 1020 (2007).
[2]Eberhardt, L., Armspach, D., Matt, D., Toupet, L., Oswald, B. *EJOC* 5395 (2007).
[3]Patureau, F.W., Kuil, M., Sandee, A.J., Reek, J.N.H. *ACIE* **47**, 3180 (2008).
[4]Smith, S.M., Thacker, N.C., Takacs, J.M. *JACS* **130**, 3734 (2008).

1,1'-Binaphthalene-2,2'-diyl phosphates and 3,3'-diaryl analogues.

Hydrogen transfer.   Using Hantzsch ester as hydrogen source imines undergo asymmetric reduction that is catalyzed by BINOL phosphates. The 3,3'-bis(9-anthracenyl)-binaphthyl phosphate *ent*-**1A** mediates the saturation of C=N bond and semihydrogenation of a conjugated triple bond.[1]

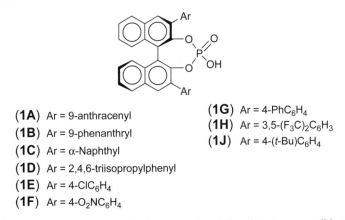

(1A) Ar = 9-anthracenyl
(1B) Ar = 9-phenanthryl
(1C) Ar = α-Naphthyl
(1D) Ar = 2,4,6-triisopropylphenyl
(1E) Ar = 4-ClC$_6$H$_4$
(1F) Ar = 4-O$_2$NC$_6$H$_4$

(1G) Ar = 4-PhC$_6$H$_4$
(1H) Ar = 3,5-(F$_3$C)$_2$C$_6$H$_3$
(1J) Ar = 4-(*t*-Bu)C$_6$H$_4$

Polyheteroaromatic systems containing a fused pyridine ring are susceptible to partial hydrogenation, with the pyridine ring the site of attack. The following examples involve **1A**[2] and 3,3'-bis(9-phenanthrenyl)-BINOL **1B**.[3]

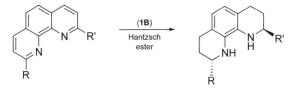

Transfer hydrogenation of quinolines has also been studied using a 3,3-linked dimeric BINOL derivative **2** in which both subunits are phosphorylated.[4]

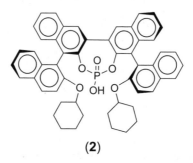

(2)

Due to intramolecular aldol reaction and the following Schiff base formation prior to transfer hydrogenation, *cis*-3-substituted cyclohexylamines are obtained from 1,5-dicarbonyl compounds and ArNH$_2$. The very crowded phosphate catalyst **1D** is used in this transformation.[5]

Addition and cyclization reactions.   Chiral propargylic amines are obtained from alkynylation of imines by catalysis of the silver salt of **1B**.[6] The enantiomer of phosphate **1D** also finds use in the addition of indole to α-acetaminostyrenes.[7] One more catalyst for intramolecular hydroamination to form pyrrolidine derivatives is the silylated **3**.[8] The reaction is conducted at 130°.

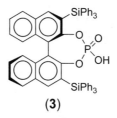

(3)

In the addition of indole to *N*-benzoyl aldimines, **3** also can be put to use.[9] The catalyst **1F** is for promoting reaction between ketene silyl acetals and *N*-(*o*-hydroxyphenyl)aldimines,[10,11] and the 3,3'-dimesityl analogue for vinylogous Mannich reaction.[12] The imines are suitably activated as shown below.

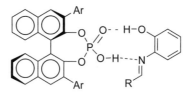

It is not surprising that there are many other studies on analogous combinations that vary in catalyst (e.g., **1C**,[13] **1D**,[14] and **1E**[15]) and substrates.

Formaldehyde hydrazones behave as nucleophiles in the reaction with aldimines. Chiral adducts are produced by conducting the reaction with octahydro-**1B**.[16]

The 3-component condensation for synthesis of 3-acyl-4-aryl-1,4-dihydropyridines from amines, β-dicarbonyl compounds and enals proceeds from enamine formation, Michael reaction and cyclodehydration is amenable to asymmetric induction, such as using *ent*-octahydro-**1B**.[17]

Three Mannich reactions occur in sequence when *N*-Boc aldimines and two equivalents of an *N*-vinylcarbamate are treated with **1G**. 2,4-Diaminopiperidines are formed.[18]

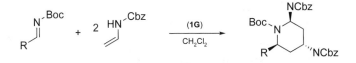

With **1B** ligating to $Rh_2(OAc)_4$ to form a chiral catalyst for inducing insertion of methyl aryl(diazo)acetate into the O-H bond of a benzylic alcohol, it also enables further addition of the product into an aldimine. Both reactions are rendered asymmetric.[19]

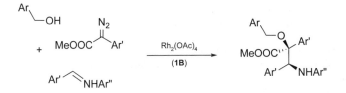

Pictet–Spengler reaction for preparation of tetrahydro-β-carbolines is rendered enantioselective by one of the BINOL-phosphate, as previously reported. A modified version describes the effectiveness of **1H** with tryptamine protected in the form of a triphenylmethanesulfenamide.[20]

Nazarov cyclization is successfully conducted with the phosphoryl triflimide derivative of **1B**.[21] Michael reaction to combine indole and β-nitrostyrene occurs on catalysis of **3** at $-35°$.[22]

Condensation reactions. In highly diastereoselective and enantioselective manner *anti*-adducts are formed when enecarbamates are brought together with ethyl glyoxylate in the presence of **1J**, which activates the formyl group by H-bonding. The adducts furnish chiral ethyl 2-hydroxy-4-oxoalkanoates on workup.[23]

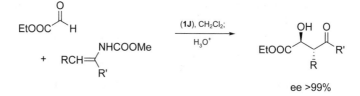

Enantioselective aza-Henry reaction can be conducted with H_8-3.[24]

Miscellaneous reactions. BINOL-phosphate **1D** has also found applications in a Pd-catalyzed allylation of aldehydes by 1-benzhydrylamino-2-alkenes,[25] and epoxidation of enals with *t*-BuOOH.[26]

Azomethine ylides are strongly H-bonded to **4**, therefore 1,3-dipolar cycloaddition is dominated by its chirality.[27]

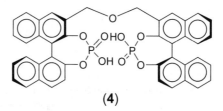

(4)

[1]Kang, Q., Zhao, Z.-A., You, S.-L. *OL* **10**, 2031 (2008).
[2]Rueping, M., Antonchick, A.P. *ACIE* **46**, 4562 (2007).
[3]Metallinos, C., Barrett, F.B., Xu, S. *SL* 720 (2008).
[4]Guo, Q.-S., Du, D.-M., Xu, J. *ACIE* **47**, 759 (2008).
[5]Zhou, J., List, B. *JACS* **129**, 7498 (2007).
[6]Rueping, M., Antonchick, A.P., Brinkmann, C. *ACIE* **46**, 6903 (2007).
[7]Jia, Y.-X., Zhong, J., Zhu, S.-F., Zhang, C.-M., Zhou, Q.-L. *ACIE* **46**, 5565 (2007).
[8]Ackermann, L., Althammer, A. *SL* 995 (2008).
[9]Rowland, G.B., Rowland, E.B., Liang, Y., Perman, J.A., Antilla, J.C. *OL* **9**, 2609 (2007).
[10]Itoh, J., Fuchibe, K., Akiyama, T. *S* 1319 (2008).
[11]Yamanaka, M., Itoh, J., Fuchibe, K., Akiyama, T. *JACS* **129**, 6756 (2007).
[12]Sickert, M., Schneider, C. *ACIE* **47**, 3631 (2008).
[13]Kang, Q., Zhao, Z.-A., You, S.-L. *JACS* **129**, 1484 (2007).
[14]Terada, M., Sorimachi, K. *JACS* **129**, 292 (2007).
[15]Guo, Q.-X., Liu, H., Guo, C., Luo, S.-W., Gu, Y., Gong, L.-Z. *JACS* **129**, 3790 (2007).
[16]Rueping, M., Sugiono, E., Theissmann, T., Kuenkel, A., Köckritz, A., Pews-Davtyan, A., Nemati, N., Beller, M. *OL* **9**, 1065 (2007).
[17]Jiang, J., Yu, J., Sun, X.-X., Rao, Q.-Q., Gong, L.-Z. *ACIE* **47**, 2458 (2008).
[18]Terada, M., Machioka, K., Sorimachi, K. *JACS* **129**, 10336 (2007).
[19]Hu, W., Xu, X., Zhou, J., Liu, W.-J., Huang, H., Hu, J., Yang, L., Gong, L.-Z. *JACS* **130**, 7782 (2008).
[20]Wanner, M.J., van der Haas, R.N.S., de Cuba, K.R., van Marseveen, J.H. *ACIE* **46**, 7485 (2007).
[21]Rueping, M., Ieawsuwan, W., Antonchick, A.P., Nachtsheim, B.J. *ACIE* **46**, 2097 (2007).
[22]Itoh, J., Fuchibe, K., Akiyama, T. *ACIE* **47**, 4016 (2008).
[23]Terada, M., Soga, K., Momiyama, N. *ACIE* **47**, 4122 (2008).

[24]Rueping, M., Antonchick, A.P. *OL* **10**, 1731 (2008).
[25]Mukherjee, S., List, B. *JACS* **129**, 11336 (2007).
[26]Wang, X., List, B. *ACIE* **47**, 1119 (2008).
[27]Chen, X.-H., Zhang, W.-Q., Gong, L.-Z. *JACS* **130**, 5652 (2008).

1,1′-Binaphthalene-2,2′-diyl phosphites.

Hydrogenation. For an effective asymmetric hydrogenation of itaconic esters, the heterocomplex with Rh(I) center associated with two BINOL-derived phosphites of opposite electron-richness (**1A, 1B**) emerges as a more active and selective catalyst.[1] Also having been examined are **2**[2] and the one bearing a carboranyl residue.[3]

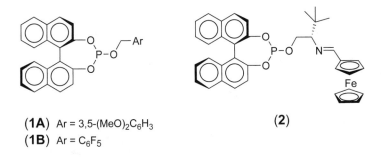

(**1A**) Ar = 3,5-(MeO)$_2$C$_6$H$_3$ (**2**)
(**1B**) Ar = C$_6$F$_5$

With Rh(I) salts heterobidentate ligands such as **3** form catalysts of greater activities than homobidentate ligands **4**, for enantioselective hydrogenation of acrylate and cinnamate esters. The findings are rationalized in terms of conformational and allosteric effects of the substrates.[4]

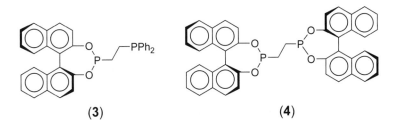

(**3**)   (**4**)

In using the complex [Ir(cod)Cl]$_2$ to conduct hydrogenation of imines, optimal diastereoselectivity is observed with mixed chiral/achiral ligands, as exemplified by BINOL phosphite and Ph$_3$P.[5]

Aminohydroxylation. Enolate anions generated from enol silyl ethers (by CsF) undergo Ag-catalyzed α-aminohydroxylation with PhN=O. An enantioselective version is readily performed in the presence of the phosphite ligand **5**.[6]

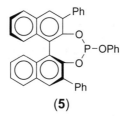

(5)

[1]Lynikaite, B., Cvengros, J., Piarulli, U., Gennari, C. *TL* **49**, 755 (2008).
[2]Gavrilov, K.N., Maksimova, M.G., Zheglov, S.V., Bondarev, O.G., Benetsky, E.B., Lyubimov, S.E., Petrovskii, P.V., Kabro, A.A., Hey-Hawkins, E., Moiseev, S.K., Kolinin, V.N., Davankov, V.A. *EJOC* 4940 (2007).
[3]Lyubimov, S.E., Tyutyunov, A.A., Kalinin, V.N., Said-Galiev, E.-E., Khokhlov, A.R., Petrovskii, P.V., Davankov, V.A. *TL* **48**, 8217 (2007).
[4]Norman, D.W., Carraz, C.A., Hyett, D.J., Pringle, P.G., Sweeney, J.B., Orpen, A.G., Phetmung, H., Wingad, R.L. *JACS* **130**, 6840 (2008).
[5]Reetz, M.T., Bondarev, O. *ACIE* **46**, 4523 (2007).
[6]Kawasaki, M., Li, P., Yamamoto, H. *ACIE* **47**, 3795 (2008).

1,1'-Binaphthalene-2-diarylphosphines.

Silylboration. Methylenecyclopropanes undergo functionalization on breaking the ring at an sp^2—sp^3 bond in a Pd-catalyzed reaction with silylboronates. The products are alkenylboronates.[1]

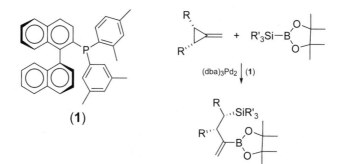

(1)

[1]Ohmura, T., Taniguchi, H., Kondo, Y., Suginome, M. *JACS* **129**, 3518 (2007).

Bis(acetonitrile)dichloropalladium(II).

Coupling reactions. Sequential and chemoselective coupling reactions have been developed for 3-chloro-4-arylthio-1,2-cyclobutenones. Thus a Stille coupling replaces the chlorine atom with a Suzuki coupling to follow. Two different Pd complexes are used.[1]

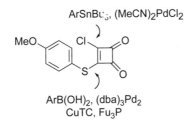

ArB(OH)$_2$, (dba)$_3$Pd$_2$
CuTC, Fu$_3$P

The Pd complex is useful for *B*-arylation of 4,4,6-trimethyl-1,3,2-dioxaborinane with ArI.[2]

Preparation of the very active catalyst (X-Phos)$_2$Pd from (MeCN)$_2$PdCl$_2$ starts by ligand exchange with TMEDA, which is followed by reaction with MeLi at 0°, and further treatment with X-Phos and 2-(*o*-chlorophenyl)ethylamine, prior to warming to room temperature to split off indoline.[3] The precatalyst (before elimination of indoline) can be used directly for *N*-arylation of arylamines.

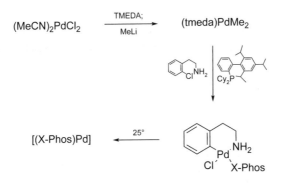

Heterocycle synthesis. Highly substituted furans are obtained from 2-alkylidene-3-alkynones via conjugate addition to generate enols that show nucleophilicity at C-3.[4]

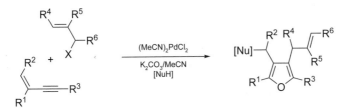

Wacker oxidation of *N*-acylpropargylamines through neighboring group participation gives 5-formyloxazoles.[5]

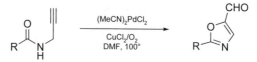

Dihydropyran formation by a 6-endo-dig cyclization from 4-alkynols is achieved with (MeCN)$_2$PdCl$_2$ as catalyst, Pd(OAc)$_2$ is much inferior for structurally complex substrates as that shown below.[6]

R = PMB   60%

Heterocycle formation proceeds from exposure of phenols, aroic acids and amides that contain an *o*-cyclopropyl substituent to (MeCN)$_2$PdCl$_2$ [and benzoquinone as reoxidant of the catalyst].[7]

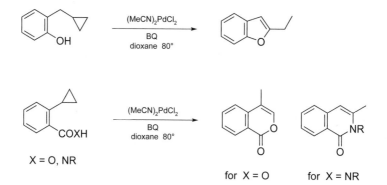

[1]Aguilar-Aguilar, A., Pena-Cabrera, L. *OL* **9**, 4163 (2007).
[2]Murata, M., Oda, T., Watanabe, S., Masuda, Y. *S* 351 (2007).
[3]Biscoe, M.R., Fors, B.P., Buchwald, S.L. *JACS* **130**, 6686 (2008).
[4]Xiao, Y., Zhang, J. *ACIE* **47**, 1903 (2008).
[5]Beccalli, E.M., Borsini, E., Broggini, G., Palmisano, G., Sottocornola, S. *JOC* **73**, 4746 (2008).
[6]Trost, B.M., Ashfeld, B.L. *OL* **10**, 1893 (2008).
[7]He, Z., Yudin, A.K. *OL* **8**, 5829 (2006).

Bis(η³-allyl)dichlorodipalladium.

Addition.   Group transfer from boronic acids to aldehydes can be carried out with the Pd complex in the presence of carbene **1**.[1]

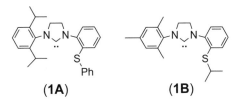

(1A)   **(1B)**

Substitution reactions.   Benzylation of phenols by benzyl methyl carbonates with Pd catalysis proceeds via transesterification and decarbonylation.[2] Triarylmethanes are obtained from a reaction of benzhydryl carbonates with arylboronic acids.[3]

1,3-Transpositional reduction, as pioneered by Tsuji, is a general method for synthesis of 3-methylenated azacycles.[4] Hydrosilanes can be used in reductive defluorination.[5]

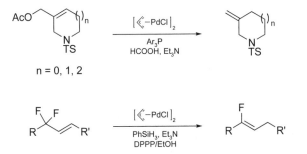

A convenient preparation of 3-amino-1-alkenes involves allylic substitution of primary carbonates with $BnONH_2$, Ph_3CCONH_2, or $Ph_2C{=}NNH_2$, followed by treatment of the products with Zn-HOAc.[6] Related work indicates the critical requirement of a base (DBU) to ensure the generation of branched isomers (conditions: $[(η^3\text{-}C_3H_5)PdCl]_2$, $(EtO)_3P$, THF).[7]

Chiral 2-vinyl-1,2,3,4-tetrahydroquinolines are accessible by a Pd-catalyzed (S_N2') cyclization in the presence of **1**.[8] A remarkable change of enantioselectivity in the Pd-catalyzed allylic substitution with amines by changing the *o*-substituent (CH_2OR to COOMe) to the diphenylphosphino group of the ferrocenyldiphosphine ligand **2**.[9] α-Arylation of aldehydes is performed with $[(η^3\text{-}C_3H_5)PdCl]_2$, in the presence of a ferrocenylphosphine ligand (**3** or **4**).[10]

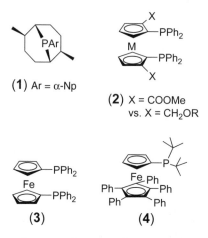

(1) Ar = α-Np

(2) X = COOMe
vs. X = CH$_2$OR

(3)   **(4)**

Coupling reactions. Recent works on coupling reactions mainly address variations of conditions, particularly new ligand and metal combinations. Suzuki coupling has now been conducted with Pd catalyst assisted by the carbene **1B**,[11] and Heck reaction in the presence of all-*cis* 1,2,3,4-tetrakis(diphenylphosphinomethyl)cyclopentane.[12,13]

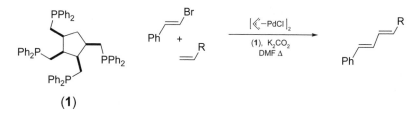

(1)

Alkenylation of benzoxazole and benzothiazole occurs at C-2 under Heck reaction conditions.[14] Cross-coupling of ArX and Ar′Si(Me)$_2$OK is improved by Ph$_3$PO, which serves as a stabilizing ligand for the Pd catalyst.[15]

[1]Kuriyama, M., Shimazawa, R., Shirai, R. *JOC* **73**, 1597 (2008).
[2]Kuwano, R., Kusano, H. *OL* **10**, 1979 (2008).
[3]Yu, J.-Y., Kuwano, R. *OL* **10**, 973 (2008).
[4]Cheng, H.-Y., Sun, C.-S., Hou, D.-R. *JOC* **72**, 2674 (2007).
[5]Narumi, T., Tomita, K., Inokuchi, E., Kobayashi, K., Oishi, S., Ohno, H., Fujii, N. *OL* **9**, 3465 (2007).
[6]Johns, A.M., Liu, Z., Hartwig, J.F. *ACIE* **46**, 7259 (2007).
[7]Dubovyk, I., Watson, I.D.G., Yudin, A.K. *JACS* **129**, 14172 (2007).
[8]Hara, O., Koshizawa, T., Makino, K., Kunimune, I., Namiki, A., Hamada, Y. *T* **63**, 6170 (2007).
[9]Xie, F., Liu, D., Zhang, W. *TL* **49**, 1012 (2008).
[10]Vo, G.D., Hartwig, J.F. *ACIE* **47**, 2127 (2008).
[11]Kuriyama, M., Shimazawa, R., Shirai, R. *T* **63**, 9393 (2007).
[12]Fall, Y., Berthiol, F., Doucet, H., Santelli, M. *S* 1683 (2007).
[13]Lemhadri, M., Battace, A., Berthiol, F., Zair, T., Doucet, H., Santelli, M. *S* 1142 (2008).
[14]Gottumukkala, A.L., Derridj, F., Djebbar, S., Doucet, H. *TL* **49**, 2926 (2008).
[15]Denmark, S.E., Smith, R.C., Tymonko, S.A. *T* **63**, 5730 (2007).

Bis[(η⁶-arene)dichlororuthenium(II)].

Arylation.   Benzalimino compounds are *o*-activated for arylation by a combination of [(η⁶-cymene)RuCl₂]₂ and mesitylenecarboxylic acid.[1]

Carbenoid insertion.[2]   The Ru complex is also effective in forming metal-carbenoids from diazoalkanes for insertion into X—H bonds, as exemplified by the formation of proline derivatives.

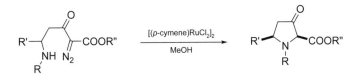

[1]Ackermann, L., Vicente, R., Althammer, A. *OL* **10**, 2299 (2008).
[2]Deng, Q.-H., Xu, H.-W., Yuen, A.W.-H., Xu, Z.-J., Che, C.-M. *OL* **10**, 1529 (2008).

Bis(benzonitrile)dichloropalladium(II).

Coupling reactions.   Under catalysis of (PhCN)₂PdCl₂ Negishi coupling performs better with diphenyl(*o*-chalconyl)phosphine, which is a π-acceptor ligand.[1]

Teraryls are readily synthesized from 1,4-bis(iodozincio)benzene by consecutive Negishi coupling reactions. Such is feasible because of the different reactivity of the two types of arylzinc reagents, products of the first coupling are less reactive toward the Pd catalyst.[2]

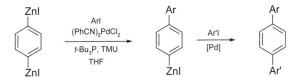

Heck reaction intermediates are intercepted as chlorides by either PhICl₂ or CuCl₂, with interesting regiochemical consequences.[3]

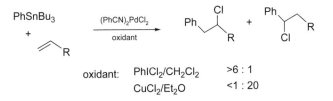

Isomerization.   Overman rearrangement of imino ethers of glycals favors the generation of β-glycosyl amides.[4] Interestingly, more ionic Pd species favor the α-isomers.

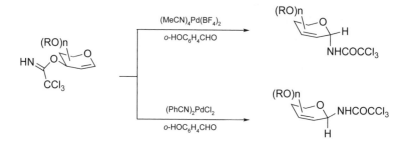

Allenyl ketones undergo cycloisomerization on exposure to $(PhCN)_2PdCl_2$. Dimeric products are produced if an α-substituent (at the allenyl group) is absent from the substrates.[5]

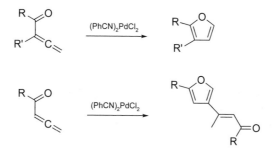

[1]Luo, X., Zhang, H., Duan, H., Liu, Q., Zhu, L., Zhang, T., Lie, A. *OL* **9**, 4571 (2007).
[2]Kawamoto, T., Ejiri, S., Kobayashi, K., Odo, S., Nishihara, Y., Takagi, K. *JOC* **73**, 1601 (2008).
[3]Kalyani, D., Sanford, M.S. *JACS* **130**, 2150 (2008).
[4]Yang, J., Mercer, G.J., Nguyen, H.M. *OL* **9**, 4231 (2007).
[5]Alcaide, B., Almandros, P., del Campo, T.M. *EJOC* 2844 (2007).

Bis[bromotricarbonyl(tetrahydrofuran)rhenium].

[2+2]Cycloadditions. Norbornene and norbornadiene undergo cycloaddition with alkynes from the *exo*-face. 2,6-Diisopropylphenyl isocyanide is provided as a ligand for the Re catalyst.[1]

[1]Kuninobu, Y., Yu, P., Takai, K. *CL* **36**, 1162 (2007).

Bis[chloro(1,5-cyclooctadiene)iridium(I)].

Addition reactions. Catalyzed by [(cod)IrCl]$_2$–BIPHEP, carboxylic acids (as cesium salts) add to the more highly substituted double bond of 1,1-dimethylallene to give α,α-dimethylallyl esters.[1]

From [(cod)IrCl]$_2$ the pincer complex **1** is prepared. It is stable to air and water, and shows catalytic activity for hydroamination (e.g., to form pyrrolidine and piperidine derivatives).[2]

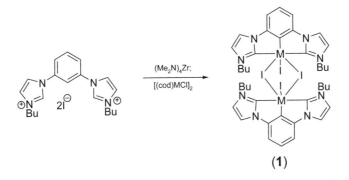

(1)

A more valuable synthetic method based on the Ir(I) complex is the stereoselective addition of alcohols to dienes to afford homoallylic alcohols.[3]

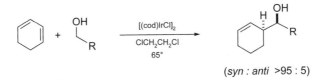

(*syn : anti* >95 : 5)

In derivatization of nucleophilic allylstannanes from allylic alcohols and SnCl$_2$ for the addition to carbonyl compounds, [(cod)IrCl]$_2$ shows superior performance than [(cod)RhCl]$_2$ and (PhCN)$_2$PdCl$_2$.[4]

Redox cyclization.   Allylic alcohol and alkyne units that are separated by several bonds undergo cyclization that involves hydrogen transfer to the triple bond and appearance of a formyl group.[5] The products also can participate in aldol-type condensation.

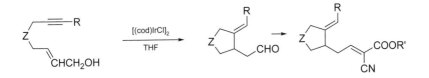

Borylation.   Directed by a silylated heteroatom the borylation opens a new trail for *o*-functionalization of arenes.[6] Allylic boranes derived from alkenes by the same protocol can be used to synthesize homoallylic alcohols and alkenylarenes.[7]

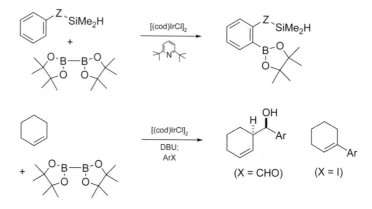

Borylation at the terminal sp^2-carbon atom of allylsilanes with bis(pinacolato)diboron furnishes a valuable reagent for homologation and/or functionalization by two different reactions.[8]

Esters.   Various esters are obtained by mixing RCHO and alcohols with [(cod)IrCl]$_2$ and K$_2$CO$_3$ at room temperature. In the case of allyl alcohol some propyl esters are also formed.[9] Primary alcohols RCH$_2$OH are oxidized to provide esters RCOOCH$_2$R on heating with [(cod)IrCl]$_2$ in open air (95°).[10]

3,5-Disubstituted benzoic esters arise from an Ir-catalyzed [2+2+2]cycloaddition involving propynoic esters and two equivalents of 1-alkynes. Aryl ethynyl sulfones also react similarly.[11]

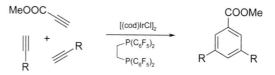

Substitution.   Secondary allylic alcohols are converted to amines by reaction with sulfamic acid, which forms the internal salt [Me$_2$N=CHOSO$_3$] and NH$_3$ to provide both activator and nucleophile.[12]

[1]Kim, I.S., Krische, M.J. *OL* **10**, 513 (2008).
[2]Bauer, E.B., Andavan, G.T.S., Hollis, T.K., Rubio, R.J., Cho, J., Kuchenbeiser, G.R., Helgert, T.R., Letko, C.S., Tham, F.S. *OL* **10**, 1175 (2008).
[3]Bower, J.F., Patman, R.L., Krische, M.J. *OL* **10**, 1033 (2008).
[4]Masuyama, Y., Marukawa, M. *TL* **48**, 5963 (2007).
[5]Kummeter, M., Ruff, C.M., Müller, T.J.J. *SL* 717 (2007).
[6]Boebel, T.A., Hartwig, J.F. *JACS* **130**, 7534 (2008).
[7]Olsson, V.J., Szabo, K.J. *ACIE* **46**, 6891 (2007).
[8]Olsson, V.J., Szabo, K.J. *OL* **10**, 3129 (2008).

[9]Kiyooka, S., Wada, Y., Ueno, M., Yokoyama, T., Yokoyama, R. *T* **63**, 12695 (2007).
[10]Izumi, A., Obora, Y., Sakaguchi, S., Ishii, Y. *TL* **47**, 9199 (2006).
[11]Onodera, G., Matsuzawa, M., Aizawa, T., Kitahara, T., Shimizu, Y., Kozuka, S., Takeuchi, R. *SL* 755 (2008).
[12]Defieber, C., Ariger, M.A., Moriel, P., Carreira, E.M. *ACIE* **46**, 3139 (2007).

Bis[chloro(1,5-cyclooctadiene)rhodium(I)].

Hydrogenation. With bis(dibenzotropyl)amine and a phosphine as coligands to modify [(cod)RhCl]₂ a hydrogenation catalyst is formed. Reduction of alkenes and ketones with this system employs EtOH as hydrogen source.[1]

Addition reactions. Alkenylsilanes[2] and alkenylboronic acids[3] are converted into Rh reagents, which add to conjugated carbonyl compounds.

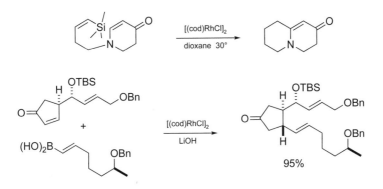

Azolecarbene complexes derived from the title reagent are found to convert alkenes into homologous saturated amines via hydroformylation and reductive amination in one operation.[4] Chloroformates and 1-alkynes combine to give (*Z*)-2-chloroalkenoic esters.[5]

Cycloisomerization. Molecules containing two allene units that are separated by four bonds undergo Rh-catalyzed cycloisomerization. Unsaturated 7-membered ring compounds with two exocyclic double bonds are produced.[6]

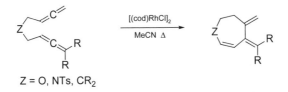

Z = O, NTs, CR₂

o-Ethynylarylamines and phenols cyclize to indoles and benzofurans, respectively, by heating with [(cod)RhCl]₂ and an Ar₃P in DMF at 85°.[7] The presence of either electron-donating or electron-withdrawing substituent(s) in the aromatic moiety has little effect.

Elimination of Ar₂CO.   β,β-Diphenyl-β-hydroxy ketones suffer cleavage to generate Rh enolates, which can be trapped in situ, for example, with RCHO.[8]

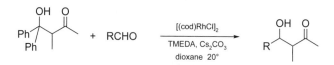

More significantly, a triarylmethanol also lose Ar₂CO and the remaining aryl group is benzannulated by reaction with two equivalents of an alkyne.[9]

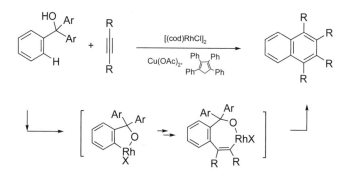

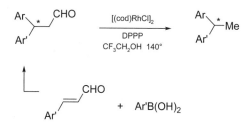

Decarbonylation and carbonylation.   1,1-Diarylethanes (of particular interest are the chiral members) are obtained from decarbonylation of 3,3-diarylpropanals. Such compounds are accessible from cinnamaldehydes in two steps, involving two different Rh-catalyzed reactions.[10]

Co-entrapment of [(cod)RhCl]₂, a sulfonated tertiary phosphine and ionic liquid in silica gel forms a hydroformylation catalyst that is shown to exhibit very high (usually >95%) selectivity for converting styrenes into α-arylacetaldehydes.[11]

Insertion of CO into the OC bond of an oxazolidine gives morpholinones.[12] Cyclopropanes are particularly susceptible to CO insertion via rhodacyclobutane intermediates. Spiro[2.2]pentanes in which the two rings have different degrees of substitution show selective transformations.[13] Cyclopropylcyclopropanes give 4-cycloheptenones.[14]

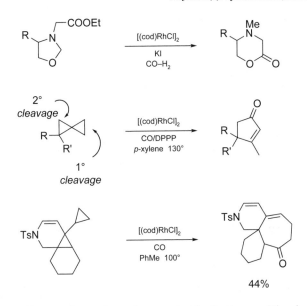

The [(cod)RhCl]₂ complex can be used as a catalyst for the Pauson–Khand reaction under CO.[15] The true catalyst may be formed by exchange of the two ligands (to CO and DPPP).[16]

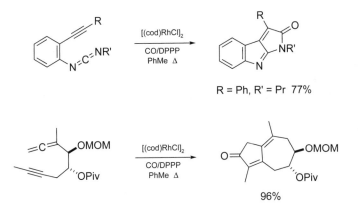

R = Ph, R' = Pr 77%

96%

Arylation of arenes.   On transforming [(cod)RhCl]₂ into a more active cationic Rh(I) species by di(2-pyridyl)aminodiphenylphosphine, reaction between ArX and Ar′H occurs in its presence to give Ar-Ar′.[17]

[1]Zweifel, T., Naubron, J.-V., Büttner, T., Ott, T., Grützmacher, H.-G. *ACIE* **47**, 3245 (2008).
[2]Furman, B., Lipner, G. *T* **64**, 3464 (2008).
[3]Wu, Y., Gao, J. *OL* **10**, 1533 (2008).
[4]Ahmed, M., Buch, C., Routaboul, L., Jackstell, R., Klein, H., Spannenberg, A., Beller, M. *CEJ* **13**, 1594 (2007).

[5]Baek, J.Y., Lee, S.I., Sim, S.H., Chung, Y.K. *SL* 551 (2008).
[6]Lu, P., Ma, S. *OL* **9**, 2095 (2007).
[7]Trost, B.M., McClory, A. *ACIE* **46**, 2074 (2007).
[8]Murakami, K., Ohmiya, H., Yorimitsu, H., Oshima, K. *TL* **49**, 2388 (2008).
[9]Uto, T., Shimizu, M., Ueura, K., Tsurugi, H., Satoh, T., Miura, M. *JOC* **73**, 298 (2008).
[10]Fessard, T.C., Andrews, S.P., Motoyoshi, H., Carreira, E.M. *ACIE* **46**, 9331 (2007).
[11]Hamza, K., Blum, J. *EJOC* 4706 (2007).
[12]Vasylyev, M., Alper, H. *OL* **10**, 1357 (2008).
[13]Matsuda, T., Tsuboi, T., Murakami, M. *JACS* **129**, 12596 (2007).
[14]Kim, S.Y., Lee, S.I., Choi, S.Y., Chung, Y.K. *ACIE* **47**, 4914 (2008).
[15]Saito, T., Sugizaki, K., Otani, T., Suyama, T. *OL* **9**, 1239 (2007).
[16]Hirose, T., Miyakoshi, N., Mukai, C. *JOC* **73**, 1061 (2008).
[17]Proch, S., Kempe, R. *ACIE* **46**, 3135 (2007).

Bis[chloro(dicyclooctene)rhodium(I)].

Coupling reactions. Alkylation of heteroaromatic compounds at a site adjacent to the heteroatom (e.g., N) by alkenes in the presence of [(coe)₂RhCl]₂ is validated for pyridines and quinolines.[1]

For more conventional arylation with ArBr the catalyst system containing a phosphepine ligand (**1**) is recommended.[2]

(1)

Cycloisomerization. Activation of a C—H bond by the Rh complex for intramolecular hydrometallation of a proximal double bond can lead to valuable cyclic products. Examples for such reactions include elaboration of dehydrobenzosuberones from *o*-formylbenzylide-necyclopropanes[3] and of cyclopentane derivatives through addition of an azadiene.[4,5]

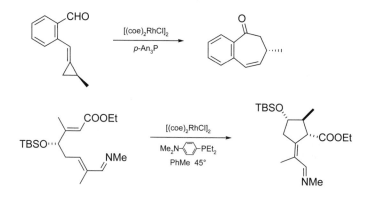

O-Silylation.   Using [(coe)$_2$RhCl]$_2$ as catalyst ROH are silylated by vinylsilanes.[5]

[1]Lewis, J.C., Bergman, R.G., Ellman, J.A. *JACS* **129**, 5332 (2007).
[2]Lewis, J.C., Berman, A.M., Bergman, R.G., Ellman, J.A. *JACS* **130**, 2493 (2008).
[3]Aissa, C., Fürstner, A. *JACS* **129**, 14836 (2007).
[4]Tsai, A.S., Bergman, R.G., Ellman, J.A. *JACS* **130**, 6316 (2008).
[5]Park, J.-W., Jun, C.-H. *OL* **9**, 4073 (2007).

Bis[chloro(diethene)rhodium(I)].

Coupling reactions.   2-Arylpyridines can be arylated by ArB(OH)$_2$, with the Rh complex and the presence of tris(*p*-trifluoromethylphenyl)phosphine and TEMPO.[1]

Heck reaction involving ArBF$_3$K with the Rh complex (and Ph$_3$P) does not require any base.[2]

Vinyl and alkenyl groups attached to the silicon atom of the *o*-hydroxyphenylsilanes are transferred to β-silyl enones on mediation of [(C$_2$H$_4$)$_2$RhCl]$_2$, and the transfer can be rendered enantioselective by adding chiral ligands such as 2,5-diphenylbicyclo[2.2.2]octa-2,5-diene.[3]

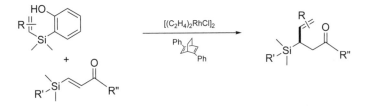

In arylation of *N*-tosylaldimines by ArB(OH)$_2$ the diene ligand is a diphenyl-tetrahydropentalene.[4]

[1]Vogler, T., Studer, A. *OL* **10**, 129 (2008).
[2]Martinez, R., Voica, F., Genet, J.-P., Darses, S. *OL* **9**, 3213 (2007).
[3]Shintani, R., Ichikawa, Y., Hayashi, T., Chen, J., Nakao, Y., Hiyama, T. *OL* **9**, 4643 (2007).
[4]Wang, Z.-Q., Feng, C.-G., Xu, M.-H., Lin, G.-Q. *JACS* **129**, 5336 (2007).

Bis[chloro(norbornadiene)rhodium(I)].

Coupling reactions.   Alkynyl epoxides react with organoboronic acids by a formal S$_N$2' process, yielding allenyl carbinols.[1]

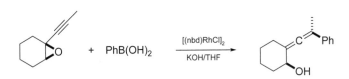

Cycloaddition. A process leading to formation of a 1,3,6-cyclooctatriene system from a conjugated diene and two alkynes is useful. A cationic Rh(I) complex fulfills the catalytic purpose.[2]

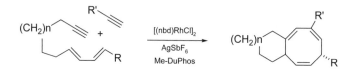

[1]Miura, T., Shimada, M., Ku, S.-Y., Tamai, T., Murakami, M. *ACIE* **46**, 7101 (2007).
[2]DeBoef, B., Counts, W.R., Gilbertson, S.R. *JOC* **72**, 799 (2007).

Bis[chloro(pentamethylcyclopentadienyl)methylthioruthenium] triflate.

Enyne synthesis. Ethynyl cyclopropyl carbinols undergo dehydrative metallation on exposure to the Ru complex, the metallocarbenoids thus formed are attacked by common nucleophile (e.g., H_2O, $ArNH_2$) at a cyclopropyl carbon.[1] The different carbenoids originated from ynoxy triflates engage in *ipso*-substitution.[2]

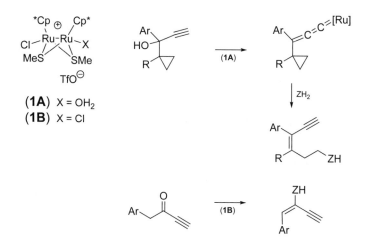

[1]Yamauchi, Y., Onodera, G., Sakata, K., Yuki, M., Miyaka, Y., Uemura, S., Nishibayashi, Y. *JACS* **129**, 5175 (2007).
[2]Yamauchi, Y., Yuki, M., Tanabe, Y., Miyaka, Y., Inada, Y., Uemura, S., Nishibayashi, Y. *JACS* **130**, 2908 (2008).

Bis[(1,5-cyclooctadiene)hydroxyiridium].

Annulation. Synthesis of 1-indanols from *o*-acylarylboronic acids and conjugated dienes involves iridium cycles. While dienes bearing electron-donating or electron-withdrawing substituent(s) are successfully used, the participating double bond is electron-richer.

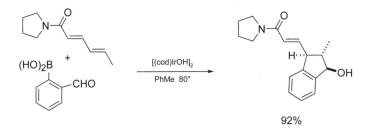

92%

Nishimura, T., Yasuhara, Y., Hayashi, T. *JACS* **129**, 7506 (2007).

Bis[(1,5-cyclooctadiene)hydroxyrhodium].

Reduction.   N-Sulfonyl imines are reduced by o-triorganosilylbenzyl alcohols, which is catalyzed by [(cod)RhOH]₂.[1]

Addition reactions.   The same reagent system is active in hydroarylation and hydroalkenylation of alkynes.[2] The arylsilanes submit the addends and thereby are converted into benzoxasiloles.[3]

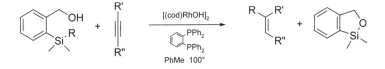

Alkynylsilanes also are active such that dienynes are formed by the reaction of bis(trimethylsilyl)ethyne with alkynes.[3]

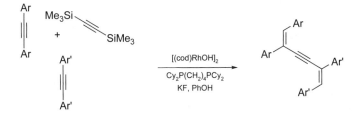

A synthesis of aroylformic esters is based on the addition of ArB(OH)₂ to cyanoformic esters, with H₃BO₃ acting as an additive for the reaction.[4] Addition reations are followed by cyclization as situation prevails, as in the case of the addition of boronic acids to 4-hydroxy-2-alkynoic esters (to give 3-substituted furanones, regiochemically differentiated from the Pd-catalyzed reaction).[5]

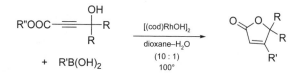

Boronic acids submit the organic groups to *o*-alkynylaryl isocyanates to afford 3-alkylideneoxindoles, the incoming group being *cis*-related to the carbonyl function.[6] Bis(pinacolato)diboron reacts similarly, and apparently the cyclic adducts are available for Suzuki coupling to generate a library of oxindoles.[7]

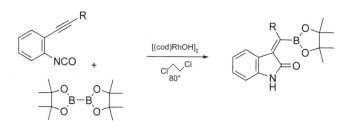

2-Alken-6-yn-1-ones react with organoboronic acids to give 3-acylmethyl-1-cyclo-pentenes containing a 2-substituent arising from the boronic acid.[8]

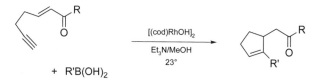

2-Alkylidene-1,3-dithiane *S*-oxides are receptive to addition of boronic acids.[9]

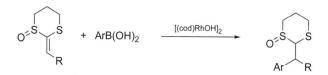

[1]Nakao, Y., Takada, M., Chen, J., Hiyama, T., Ichikawa, Y., Shintani, R., Hayashi, T. *CL* **37**, 290 (2008).
[2]Nakao, Y., Takeda, M., Chen, J., Hiyama, T. *SL* 774 (2008).
[3]Horita, A., Tsurugi, H., Satoh, T., Miura, M. *OL* **10**, 1751 (2008).
[4]Shimizu, H., Murakami, M. *CC* 2855 (2007).
[5]Alfonsi, M., Arcadi, A., Chiarini, M., Marinelli, F. *JOC* **72**, 9510 (2007).
[6]Miura, T., Takahashi, Y., Murakami, M. *OL* **9**, 5075 (2007).
[7]Miura, T., Takahashi, Y., Murakami, M. *OL* **10**, 1743 (2008).
[8]Chen, Y., Lee, C. *JACS* **128**, 15598 (2006).
[9]Yoshida, S., Yorimitsu, H., Oshima, K. *SL* 1622 (2007).

Bis[(1,5-cyclooctadiene)methoxyiridium(I)].

Borylation.   Arenes[1] (including thiophene[2]) are borylated by pinacolatoborane using [(cod)IrOMe]$_2$ as catalyst. The remarkable feature of this reaction is *m*-substitution, through such unusually patterned aromatic compounds become available.[3,4]

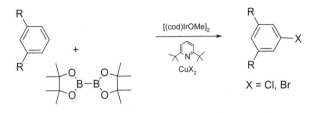

[1]Kikuchi, T., Nobuta, Y., Umeda, J., Yamamoto, Y., Ishiyama, T., Miyaura, N. *T* **64**, 4967 (2008).
[2]Chotana, G.A., Kallepalli, V.A., Maleczka Jr, R.E., Smith III, M.R. *T* **64**, 6103 (2008).
[3]Murphy, J.M., Liao, X., Hartwig, J.F. *JACS* **129**, 15434 (2007).
[4]Murphy, J.M., Tzschucke, C.C., Hartwig, J.F. *OL* **9**, 757 (2007).

Bis[(1,5-cyclooctadiene)methoxyrhodium(I)].

Hydration.   Nitriles are converted to amides at room temperature with aqueous NaOH and catalytic amounts of [(cod)RhOMe]$_2$–Cy$_3$P.[1]

Condensation reactions.   Nitriles activated through coordination to Rh become nucleophilic toward aldehydes in DMSO such that β-hydroxy alkanitriles are formed at room temperature.[2] The Rh complex also promotes transfer reaction of an organoborane to conjugated esters, and those with additional bonding opportunities cyclic structures may be erected.[3]

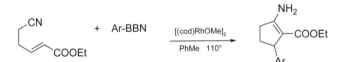

[1]Goto, A., Endo, K., Saito, S. *ACIE* **47**, 3607 (2008).
[2]Goto, A., Endo, K., Ukai, Y., Irle, S., Saito, S. 2212 (2008).
[3]Miura, T., Harumashi, T., Murakami, M. *OL* **9**, 741 (2007).

Bis(1,5-cyclooctadiene)nickel(0).

Addition to C=O bond.   Excellent regioselective addition of organozinc reagents to one of the C=O group of a cyclic anhydride (see equation below) can be attributed to precoordination to the electronically more favorable double bond.[1]

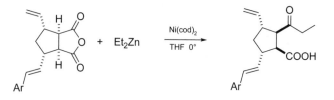

In the reductive aldol reaction *t*-butyl acrylate is formally transformed into an enolate of the propanoate ester. Such a reaction requires PhI in addition to Ni(cod)$_2$ and Et$_3$B.[2]

Unactivated conjugated dienes also undergo reduction with a hydrosilane in situ to form allylating nucleophiles.[3] The double bond of the allyl residue has a (*Z*)-configuration.

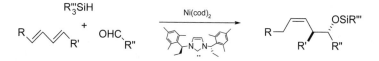

Activation of the C-2 of 1-alkenes with the Ni complex and an azolecarbene enables preparation of α-substituted acrylamides by adding to isocyanate esters.[4] A similar addition of 1-alkenes to ArCHO in the presence of Et$_3$SiOTf is also reported.[5]

Involvement of either a double bond or a triple bond is shown to depend on the substitution status of the alkyne unit, to result in the formation of a common ring or macrocycle, due to preference of activation.[6]

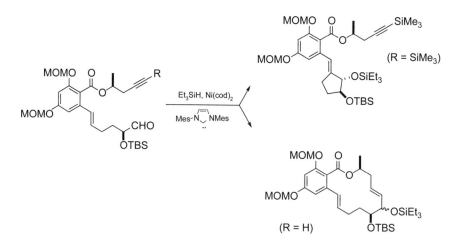

Addition to CC multiple bonds. Allenes are carboxylated at the central carbon by the Ni-catalyzed reaction with CO_2, hydroxylation follows on subsequent exposure to oxygen.[7]

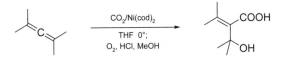

Conjugate addition to α,β-unsaturated ketones[8] and esters[9] by organoboron reagents is accomplished with intervention of Ni(cod)$_2$. Such processes are also subject to asymmetric induction.[7]

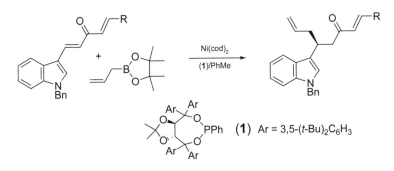

Selective addition to one double bond of a cross-conjugated dienone is attributed to formation of the intermediate with the metal binding to both a η^3-boroxyallyl ligand and a η^1-allyl ligand prior to the allyl group transfer.[10]

The conjugate addition can use bis(pinacolato)diboron to prepare β-boryl esters and amides.[11]

1-Alkenes combine with conjugated aldehydes and ketones in the Michael reaction style when a silyl triflate is present to polarize the acceptors.[12] (Note styrenes are activated at the β-carbon.)

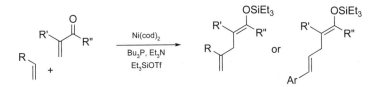

Reductive coupling involving alkynes and conjugated carbonyl compounds generate γ,δ-unsaturated carbonyl compounds.[13] Under somewhat different reaction conditions (mainly with respect to ligand) acrolein participates in the reaction and its formyl group becomes oxidized.[14]

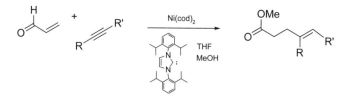

Alkenes such as norbornene and styrene, and also 1,3-dienes add alkynylsilanes, the latter at the terminal double bond to afford branched skipped enynes.[15] Cyanoalkynes split and add to alkynes and allenes to generate conjugated enynes.[16]

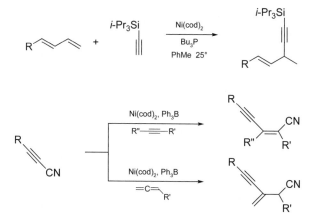

A highly efficient preparation of crotonitrile is by HCN addition to 1,3-butadiene, catalyzed by Ni(cod)₂ in the presence of the bis(diphenylphosphino)triptycene ligand **1**.[17]

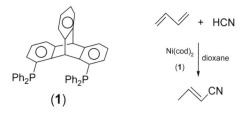

The Ni(cod)₂ – Me₃P reagent is able to split R-CN (alkyl, alkenyl, and aryl nitriles) and deliver the two components to a triple bond. A Lewis acid facilitates the initial process by coordinating to the nitrogen atom of the nitrile.[18]

From allyl sulfides the formation of π-allylnickel species determines the reaction course with alkynes.[19]

Coupling reactions.   In the presence of Ni(cod)₂ – Cy₃P and CsF, cross-coupling between boronic esters and ArOMe can be achieved.[20]

The nickel complex supported by an azolecarbene under basic conditions smooths the preparation of ArSR from ArBr and RSH.[21] C-Arylation of ketones catalyzed by

Ni(cod)₂ – Difluorphos affords much better enantiomer ratios than the reaction using (dba)₂Pd.[22]

The Ni-catalyzed coupling is a superior method for the preparation of unsymmetrical 1,5-dienes, an allylic nucleophile being generated from fragmentation of a tertiary homo-allylic alcohol.[23]

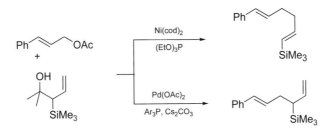

Phthalimides undergo decarbonylative incorporation of alkynes to give isoquinolones.[24] Pyridine *N*-oxides couple with alkynes to provide 2-alkenyl derivatives.[25] With pyridines reaction also occurs but it requires a Lewis acid.[26]

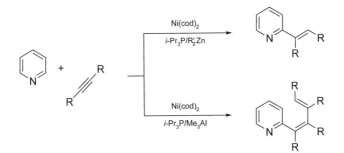

Also catalyzed by Ni(cod)₂ is the coupling reaction involving arynes, alkenes, and boronic acids.[27]

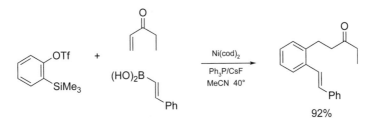

Silacyclobutanes are alkenylated with ring opening, on treatment with 1-alkenes in the presence of Ni(cod)₂ – Cy₃P.[28]

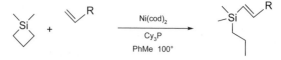

The modified 2:1 Ni-COD complex in which the metal also binds to an imidazolidene unit shows superior selectivity in the Suzuki coupling of polyfluoroarenes. For example, reaction of perfluorotoluene occurs at the *p*-position of the trifluoromethyl group.[29]

Cycloaddition. A notable application of the [2+2+2]cycloaddition of alkynes to form a benzene ring is the preparation of tetrahydrohexahelicenes.[30]

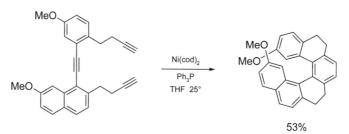

53%

Other useful cycloadditions based on catalysis of the Ni(0) complex include [3+2] and [3+2+2] versions, which produce cyclopentanes,[31] 4-alkenylimidazolidinones,[32] and 5-alkylidene-1,3-cycloheptadienes,[33] respectively.

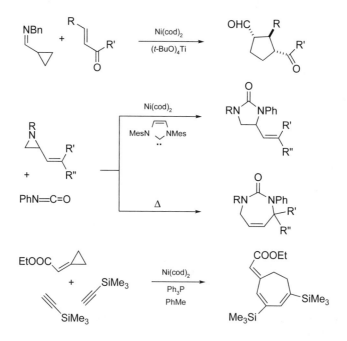

1,2-Dihydropyridines are formed by the Ni-catalyzed cycloaddition, each product being derived from two molecules of alkynes and an *N*-sulfonylaldimine.[34]

[1]Rogers, R.L., Moore, J.L., Rovis, T. *ACIE* **46**, 9301 (2007).
[2]Chrovian, C.C., Montgomery, J. *OL* **9**, 537 (2007).
[3]Sato, Y., Hinata, Y., Seki, R., Oonishi, Y., Saito, N. *OL* **9**, 5597 (2007).
[4]Schleicher, K.D., Jamison, T.F. *OL* **9**, 875 (2007).
[5]Ho, C.-Y., Jamison, T.F. *ACIE* **46**, 782 (2007).
[6]Chrovian, C.C., Knapp-Reed, B., Montgomery, J. *OL* **10**, 811 (2008).
[7]Aoki, M., Izumi, S., Kaneko, M., Ukai, K., Takaya, J., Iwasawa, N. *OL* **9**, 1251 (2007).
[8]Sieber, J.D., Morken, J.P. *JACS* **130**, 4978 (2008).
[9]Hirano, K., Yorimitsu, H., Oshima, K. *OL* **9**, 1541 (2007).
[10]Sieber, J.D., Liu, S., Morken, J.P. *JACS* **129**, 2214 (2007).
[11]Hirano, K., Yorimitsu, H., Oshima, K. *OL* **9**, 5031 (2007).
[12]Ho, C.-Y., Ohmiya, H., Jamison, T.F. *ACIE* **47**, 1893 (2008).
[13]Herath, A., Thompson, B.B., Montgomery, J. *JACS* **129**, 8712 (2007).
[14]Herath, A., Li, W., Montgomery, J. *JACS* **130**, 469 (2008).
[15]Shirakura, M., Suginome, M. *JACS* **130**, 5410 (2008).
[16]Nakao, Y., Hirata, Y., Tanaka, M., Hiyama, T. *ACIE* **47**, 385 (2008).
[17]Bini, L., Muller, C., Wilting, J., von Chrzanowski, L., Spek, A.L., Vogt, D. *JACS* **129**, 12622 (2007).
[18]Nakao, Y., Yada, A., Ebata, S., Hiyama, T. *JACS* **129**, 2428 (2007).
[19]Hua, R., Takeda, H., Onozawa, S., Abe, Y., Tanaka, M. *OL* **9**, 263 (2007).
[20]Tobisu, M., Shimasaki, T., Chatani, N. *ACIE* **47**, 4866 (2008).
[21]Zhang, Y., Ngeow, K.C., Ying, J.Y. *OL* **9**, 3495 (2007).
[22]Liao, X., Weng, Z., Hartwig, J.F. *JACS* **130**, 195 (2008).
[23]Sumida, Y., Hayashi, S., Hirano, K., Yorimitsu, H., Oshima, K. *OL* **10**, 1629 (2008).
[24]Kajita, Y., Matsubara, S., Kurahashi, T. *JACS* **130**, 6058 (2008).
[25]Kanyiva, K.S., Nakao, Y., Hiyama, T. *ACIE* **46**, 8872 (2007).
[26]Nakao, Y., Kanyiva, K.S., Hiyama, T. *JACS* **130**, 2448 (2008).
[27]Jayanth, T.T., Cheng, C.-H. *ACIE* **46**, 5921 (2007).
[28]Hirano, K., Yorimitsu, H., Oshima, K. *JACS* **129**, 6094 (2007).
[29]Schaub, T., Backes, M., Radius, U. *JACS* **128**, 15964 (2006).
[30]Tepley, F., Stara, I.G., Stary, I., Kollarovic, A., Lustinec, D., Krausova, Z., Fiedler, P. *EJOC* 4244 (2007).
[31]Liu, L., Montgomery, J. *OL* **9**, 3885 (2007).
[32]Zhang, K., Chopade, P.R., Louie, J. *TL* **49**, 4306 (2008).
[33]Saito, S., Komagawa, S., Azumaya, I., Masuda, M. *JOC* **72**, 9114 (2007).
[34]Ogoshi, S., Ikeda, H., Kurosawa, H. *ACIE* **46**, 4930 (2007).

Bis(1,5-cyclooctadiene)rhodium(I) salts.

Hydrogenation. Catalyst derived from (cod)$_2$RhOTf and the SEGPHOS analogue **1** is instrumental for hydrogenation of piperitenone to (-)-menthol via pulegone.[1]

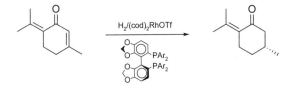

Addition reactions. Hydroboration of styrenes with pinacolatoborane catalyzed by (cod)$_2$RhBF$_4$ furnishes benzylic boranes. The rate of addition is influenced by electronic effects of the nuclear substituents.[2]

Intramolecular hydroamination to afford 5- and 6-membered cyclic amines is also assisted by (cod)$_2$RhBF$_4$, together with a bidentate ligand such as **2** or **3**.[3]

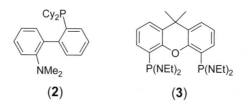

(2)   **(3)**

Reductive hydroxyalkylation starts with enynes under hydrogen (1 atm.) to give dienyl carbinols with a glyoxylic ester.[4] Ethyne forms a rhodacyclopentadiene which serves as 1,3-butadienylating agent for *N*-sulfonyl imines.[5]

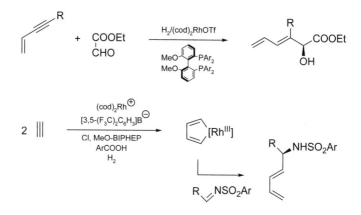

Indanes are formed in a Rh-catalyzed reaction of *o*-bis(3-oxoalkenyl)arenes with RB(OH)$_2$.[6] Formally, there is a conjugate transfer of the R group to one of the enone unit with an intramolecular Michael reaction to follow.

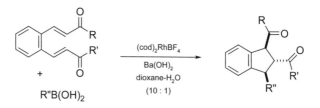

Cycloaddition. A synthesis of phthalides is based on a [2+2+2]cycloaddition of alkynes. The ligand (*R*)-Solphos is used in this case to complement (cod)$_2$RhBF$_4$.[7] The elaboration of dioxotetrahydro[7]helicenes from bis-[2,2′-(propargyloxy)naphthyl]ethyne is also remarkable.[8]

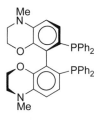

(*R*)-Solphos

Cyclization. Another pattern of cyclization for 1,5-diynes is revealed for the Rh-catalyzed process. The products have a five-membered ring adorned with the cross-conjugated triene system.[9]

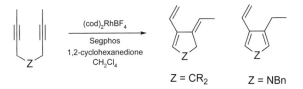

$$Z = CR_2 \qquad Z = NBn$$

N-Propargylarylamines are transformed into 2-methylindoles via a Claisen rearrangement which is followed by an intramolecular hydroamination.[10]

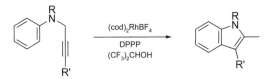

Coupling reactions. Quaternary salts of gramine couple with organoboronic acids to provide indoles with a 3-benzyl or a 3-allyl group.[11]

[1]Ohshima, T., Tadaoka, H., Hori, K., Sayo, N., Mashima, K. *CEJ* **14**, 2060 (2008).
[2]Edwards, D.R., Hleba, Y.B., Lata, C.J., Calhoun, L.A., Crudden, C.M. *ACIE* **46**, 7799 (2007).
[3]Liu, Z., Hartwig, J.F. *JACS* **130**, 1570 (2008).
[4]Hong, Y.-T., Cho, C.-W., Skucas, E., Krische, M.J. *OL* **9**, 3745 (2007).
[5]Skucas, E., Kong, J.R., Krische, M.J. *JACS* **129**, 7242 (2007).
[6]Navarro, C., Csaky, A.G. *OL* **10**, 217 (2008).
[7]Tanaka, K., Osaka, T., Noguchi, K., Hirano, M. *OL* **9**, 1307 (2007).
[8]Tanaka, K., Kamisawa, A., Suda, T., Noguchi, K., Hirano, M. *JACS* **129**, 12078 (2007).

[9]Tanaka, K., Otake, Y., Hirano, M. *OL* **9**, 3953 (2007).
[10]Saito, A., Kanno, A., Hanzawa, Y. *ACIE* **46**, 3931 (2007).
[11]de la Herran, G., Segura, A., Csaky, A.G. *OL* **9**, 961 (2007).

Bis(dibenzylideneacetone)palladium(0).

Arylation.   Oxindole and ester enolates are arylated by ArX, with (dba)$_2$Pd and a bulky phosphine ligand present,[1,2] although in the case dealing with the esters [t-Bu$_3$P · PdBr]$_2$ is equally effective.[3]

An intramolecular version of such arylation pertains to formation of oxindoles.[4]

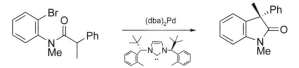

Coupling reactions.   Preparation of oxindoles in which C-3 is fully substituted can be achieved by a Heck reaction, if the neopentyl σ-palladium intermediates are coerced into another coupling reaction. In the context of a synthetic approach to physostigmine and related alkaloids it requires only to supply a cyanide source to complete the task.[5]

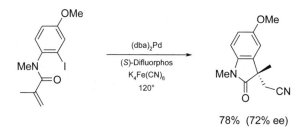

78% (72% ee)

Conversion of ArX to styrenes using the inexpensive divinyltetramethyldisiloxane an activator (KOSiMe$_3$) is added to facilitate the Pd-catalyzed coupling.[6] A procedure of Suzuki coupling in the presence of (dba)$_2$Pd also prescribes the ruthenocene ligand **1**.[7]

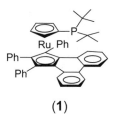

(1)

Coupling of arynes with either simple ArI or 2-iodobiaryls leads to triphenylenes. The two different situations differ in terms of stoichiometry of the reactants (2 : 1 and 1 : 1, respectively).[8]

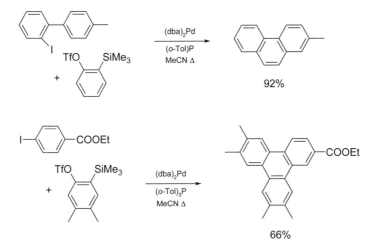

92%

66%

Two alkynes molecules are gathered by the Pd catalyst to react with dialkylamino-(pinacolatoboryl)silanes, consequently 2,4-disubstituted siloles are produced.[9]

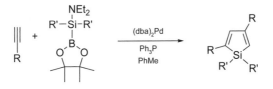

α-(o-Nitroaryl)acrylic esters undergo reductive coupling, in the presence of CO, to afford 3-indolecarboxylic esters.[10]

[1]Durbin, M.J., Willis, M.C. *OL* **10**, 1413 (2008).
[2]Hama, T., Hartwig, J.F. *OL* **10**, 1549 (2008).
[3]Hama, T., Hartwig, J.F. *OL* **10**, 1545 (2008).
[4]Kündig, E.P., Seidel, T.M., Jia, Y., Bernardinelli, G. *ACIE* **46**, 8484 (2007).
[5]Pinto, A., Jia, Y., Neuville, L., Zhu, J. *CEJ* **13**, 961 (2007).
[6]Denmark, S.E., Butler, C.R. *JACS* **130**, 3690 (2008).
[7]Hoshi, T., Nakazawa, T., Saitoh, I., Mori, A., Suzuki, T., Sakai, J., Hagiwara, H. *OL* **10**, 2063 (2008).
[8]Liu, Z., Larock, R.C. *JOC* **72**, 223 (2007).
[9]Ohmura, T., Masuda, K., Suginome, M. *JACS* **130**, 1526 (2008).
[10]Söderberg, B.C.G., Banini, S.R., Turner, M.R., Minter, A.R., Arrington, A.K. *S* 903 (2008).

Bis[dicarbonylchlororhodium(I)].

Coupling reactions.   Electron-rich heteroarenes (furan, thiophene, indole, . . .) couple with ArI using a catalyst derived from [Rh(CO)$_2$Cl]$_2$ and [(CF$_3$)$_2$CHO]$_3$P and Ag$_2$CO$_3$.[1]

Cycloadditions. Many types of cycloaddition are found to be catalyzed by [Rh(CO)$_2$Cl]$_2$. The divergent reaction courses of 2-vinylcyclopropylalkenes due to stereochemical differences are synthetically significant.[2]

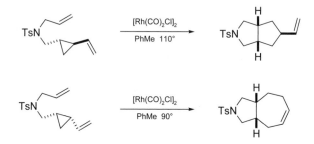

A unique reorganization of the bicyclo[1.1.0]butane unit during its participation in an intramolecular cycloaddition has been recognized.[3]

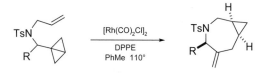

Alkenylcyclopropanes and ethyne combine to give 1,4-cycloheptadienes. The rate of this [5+2]cycloaddition is enhanced by a substituent at C-1, especially a heteroatomic group.[4]

The commonly employed reagent for carbonylative Pauson–Khand reaction is Co$_2$(CO)$_8$, but [Rh(CO)$_2$Cl]$_2$ is a valuable catalyst. When a paste made of powdered 4A-molecular sieves and *t*-BuOH is added to absorb CO, conversion of the substrates is increased.[5]

A cycloctenone synthesis is based on the [5+2+1]cycloaddition in which alkenylcyclopropane, alkene, and CO are the participants.[6] The reaction is carried out under CO and N$_2$ (0.2 and 0.8 atm., respectively). Its synthetic potential is illustrated in an approach to hirsutene.[7]

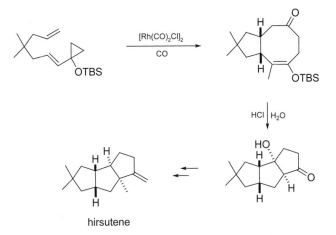

hirsutene

Pauson–Khand reaction of alkynyl ketones in which an allenyl group is extended further from the α'-position is intriguing. It has been found that one of the double bonds of the allene unit can be selected to participate by using certain transition metal catalysts besides modification of the substrates.[8]

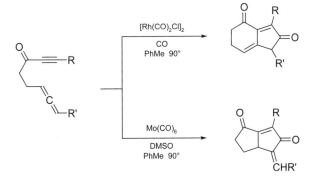

Cycloisomerization. Conjugated alkynones bearing at the α'-position an allenyl substituent (α'-carbon usually quaternary) undergo cyclization to afford cyclopentenones.[9]

[1]Yanagisawa, S., Sudo, T., Noyori, R., Itami, K. *T* **64**, 6073 (2008).
[2]Jiao, L., Ye, S., Yu, Z.-X. *JACS* **130**, 7178 (2008).
[3]Walczak, M.A.A., Wipf, P. *JACS* **130**, 6924 (2008).
[4]Liu, P., Cheong, P.H.-Y., Yu, Z.-X., Wender, P.A., Houk, K.N. *ACIE* **47**, 3939 (2008).

[5]Blanco-Urgoiti, J., Abdi, D., Dominguez, G., Perez-Castells, J. *T* **64**, 67 (2008).
[6]Wang, Y., Wang, J., Su, J., Huang, F., Jiao, L., Liang, Y., Yang, D., Zhang, S., Wender, P.A., Yu, Z.-X. *JACS* **129**, 10060 (2007).
[7]Jiao, L., Yuan, C., Yu, Z.-X. *JACS* **130**, 4421 (2008).
[8]Brummond, K.M., Chen, D. *OL* **10**, 705 (2008).
[9]Brummond, K.M., Chen, D., Painter, T.O., Mao, S., Seifried, D.D. *SL* 759 (2008).

Bis[dicarbonyl(cyclopentadienyl)iron].

Carbodiimide formation. Deoxygenative dimerization of isocyanate esters occurs on heating with [CpFe(CO)$_2$]$_2$ in xylene.[1]

[1]Rahman, A.K.F., Nicholas, K.M. *TL* **48**, 6002 (2007).

Bis[dichloro(1,5-cyclooctadiene)hydridoiridium(II)].

Mannich reaction. Synthesis of β-amino ketones involving ArNH$_2$ is accomplished in DMSO at room temperature in using the Ir complex as catalyst.[1] When β-amino ketones are desired Mannich adducts should be formed from *o*-anisylamine, as they can be dearylated by oxidation with CAN.

[1]Sueki, S., Igarashi, T., Nakajima, T., Shimizu, I. *CL* **35**, 682 (2006).

Bis[dichloro(*p*-cymene)ruthenium(II)].

Cyclization. Diynes such as 1,6-diynes undergo cyclization with incorporation of a RCOOH molecule on warming with the Ru complex and a phosphine ligand.[1]

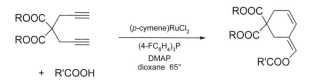

N-Alkylation. The Ru complex turns alcohols into alkylating agents for amines. The reaction of diols such as 1,5-pentanediol gives cyclic amines.[2,3]

[1]Kim, H., Goble, S.D., Lee, C. *JACS* **129**, 1030 (2007).
[2]Hamid, M.H.S.A., Williams, J.M.J. *TL* **48**, 8263 (2007).
[3]Hamid, M.H.S.A., Williams, J.M.J. *CC* 725 (2007).

Bis[dichloro(pentamethylcyclopentadienyl)iridium(II)].

Substitution. Alcohols are transformed into secondary and tertiary amines in the Ir-catalyzed reaction with an ammonium salt. Remarkably, the counter-anion of the ammonium salt determines the extent of *N*-alkylation.[1]

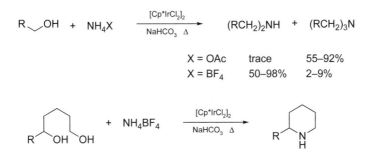

$$X = OAc \quad trace \quad 55-92\%$$
$$X = BF_4 \quad 50-98\% \quad 2-9\%$$

N-Alkylation of primary and secondary amines is also accomplished.[2]

Oxidative amination.[3] A different reaction pathway is adopted in the reaction of primary alcohols with hydroxylamine hydrochloride under the influence of [Cp*IrCl₂]₂. Dehydrogenation of the alcohols (to form RCHO) and oximation are followed by a rearrangement step, which leads to RCONH₂.

Annulation. Aroic acids with a free *o*-position incorporate two equivalents of an alkynes to form a benzene ring. Decarboxylation is arrested if the catalyst is switched from [Cp*IrCl₂]₂ to [Cp*RhCl₂]₂ (also there is a change of the auxiliary metal salt).[4]

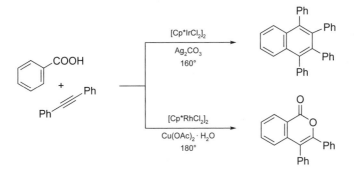

2,3-Disubstituted indoles are obtained from a reaction of *o*-aminobenzyl carbinols or *o*-nitrobenzyl carbinols and a primary (preferably benzyl) alcohol.[5] Redox transformation of various functional groups and proper condensation thereof lead to the results.

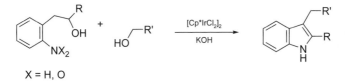

X = H, O

[1]Yamaguchi, R., Kawagoe, S., Asai, C., Fujita, K. *OL* **10**, 181 (2008).
[2]Fujita, K., Enoki, Y., Yamaguchi, R. *T* **64**, 1943 (2008).
[3]Owston, N.A., Parker, A.J., Williams, J.M.J. *OL* **9**, 73 (2007).
[4]Ueura, K., Satoh, T., Miura, M. *JOC* **72**, 5362 (2007).
[5]Whitney, S., Grigg, R., Derrick, A., Keep, A. *OL* **9**, 3299 (2007).

Bis[dichloro(pentamethylcyclopentadienyl)rhodium(II)].

Annulation. Starting from heteroatom-directed *o*-metallation, two molecules of alkynes are incorporation into the benzene ring to form naphthalenes.[1]

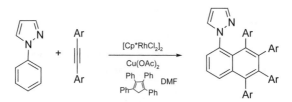

[1]Umeda, N., Tsurugi, H., Satoh, T., Miura, M. *ACIE* **47**, 4109 (2008).

2,2′-Bis(diphenylphosphino)-1,1′-binaphthyl and analogues.

Copper complexes.

Addition. In the presence of a chiral BINAP to coordinate with Cu(OTf)$_2$ a useful catalyst for the addition of diorganozincs to *N*-(2-pyridinesulfonyl) aldimines is achieved.[1]

Grignard reagents perform enantioselective conjugate addition to α,β-unsaturated esters in the presence of a CuI complex of Tol-BINAP.[2]

[1]Desrosiers, J.-N., Bechara, W.S., Charette, A.B. *OL* **10**, 2315 (2008).
[2]Wang, S.-Y., Ji, S.-J., Loh, T.-P. *JACS* **129**, 276 (2007).

Gold complexes.

Cycloisomerization.[1] Gold salts and complexes are popular catalysts for organic transformations because it is found that the metal has high affinity to allenes and alkynes. A gold ion usually requires stabilization of a phosphine. As shown by the cyclization of 1,2,7-alkatrienes, BINAP and its congners are adequate ligands.

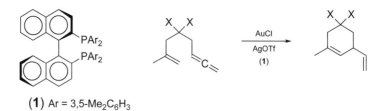

(**1**) Ar = 3,5-Me$_2$C$_6$H$_3$

Hydroamination.[2]   On complexing to (R)-xylyl-BINAP gold p-nitrobenzoate activates a double bond of an allene moiety to allow intramolecular attack by an amino group, asymmetrically.

[1]Tasselli, M.A., Chianese, A.R., Lee, S.J., Gagne, M.R. *ACIE* **46**, 6670 (2007).
[2]LaLonde, R.L., Sherry, B.D., Kang, E.J., Toste, F.D. *JACS* **129**, 2452 (2007).

Iridium complexes.

Hydrogenation.   For asymmetric hydrogenation of 2-substituted quinolines to give the tetrahydro derivatives a catalyst is created from [(cod)IrCl]$_2$ and dendrimers with a 5,5′-carboxamido-BINAP core for enhanced activity.[1]

Coupling reactions.   The [(cod)IrCl]$_2$ – BINAP specimen transforms allyl acetate into a π-allyliridium complex. Reaction with a primary alcohol or an aldehyde affords homoallylic alcohol in chiral form.[2]

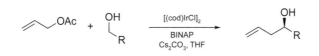

The ability of Ir complexes in performing dehydrogenation/hydrogenation is exploitable in that a primary alcohol acts as an alkylating agent for certain Wittig reagents.[3] An emerging aldehyde is intercepted by the Wittig reagent and hydrogenation of the resulting alkene completes the process. With a chiral BINAP ligand the iridium complex mediates hydrogen transfer while the hydrogenation step is rendered enantioselective.

$$Ph\!\!\nearrow\!\!OH \; + \; Ph_3P\!=\!\!\!\!<\!\!^{COOEt} \xrightarrow[\substack{(S)\text{-BINAP} \\ PhMe \, \Delta}]{[(cod)IrCl]_2} Ph\!\!\diagdown\!\!\diagup\!\!COOET$$

(87% ee)

[1]Wang, Z.-J., Deng, G.-J., Li, Y., He, Y.-M., Tang, W.-J., Fan, Q.-H. *OL* **9**, 1243 (2007).
[2]Kim, I.S., Ngai, M.-Y., Krische, M.J. *JACS* **130**, 6340 (2008).
[3]Shermer, D.J., Slatford, P.A., Edney, D.D., Williams, J.M.J. *TA* **18**, 2845 (2007).

Palladium complexes.

Alkylation.   Acetals of enals serve as alkylating agents for t-butyl β-keto esters using a Pd complex of BINAP.[1] With a cationic Pd complex of (R)-BINAP intramolecular addition of arylboronic acid moiety to a sidechain ketone leads to chiral, tertiary benzylic alcohols.[2]

Heck reaction.   Cyclization with desymmetrization is shown to proceed in excellent yields and ee by the formation of tetralin derivatives.[3]

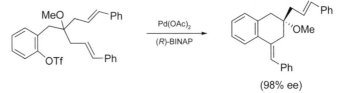

(98% ee)

With Ag₃PO₄ as additive for an intramolecular Heck reaction to form 3,3-disubstituted oxindoles considerable variation of enantioselectivity and direction of asymmetric induction is observed.[4]

[1]Umebayashi, N., Hamashima, Y., Hashizume, D., Sodeoka, M. *ACIE* **47**, 4196 (2008).
[2]Liu, G., Lu, X. *JACS* **128**, 16504 (2006).
[3]Machotta, A.B., Straub, B.F., Oestreich, M. *JACS* **129**, 13455 (2007).
[4]McDermott, M.C., Stephenson, G.R., Walkington, A.J. *SL* 51 (2007).

Platinum complexes.

Aldol reaction.[1]   A cationic Pt(II) salt complexed to BINAP is active in catalyzing the Mukaiyama aldol reaction in DMF. However, ee are not as high as desired.

[1]Kiyooka, S., Matsumoto, S., Kojima, M., Sakonaka, K., Maeda, H. *TL* **49**, 1589 (2008).

Rhodium complexes.

Kinetic resolution.   A method for kinetic resolution of *t*-homoallylic alcohols is hinged on selective cleavage of one enantiomeric series of compounds by a Rh complex of chiral octahydro-BINAP.[1]

Isomerization.   Heating *N*-allylaziridines with (cod)₂RhOTf and *rac*-BINAP causes migration of the double bond to produce (*Z*)-propenylaziridines.[2]

Addition reactions.   By intramolecular hydroacylation in an ionic liquid, indanones are prepared in a Rh-catalyzed reaction, the metal ion in use is ligated to (*R*)-BINAP.[3]

The influence of a ligand on the Rh-catalyzed addition of RAlMe₂ to 2-cycloalkenones can be quite profound. The 1,2-addition in the presence of BINAP is switched over to the 1,4-addition mode when the ligand is omitted.[4]

Chiral dibenzylacetic esters are accessible from conjugate addition of ArB(OH)₂ to *t*-butyl α-benzylacrylates when protonation with B(OH)₃ is rendered enantioselective, by the presence of a chiral BINAP.[5]

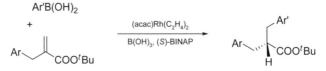

A similar strategy of enantioselective protonation affords α-amino esters from synthesis involving conjugate addition of RBF₃K to N-protected α-aminoacrylic esters.[6]

Asymmetry is directly established at the β-carbon of an α,β-unsaturated carbonyl compound during silyl group transfer from a (pinacolatoboryl)silane, which is catalyzed by a Rh-BINAP complex.[7]

Cycloaddition.   Different versions of [2+2+2]cycloaddition are known to be induced by cationic Rh(I) salts with support of BINAP ligands.[8,9] Diynes combining with enol ethers lead to products containing a new benzene ring,[10] and ring fused dihydropyrans are formed from enynes and α-dicarbonyl compounds.[11]

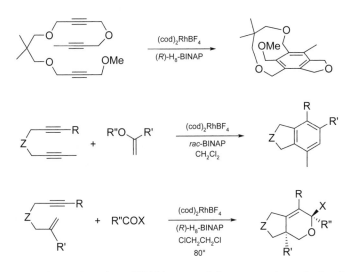

Reorganizations.   Insertion of [Rh] into a cyclobutanone unit can lead to interesting consequences. Dihydrocoumarins are obtained from 3-(o-hydroxyaryl)cyclobutanones.[12] As metal migration occurs from the rhodacyclopentane intermediates to the aromatic ring, site-selective functionalization is achieved.

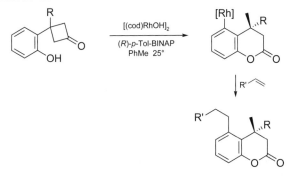

Asymmetric rearrangement of alkenyl alkynyl carbinols with a chiral Rh-BINAP catalyst furnishes β-alkynyl ketones.[13] The transformation is synthetically equivalent to enantioselective conjugate addition of the alkynyl unit, saving the asymmetric induction delegated to a different operation.

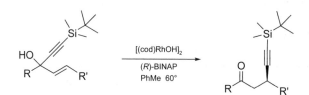

Rearrangement of dialkynylbenzyl alcohols gives 3-alkynylindanones.[14]

Coupling reactions.   Cyclization of 1,6-enynes and diynes with concomitant aryl-coupling is realized using diaryl ketones (possessing a free *o*-CH).[15] *o*-Acylarylrhodium hydride species are formed to initiate hydrometallation at the triple bond.

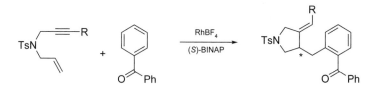

Cyclization involving sp^2-sp coupling from an allylic alcohol and an ynoate segments that forms a cyclopentane ring substituted by two functional sidechains in adjacent positions is a key step in a synthesis of (-)-platensimycin.[16]

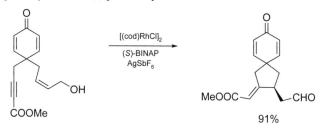

[1]Shintani, R., Takatsu, K., Hayashi, T. *OL* **10**, 1191 (2008).
[2]Tsang, D.S., Yang, S., Alphonse, F.-A., Yudin, A.K. *CEJ* **14**, 886 (2008).
[3]Oonishi, Y., Ogura, J., Sato, Y. *TL* **48**, 7505 (2007).
[4]Siewert, J., Sandmann, R., von Zezschwitz, P. *ACIE* **46**, 7122 (2007).

[5]Frost, C.G., Penrose, S.D., Lambshead, K., Raithby, P.R., Warren, J.E., Gleave, R. *OL* **9**, 2119 (2007).
[6]Navarre, L., Martinez, R., Genet, J.-P., Darses, S. *JACS* **130**, 6159 (2008).
[7]Walter, C., Oestreich, M. *ACIE* **47**, 3818 (2008).
[8]Tanaka, K. *SL* 1977 (2007).
[9]Tanaka, K., Sagae, H., Toyoda, K., Noguchi, K., Hirano, M. *JACS* **129**, 1522 (2007).
[10]Hara, H., Hirano, M., Tanaka, K. *OL* **10**, 2537 (2008).
[11]Tanaka, K., Otake, Y., Sagae, H., Noguchi, K., Hirano, M. *ACIE* **47**, 1312 (2008).
[12]Matsuda, T., Shigeno, M., Murakami, M. *JACS* **129**, 12086 (2007).
[13]Nishimura, T., Katoh, T., Takatsu, K., Shintani, R., Hayashi, T. *JACS* **129**, 14158 (2007).
[14]Shintani, R., Takatsu, K., Katoh, T., Nishimura, Y., Hayashi, T. *ACIE* **47**, 1447 (2008).
[15]Tsuchikama, K., Kuwata, Y., Tahara, Y., Yoshinami, Y., Shibatas, T. *OL* **9**, 3097 (2007).
[16]Nicolaou, K.C., Edmonds, D.J., Li, A., Tria, G.S. *ACIE* **46**, 3942 (2007).

Ruthenium complexes.

Asymmetric hydrogenation. Hydrogenation of *t*-butyl β-ketoalkanoates with (binap)$_2$RuCl$_2$ is selective in the presence of the corresponding hexafluoroisopropyl esters (which is practically unreduced).[1] The turnover rates for the hydrogenation of several β-keto-carboxylic acid derivatives have been determined, hydrogenation of amides (pyrrolidine and piperidine > diethylamine) is generally more facile than esters.[2]

Fluorous BINAP ligands such as **1**, prepared from the bromo-BINOL precursor(s), coordinate with RuCl$_2$ to form reusable catalysts that have shown activities in hydrogenation of α-substituted acrylic esters and dehydroamino esters.[3]

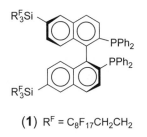

(1) R^F = C$_8$F$_{17}$CH$_2$CH$_2$

In the presence of **2**, aryl ketones in which the α-carbon carries a heteroatom substituent undergo hydrogenation diastereoselectively and enantioselectively.[4]

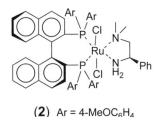

(2) Ar = 4-MeOC$_6$H$_4$

Catalysts constituting a C_2-symmetric 1,2-diamine have been used to hydrogenate α-aryl aldehydes to yield chiral alcohols, under dynamic kinetic resolution conditions.[5] Hydrogenation of the carbonyl group of acylsilanes with **3** (presence of *t*-AmOK or NaBH$_4$ as activator) is applicable to acquisition of α-silyl allylc alcohols from conjugated acylsilanes.[6]

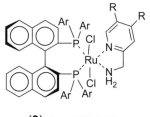

(3) Ar = 4-MeC$_6$H$_4$

[1]Kramer, R., Brückner, R. *ACIE* **46**, 6537 (2007).
[2]Kramer, R., Brückner, R. *CEJ* **13**, 9076 (2007).
[3]Horn, J., Bannwarth, W. *EJOC* 2058 (2007).
[4]Arai, N., Ooka, H., Azuma, K., Yabuuchi, T., Kurono, N., Inoue, T. *OL* **9**, 939 (2007).
[5]Li, X., List, B. *CC* 1739 (2007).
[6]Arai, N., Suzuki, K., Sugizaki, S., Sorimachi, H., Ohkuma, T. *ACIE* **47**, 1770 (2008).

Silver complexes.

Hydroxyalkylation.   The complex of AgOTf with (*S*)-BINAP is used in enantioselective reaction of 3-trimethylsilyl-1,4-cyclohexadiene with ArCHO.[1] It is important to note the regiochemical aspect in its application to unsymmetrical pronucleophiles. The products are converted into chiral benzhydrols on dehydrogenation with DDQ.

A review on the use of AgX – BINAP complexes in synthesis has been written.[2]

[1]Umeda, R., Studer, A. *OL* **10**, 993 (2008).
[2]Yanagisawa, A., Arai, T. *CC* 1165 (2008).

[Bis(*o*-diphenylphosphinobenzylidene)ethanediamine]dichlororuthenium(II).

Reduction.[1]   Under hydrogenation conditions the title ruthenium complex reduces an ester to a primary alcohol without affecting a double bond.

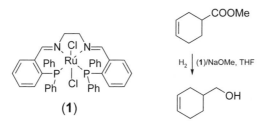

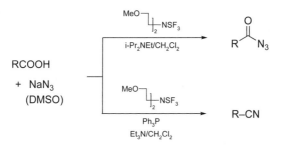

[1]Saudan, L.A., Saudan, C.M., Debieux, C., Wyss, P. *ACIE* **46**, 7473 (2007).

Bis(ethene)trispyrazolylboratoruthenium.

Hydroamination.[1] Derivatization of 1-alkynes into either imines or enamines by RNH_2 and R_2NH, respectively, in the anti-Markovnikov sense, is accomplished by heating the mixtures with $TpRu(C_2H_4)_2$ and Ph_3P in toluene at $100°$.

[1]Fukumoto, Y., Asai, H., Shimizu, M., Chatani, N. *JACS* **129**, 13792 (2007).

Bis(iodozincio)methane.

Homoenolate ions.[1] Cyclopropyloxyzinc iodides are generated from α-sulfonyloxy carbonyl compounds on reaction with $CH_2(ZnI)_2$.

[1]Nomura, K., Matsubara, S. *CL* **36**, 164 (2007).

Bis(2-methoxyethyl)aminosulfur trifluoride, Deoxo-Fluor.

Azides and nitriles. Carboxylic acids are activated (to form RCOF) by Deoxo-Fluor for conversion into acyl azides on reaction with NaN_3. Nitriles are formed by slight variation of conditions. Usually DAST can be used but the latter reagent is thermally less stable.[1]

Of particular interest is the reaction profile of arylacetic acids that is dependent on an additive.[2]

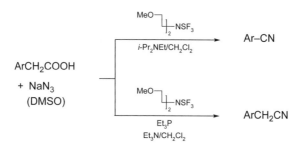

[1]Kangani, C.O., Day, B.W., Kelley, D.E. *TL* **48**, 5933 (2007).
[2]Kangani, C.O., Day, B.W., Kelley, D.E. *TL* **49**, 914 (2008).

Bis(4-methoxyphenyl)-1,3-dithia-2,4-diphosphetane-2,4-disulfide, Lawesson's reagent.

Benzothiazoles.   Heating *o*-halobenzanilides with Lawesson's reagent and Cs$_2$CO$_3$ in xylene leads to the formation of benzothiazoles.[1]

[1]Bernardi, D., Ba, L.A., Kirsch, G. *SL* 2121 (2007).

Bismuth(III) sulfate.

Friedel–Crafts reaction.   Active arenes (phenols, aryl ethers, aryl sulfides, . . .) are alkylated by *N*-tosyl aldimines (and benzylamines) such that 1,1-diarylalkanes result, with promotion by Bi$_2$(SO$_4$)$_3$–Me$_3$SiCl.[1]

[1]Liu, C.-R., Li, M.-B., Yang, C.-F., Tian, S.-K. *CC* 1249 (2008).

Bismuth(III) triflate.

Rearrangement.[1]   Acetates of Baylis–Hillman adducts undergo 1,3-migration of the acetoxy group on heating with Bi(OTf)$_3$ · 4H$_2$O in MeCN.

Substitution.   The Lewis acidity of Bi(OTf)$_3$ caters to use in activating electron-rich benzyl ethers and acetates to react with enol silyl ethers.[2] By the same token, only the alcohols (instead of halides) are needed in allylation and benzylation of 1,3-dicarbonyl compounds.[3]

N-Alkylation of sulfonamides occurs at room temperature with benzylic, allylic and propargylic alcohols in the presence of Bi(OTf)$_3$ and KPF$_6$.[4]

Addition.   Assistance is rendered by Bi(OTf)$_3$ · 4H$_2$O to aldehydes to convert them into homoallylic alcohols with allyltributylstannane, under microwave irradiation.[5]

Cyclization.   Ionization of an allylic bromide in the presence of Bi(OTf)$_3$, to trigger π-participation, leading to a cyclized product is expected.[6]

Cyclization of 4-alkynoic acids to give enol lactones shows effects of a terminal substituent.[7]

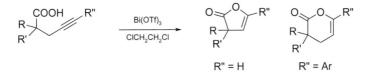

[1]Olleivier, T., Mwene-Mbeja, T.M. *T* **64**, 5150 (2008).
[2]Rubenbauer, P., Bach, T. *TL* **49**, 1305 (2008).
[3]Rueping, M., Nachtsheim, B.J., Kuenkel, A. *OL* **9**, 825 (2007).
[4]Qin, H., Yamagiwa, N., Matsunaga, S., Shibasaki, M. *ACIE* **46**, 409 (2007).
[5]Ollevier, T., Li, Z. *EJOC* 5665 (2007).
[6]Hayashi, R., Cook, G.R. *TL* **49**, 3888 (2008).
[7]Komeyama, K., Takahashi, K., Takaki, K. *CL* **37**, 602 (2008).

Bis(naphtho[2,1-*c*])azepines.

Aldol reaction. Aldol reaction in the presence of the chiral *N*-prolylbis(naphtho[2,1-*c*])azepine **1** is benefited by high diastereoselectivity and enantioselectivity.[1] Formation of 2-substituted 3-aryl-3-hydroxypropanals from aliphatic and aromatic aldehydes proceeds well when effectuated by **2**.[2] For accomplishing enantioselective nitroaldol reaction at room temperature the use of a complex derived from Cu(OAc)$_2$ and the diamine **3** can be used.[3]

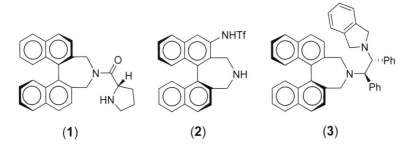

(1)   (2)   (3)

Substitutions. Bis(naphtho[2,1-c])azepine **4** containing two bulky substituents directs enantioselective α-iodination of RCH$_2$CHO with NIS.[4] Quaternary ammonium salt **5** is a chiral phase-transfer agent with proven utility in the alkylation of glycine derivatives.[5]

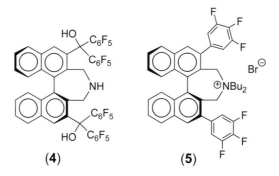

(4)   (5)

Addition reactions. Another quaternary ammonium salt **6** promotes asymmetric replacement of the tosyl residue from α-aminoalkyl *p*-tolyl sulfones by a cyano group, via addition of KCN to the imines generated in situ.[6]

Michael reaction of α-substituted *t*-butyl cyanoacetates to *t*-butyl propynoate establishes a quaternary carbon center in the adducts. Excellent asymmetric induction is achieved by much more bulky ammonium salt.[7]

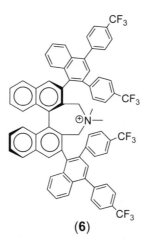

(6)

Hydrazone **7** behaves as a bidentate ligand for zinc species. However, the steric effect of the BINAP moiety on organozinc addition to ArCHO is not sufficient to give good ee of the adducts.[8]

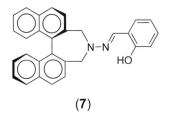

(7)

[1]Li, X.-J., Zhang, G.-W., Wang, L., Hua, M.-Q., Ma, J.-A. *SL* 1255 (2008).
[2]Kano, T., Yamaguchi, Y., Tanaka, Y., Maruoka, K. *ACIE* **46**, 1738 (2007).
[3]Arai, T., Watanabe, M., Yanagisawa, A. *OL* **9**, 3595 (2007).
[4]Kano, T., Ueda, M., Maruoka, K. *JACS* **130**, 3728 (2008).
[5]Kitamura, M., Arimura, Y., Shirakawa, S., Maruoka, K. *TL* **49**, 2026 (2008).
[6]Ooi, T., Uematsu, Y., Fujimoto, J., Fukumoto, K., Maruoka, K. *TL* **48**, 1337 (2007).
[7]Wang, M., Kitamura, M., Maruoka, K. *JACS* **129**, 1038 (2007).
[8]Arai, T., Endo, Y., Yanagisawa, A. *TA* **18**, 165 (2007).

Bis(naphtho[2,1-c])phosphepins.

Hydrogenation.   The phosphepin ligands **1** and **2**, constituting with Rh and Pd respectively, mediate asymmetric hydrogenation of enol carbamates[1] and cyclic sulfamidates.[2]

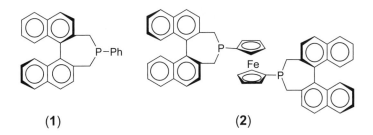

(1)   **(2)**

The 2-pyridone unit in ligand **3** has a high tendency to associate through intermolecular hydrogen bonding. To this dimeric structure the binding of Rh creates a hydrogenation catalyst that performs well in propylene carbonate.[3]

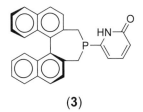

(3)

[1]Enthaler, S., Erre, G., Junge, K., Michalik, D., Spaqnnenberg, A., Marras, F., Gladiali, S., Beller, M. *TA* **18**, 1288 (2007).
[2]Wang, Y.-Q., Yu, C.-B., Wang, D.-W., Wang, X.-B., Zhou, Y.-G. *OL* **10**, 2071 (2008).
[3]Schäffner, B., Hotz, J., Verevkin, S.P., Börner, A. *TL* **49**, 768 (2008).

Bis(pentafluorophenyl)[4-dimesitylphosphino-2,3,5,6-tetrafluorophenyl]borane.

Hydrogenation.[1] The title reagent **1** activates molecular hydrogen such that imines and nitriles are saturated.

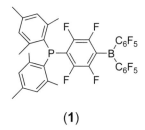

(1)

[1]Chase, P.A., Welch, G.C., Jurca, T., Stephan, D.W. *ACIE* **46**, 8050 (2007).

1,1′;3,3′-Bispropanediyl-2,2′-diimidazolylidene.

Reduction. The superelectron donor **1**, readily prepared from imidazole and 1,3-diiodopropane, reduces aryl halides[1] and promotes desulfonylation[2] in DMF.

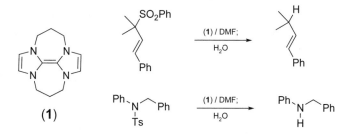

(1)

[1]Murphy, J.A., Zhou, S., Thomson, D.W., Schoenebeck, F., Mahesh, M., Park, S.R., Tuttle, T., Berlouis, L.E.A. *ACIE* **46**, 5178 (2007).
[2]Schoenebeck, F., Murphy, J.A., Zhou, S., Uenoyama, Y., Miclo, Y., Tuttle, T. *JACS* **129**, 13368 (2007).

Bis(trialkylphosphine)palladium.

Coupling reaction. Arylation on the silicon atom of a hydrosilane is shown to be catalyzed by $(t\text{-}Bu_3P)_2Pd$.[1]

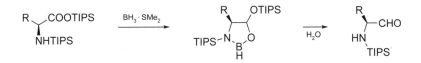

[1]Yamanoi, Y., Taira, T., Sato, J., Nakamula, I., Nishihara, H. *OL* **9**, 4543 (2007).

Borane-sulfides.

Reduction.[1]   Reduction of the carboxyl group of an α-amino acid without racemization can be carried out by the BH_3-Me_2S complex on its TIPS derivative.

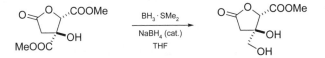

Selective reduction of an ester group adjacent to an alcohol is achieved.[2]

Amide synthesis.[3]   Limited amounts (0.35 equivalent) of the borane-dimethyl sulfide or THF complex promote condensation of carboxylic acids and amines.

Hydroboration.[4]   Synthesis of of *B*-alkenylpinacolatoborons is achieved via reaction of 1-alkynes with the new complex obtained from admixture of $BH_3 \cdot Me_2S$ with $(C_6F_5)_3B$, followed by treatment with pinacolborane.

[1]Soto-Cairoli, B., de Pomar, J.J., Soderquist, J.A. *OL* **10**, 333 (2008).
[2]Varugese, S., Thomas, S., Haleema, S., Puthiaparambil, T.T., Ibnusaud, I. *TL* **48**, 8209 (2007).
[3]Huang, Z., Reilly, J.E., Buckle, R.N. *SL* 1026 (2007).
[4]Hoshi, M., Shirakawa, K., Okimoto, M. *TL* **48**, 8475 (2007).

Boric acid.

Amidation.   Boric acid promotes condensation of carboxylic acids with amines by heating in toluene.[1]

[1]Barajas, J.G.H., Mendez, L.Y.V., Kouznetsov, V.V., Stashenko, E.E. *S* 377 (2008).

Boron tribromide.

Rearrangement.   Tertiary *N*-benzylglycinamides are subject to [1,2]-rearrangement by consecutive treatment with BBr_3 and Et_3N to afford phenylalaninamide analogues.[1] The transformation cannot be effected by replacing boron tribromide with boron trifluoride.

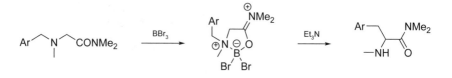

[1]Tuzina, P., Somfai, P. *TL* **48**, 4947 (2007).

Boron trifluoride etherate.

Substitution. With $BF_3 \cdot OEt_2$ and CF_3COOH as activators, aryltriazenes are converted to aryl azides.[1] Friedel–Crafts propargylation of arenes can be accomplished with propargyl trichloroacetimidates in the presence of $BF_3 \cdot OEt_2$ at room temperature.[2]

α-Acetoxyalkyl carboxamides/lactams form acyliminium species in contact with $BF_3 \cdot OEt_2$, and they are intercepted by potassium organotrifluoroborates.[3]

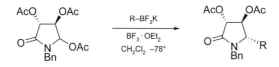

Reaction of trisubstituted epoxides with alkynylalanes in the presence of $BF_3 \cdot OEt_2$ occurs at the quaternary carbon site. Without the Lewis acid much more side products appear.[4]

Addition reactions. 2-Aryl-3-vinyl-2,3-dihydrobenzofurans are obtained from reaction of benzoxasilepins and ArCHO. Substituent effects of ArCHO set the trend for *cis/trans*-isomer variation of the products.[5]

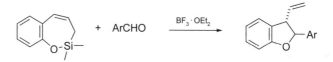

Also in the alkenyl group transfer from the trifluoroborate salts to dichloroalkyl aldimines $BF_3 \cdot OEt_2$ plays an activating role.[6]

Cyclization and cycloreversion.   Despite its adjacency to a carbonyl group a tertiary oxy substituent at C-3 of oxindoles undergoes ionization on exposure to $BF_3 \cdot OEt_2$, and that interaction with an alkenyl chain leads to spirocyclic products.[7]

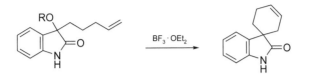

N-Protected tetrahydroisoquinolines are formed from N-acylcarbamates of 2-arylethyl-amines in two steps: Reduction with Dibal-H and cyclization by $BF_3 \cdot OEt_2$.[8]

A tandem reaction sequence involving electrocyclization, azide capture, and Schmidt rearrangement is observed when certain cross-conjugated dienones and alkyl azides are treated with $BF_3 \cdot OEt_2$.[9]

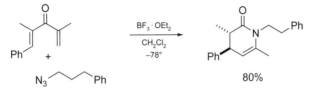

Cycloadducts of 4-hydroxy-2-isoxazoline N-oxides with alkenes are decomposed by $BF_3 \cdot OEt_2$. It takes much longer if silica gel is used to effect the cycloreversion (48 hr. vs. 1 hr.)[10]

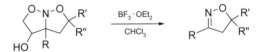

The exposure of 2-Boc-aminomethylaziridines to $BF_3 \cdot OEt_2$ results in 5-aminomethyl-2-oxazolidinones. The three-membered heterocycle is replaced by a 5-membered and there is a loss of the t-butyl group.[11]

Intramolecular Friedel–Crafts alkylation forming a seven-membered ring by a formal Michael addition at the terminus of a 3-ethynyl-2-benzyl-2-cyclohexenone unit is made sterically possible by the presence of EtSH.[12]

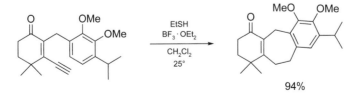

94%

4,4-Difluoro-3-buten-1-ylarenes afford α-tetralones on exposure to BF$_3$ · OEt$_2$ (1 equiv.) and catalytic quantitites of (MeCN)$_4$Pd(BF$_4$)$_2$, with hydrolytic work up.[13]

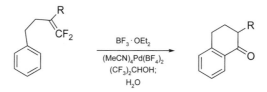

Vinyl sulfides are stable to BF$_3$ · OEt$_2$ therefore Nazarov cyclization intermediates can be trapped by them to produce 2-organothio-bicyclo[2.2.1]heptan-7-ones.[14]

Rearrangements.   Under microwave irradiation the Claisen rearrangement of *N*-allyl-anilines is effectively catalyzed by BF$_3$ · OEt$_2$. Other Lewis acids are much inferior catalysts under the same conditions.[15]

A significantly different reaction profile for isomeric diketo epoxides (as shown below) has been discovered.[16]

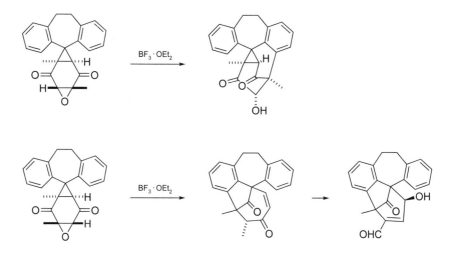

Alkenylcyclopropenes pursue a reaction pathway on exposure to BF$_3$ · OEt$_2$ different from that observed when they are treated with Cu(OTf)$_2$.[17]

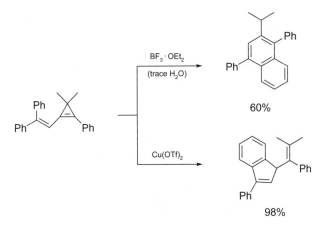

1,3-Oxy migration of propargylic alcohols and stabilization of the allenyl products by carbamoylation are performed by heating the alcohols and RNCO with a ruthenium complex (**1**) and $BF_3 \cdot OEt_2$.[18]

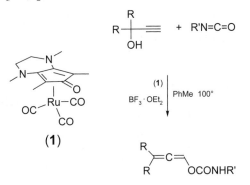

[1]Liu, C.-Y., Knochel, P. *JOC* **72**, 7106 (2007).

[2]Li, C., Wang, J. *JOC* **72**, 7431 (2007).

[3]Vieira, A.S., Ferreira, F.P., Fiorante, P.F., Guadagnin, R.C., Stefani, H.A. *T* **64**, 3306 (2008).

[4]Zhao, H., Engers, D.W., Morales, C.L., Pagenkopf, B.L. *T* **63**, 8774 (2007).

[5]Jimenez-Gonzalez, L., Garcia-Munoz, S., Alvarez-Corral, M., Munoz-Dorado, M., Rodriguez-Garcia, I. *CEJ* **13**, 557 (2007).

[6]Stas, S., Tehrani, K.A. *T* **63**, 8921 (2007).

[7]England, D.B., Merey, G., Padwa, A. *OL* **9**, 3805 (2007).

[8]Kuhakarn, C., Panyachariwat, N., Ruchirawat, S. *TL* **48**, 8182 (2007).

[9]Song, D., Rostami, A., West, F.G. *JACS* **129**, 12019 (2007).

[10]Nishiuchi, M., Sato, H., Ohmura, H. *CL* **37**, 144 (2008).

[11]Moran-Ramallal, R., Liz, R., Gotor, V. *OL* **10**, 1935 (2008).

[12]Majetich, G., Zou, G., Grove, J. *OL* **10**, 85 (2008).
[13]Yokota, M., Fujita, D., Ichikawa, J. *OL* **9**, 4639 (2007).
[14]Mahmoud, B., West, F.G. *TL* **48**, 5091 (2007).
[15]Gonzalez, I., Bellas, I., Souto, A., Rodriguez, R., Cruces, J. *TL* **49**, 2002 (2008).
[16]Asahara, H., Kubo, E., Togaya, K., Koizumi, T., Mochizuki, E., Oshima, T. *OL* **9**, 3421 (2007).
[17]Shao, L.-X., Zhang, Y.-P., Qi, M.-H., Shi, M. *OL* **9**, 117 (2007).
[18]Haak, E. *EJOC* 788 (2008).

Bromine.

Aza-transfer.   Certain unsaturated *t*-butylaldimines are found to transfer the nitrogen moiety to the double bond on consecutive treatment with bromine and a base.[1]

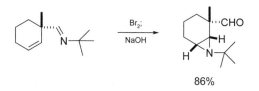

86%

Von Braun degradation.   The triphenyl phosphite complex of bromine (and other halogens) converts tertiary amides into nitriles.[2]

[1]D'hooghe, M., Boelens, M., Piqueur, J., De Kimpe, N. *CC* 1927 (2007).
[2]Vaccari, D., Davoli, P., Spaggiari, A., Prati, F. *SL* 1317 (2008).

Bromopentacarbonylmanganese.

Coupling reactions.   Coordinative stabilization of the arylmanganese species by an imino nitrogen atom greatly facilitates their formation via metal insertion of an *o*-C-H bond. *o*-Functionalization of the aryl residue in a 2-arylimidazole by reaction with electrophiles such as PhCHO after treatment with Mn(CO)$_5$Br meets expectation.[1]

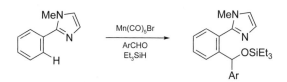

Cycloaddition.[2]   β-Keto esters are found to participate in [2+2+2]cycloaddition in the enolic form with two equivalents of alkynes, using Mn(CO)$_5$Br as catalyst.

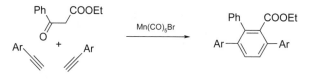

[1]Kuninobu, Y., Nishina, Y., Takeuchi, T., Takai, K. *ACIE* **46**, 6518 (2007).
[2]Tsuji, H., Yamagata, K., Fujimoto, T., Nakamura, E. *JACS* **130**, 7792 (2008).

N-Bromosuccinimide, NBS.

Debenzylation.   A benzyl group attached to the nitrogen atom of a carboxamide is subject to removal by NBS at room temperature.[1]

Bromodesilylation.   Alkenylsilanes are converted into the corresponding bromo-alkenes by NBS.[2]

66–90%

Bromination.   In bromolactamization of unsaturated *N*-Boc amides with NBS in THF, adding the base *t*-BuOLi is recommended.[3] α-Bromo enones are formed by treatment of enol silyl ethers with NBS under uv irradiation and in the presence of a free radical initiator (AIBN).[4] The allylic position of an ordinary double bond, being less reactive, remains unaffected.

75%

Another protocol for α-bromination of enals and cyclic enones using NBS indicates pyridine *N*-oxide to be a beneficial additive.[5]

Amination.   In a single step benzylic amination is achieved with NBS with catalytic amounts of $FeCl_2$. Aminating agents include carboxamides and sulfonamides.[6] *N*-Cyano sulfilimines are formed by reaction of sulfides with cyanamide in the presence of NBS, probably involving *S*-bromination.[7]

[1]Kuang, L., Zhou, J., Chen, S., Ding, K. *S* 3129 (2007).
[2]Pawluc, P., Hreczycho, G., Walkowiak, J., Marciniec, B. *SL* 2061 (2007).
[3]Yeung, Y.-Y., Corey, E.J. *TL* **48**, 7567 (2007).
[4]Kraus, G.A., Jeon, I. *TL* **49**, 286 (2008).
[5]Bovonsombat, P., Rujiwarangkul, R., Bowornkiengkai, T., Leykajarakul, J. *TL* **48**, 8607 (2007).
[6]Wang, Z., Zhang, Y., Fu, H., Jiang, Y., Zhao, Y. *OL* **10**, 1863 (2008).
[7]Mancheno, O.G., Bistri, O., Bolm, C. *OL* **9**, 3809 (2007).

t-Butanesulfinamide.

Nitriles.   Reaction of aldehydes with *t*-BuSONH$_2$ in the presence of $(EtO)_4Ti$ and with microwave irradiation leads to nitriles.[1]

Resolution of 2,2′-diformylbiphenyls.   The chiral sulfinamide forms separable diastereomeric imines with the biphenyldialdehyde, wherefrom optically active aldehydes are obtained.[2]

[1]Tanuwidjaja, J., Peltier, H.M., Lewis, J.C., Schenkel, L.B., Ellman, J.A., *S* 3385 (2007).
[2]Zhu, C., Shi, Y., Xu, M.-H., Lin, G.-Q. *OL* **10**, 1243 (2008).

t-Butyl hydroperoxide.

Carboxamides.   Because of the facility in benzylic oxidation for the adducts of ArCHO and secondary amines, their rapid conversion into ArCONR$_2$ by *t*-BuOOH (metal-free conditions) is observed.[1]

[1]Ekoue-Kovi, K., Wolf, C. *OL* **9**, 3429 (2007).

t-Butyl hydroperoxide – metal salts.

Oxidation.   Oxidation of aldehydes to carboxylic acids by *t*-BuOOH (70%) with CuCl as catalyst occurs at room temperature.[1] Using anhydrous *t*-BuOOH in decane, primary alcohols are also converted to the same products and secondary alcohols to ketones.[2] More unusual is the oxidation of aldehydes in the presence of an alcohol to provide esters by *t*-BuOOH with both Cu(ClO$_4$)$_2$ and InBr$_3$ as catalyst.[3]

The combination of TiCl$_4$ and *t*-BuOOH constitutes an epoxidizing system (for allylic alcohols) at low temperature. However, it oxidizes the same substrates to enones in refluxing CHCl$_3$; and ordinary secondary alcohols to ketones, naturally.[4]

Another oxidizing system contains Rh$_2$(cap)$_4$,[5] which also finds use in dehydrogenating secondary amines to form imines.[6]

A Pd-catalyzed oxidation to bring about ring closure is synthetically viable. Also note-worthy is that products containing different functional groups arise from a change of reaction conditions.[7]

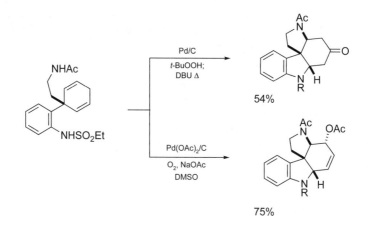

Epoxidation.   New metal specimens for activating *t*-BuOOH to epoxidize alkenes include molybdenum oxide (with R$_3$PO),[8] and nanoparticles thereof.[9] They have the advantage of being paramagnetic.

The asymmetric epoxidation of allylic alcohols based on vanadyl isopropoxide and *t*-BuOOH has been reexamined with a series of chiral hydroamic acids (**1**).[10]

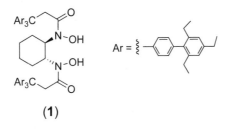

(1)

γ-(*N*-Sulfonylamino)propargylic alcohols undergo epoxidation but isomerization of the products to carbene species is rapid.[11]

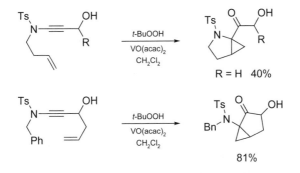

R = H 40%

81%

Cleavage of multiple CC bonds. Both alkenes and alkynes are cleaved by InCl₃– *t*-BuOOH in water.[12]

[1]Mannam, S., Sekar, G. *TL* **49**, 1083 (2008).
[2]Mannam, S., Sekar, G. *TL* **48**, 2457 (2007).
[3]Yoo, W.-J., Li, C.-J. *TL* **48**, 1033 (2007).
[4]Shei, C.-T., Chien, H.-L., Sung, K. *SL* 1021 (2008).
[5]Choi, H., Doyle, M.P. *OL* **9**, 5349 (2007).
[6]Choi, H., Doyle, M.P. *CC* 745 (2007).
[7]Beniazza, R., Dunet, J., Robert, F., Schenk, K., Landais, Y. *OL* **9**, 3913 (2007).
[8]Kiraz, C.I.A., Mora, L., Jimenez, L.S. *S* 92 (2007).
[9]Shokouhimehr, M., Piao, Y., Kim, J., Jang, Y., Hyeon, T. *ACIE* **46**, 7039 (2007).
[10]Barlan, A.U., Zhang, W., Yamamoto, H. *T* **63**, 6075 (2007).
[11]Couty, S., Meyer, C., Cossy, J. *SL* 2819 (2007).
[12]Ranu, B.C., Bhadra, S., Adak, L. *TL* **49**, 2588 (2008).

Butyllithium.

X/Li Exchange.   Different reactivities of polyhalogen substituents in an aromatic ring permit orderly exchange to form aryllithium species for stepwise functionalization.[1]

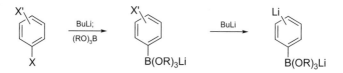

Deprotonation.   N'-Boc arylhydrazines are deprotonated initially at the nitrogen atom of the carbamate group, but dianions are generated on treatment with 2 equivalents of BuLi. Selective dialkylation of the dianions can be accomplished.[2]

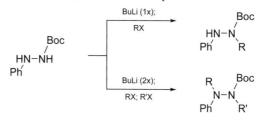

4-Alken-6-ynamines undergo deprotonation; an ensuing cyclization affords 2-allenylpyrrolidines.[3]

Arsenic analogue of the Wittig reaction operates on diallylarsonium salts delivers 1,3-dienes. The internal double bond of the major products (from RCHO) has an (E)-configuration.[4]

1,3,5-Trimethylperhydro-1,3,5-triazine is readily deprotonated to provide a synthetic equivalent of formyl anion. After addition of the lithiated species to carbonyl compounds a workup with HCl gives α-hydroxy aldehydes.[5]

Elimination.   (*E,E*)-1,3-Dienamines are obtained from (*Z*)-4-methoxy-2-alkenyl-amines by treatment with BuLi (or NaHMDS). Elimination of MeOH is stereoselective.[6]

A preparation of trifluoromethylallenes from 1,1-dichloro-3,3,3-trifluoropropen-2-yl tosylate involves treatment with BuLi to generate lithium 3,3,3-trifluoropropynide, which is used for reaction with carbonyl compounds and then Negishi coupling.[7]

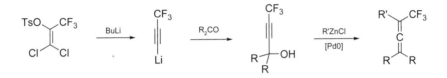

Dilithioethyne used for a synthesis of bis(pinacolato)ethyne is generated from trichloro-ethene with BuLi. The boronate has many synthetic applications.[8]

Brook rearrangement.   α-Trimethylsilylpropargyl alcohols undergo Brook rearrange-ment to afford allenyl silyl ethers, which can be used to condense with aldehydes.[9] Silyl group transfer from the ether four bonds apart is preferred after the Sn/Li exchange from silyl ethers of 1-tributylstannyl-1,3-alkanediols.[10]

Addition.   Hydroamination of cinnamyl alcohol occurs on exposure to amines that are deprotonated by BuLi. The products are vicinal amino alcohols.[11]

[1]Kurach, P., Lulinski, S., Serwatowski, J. *EJOC* 3171 (2008).
[2]Bredihhin, A., Groth, U.M., Mäeorg, U. *OL* **9**, 1097 (2007).
[3]Zhang, W., Werness, J.B., Tang, W. *OL* **10**, 2023 (2008).
[4]Habrant, D., Stengel, B., Meunier, S., Mioskowski, C. *CEJ* **13**, 5433 (2007).
[5]Bojer, D., Kamps, I., Tian, X., Hepp, A., Pape, T., Fröhlich, R., Mitzel, N.W. *ACIE* **46**, 4175 (2007).
[6]Tayama, E., Sugai, S. *TL* **48**, 6163 (2007).
[7]Shimizu, M., Higashi, M., Takeda, Y., Jiang, G., Murai, M., Hiyama, T. *SL* 1163 (2007).
[8]Kang, Y.K., Deria, P., Carroll, P.J., Therien, M.J. *OL* **10**, 1341 (2008).
[9]Reynolds, T.E., Scheidt, K.A. *ACIE* **46**, 7806 (2007).
[10]Mori, Y., Futamura, Y., Horisaki, K. *ACIE* **47**, 1091 (2008).
[11]Barry, C.S., Simpkins, N.S. *TL* **48**, 8192 (2007).

Butyllithium – (-)-sparteine.

Amines -> alcohols.   A synthesis of secondary alcohols in chiral form from *N*-Boc secondary amines involves lithiation with the BuLi – (-)-sparteine complex, quenching with R_3B, and decomposing the reaction mixture with $NaOH–H_2O_2$.[1]

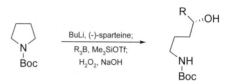

[1]Coldham, I., Patel, J.J., Raimbault, S., Whittaker, D.T.E., Adams, H., Fang, G.Y., Aggarwal, V.K. *OL* **10**, 141 (2008).

s-Butyllithium.

Lithiation.   A synthesis of 1-biarylcarboxylic acids from aroic acids (selective arylation) is based on *o*-lithiation by *s*-BuLi, Li/Sn exchange, and Stille coupling.[1] Selective functionalization at the methyl group of *o*-tolylmethanol via lateral lithiation is facilitated by derivatizing the hydroxyl group to form a methoxyethyl ether or dimethylaminoethyl ether.[2]

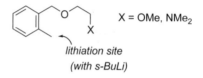

The benzylic position of 2-arylaziridines is readily lithiated. *N*-Substituted *trans*-2,3-diphenylaziridines afford diastereomers according to the solvent used (whether HMPA is present).[3] More dramatic differences are observed in the case of aziridines bearing an oxazoline substituent at C-2.[4]

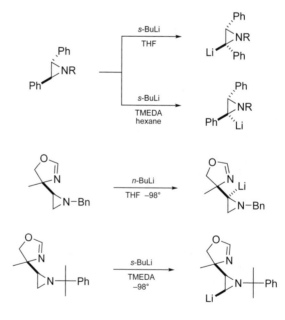

Rearrangement. Ureas containing aryl and benzyl units on different nitrogen atoms undergo N -> C aryl shift, as a result of benzylic lithiation and addition to the distal aryl group.[5]

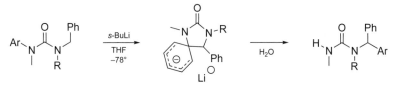

[1]Castanet, A.-S., Tilly, D., Veron, J.-B., Samanta, S.S., De, A., Ganguly, T., Mortier, J. *T* **64**, 3331 (2008).
[2]Wilkinson, J.A., Raiber, E.A., Ducki, S. *TL* **48**, 6434 (2007).
[3]Luisi, R., Capriati, V., Florio, S., Musio, B. *OL* **9**, 1263 (2007).
[4]Luisi, R., Capriati, V., DiCunto, P., Florio, S., Mansueto, R. *OL* **9**, 3295 (2007).
[5]Clayden, J., Dufour, J., Grainger, D.M., Helliwell, M. *JACS* **129**, 7488 (2007).

t-Butyllithium.

X/Li exchange. A synthetic approach to 4-hydroxycycloheptanones from γ-(3-iodopropyl)butyrolactones is proven successful via I/Li exchange and intramolecular acylation. At least in the example shown below, cyclization induced by SmI_2 is not a viable alternative.[1]

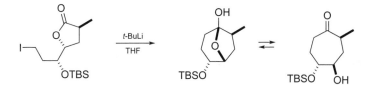

The unique quality of Br/Li exchange with *t*-BuLi enables the synthesis of an α-hydroxy(3-furanylmethyl)-2-butenolide, whereas many other coupling method fail.[2]

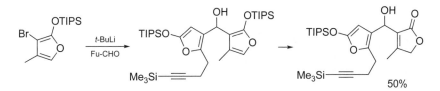

A method for the preparation of $ArBF_3K$ from ArX calls for treatment with *t*-BuLi and subsequent reaction with $(i\text{-PrO})_3B$, and finally, with KHF_2.[3]

Lithiation. A molecule featuring both a fluoroarene and a bromoarene subunit gives rise to aryne and lithioarene moieties on the treatment with *t*-BuLi. If such reactive components are spatially interactable, intramolecular addition can occur.[4]

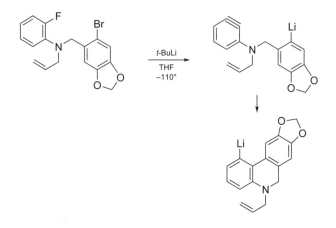

Reduction.   A rare showing of the reductive potential of *t*-BuLi is in its reaction with 1-bromo-1-dicyclohexylborylalkenes. Hydride transfer from *t*-BuLi to the boron atom triggers a debrominative rearrangement.[5]

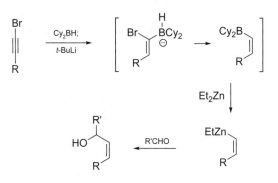

[1]Ohtsuki, K., Matsuo, K., Yoshikawa, T., Moriya, C., Tomita-Yokotani, K., Shishido, K., Shindo, M. *OL* **10**, 1247 (2008).

[2]He, W., Huang, J., Sun, X., Frontier, A.J. *JACS* **130**, 300 (2008).

[3]Park, Y.H., Ahn, H.R., Canturk, B., Jeon, S.I., Lee, S., Kang, H., Molander, G.A., Ham, J. *OL* **10**, 1215 (2008).

[4]Sanz, R., Fernandez, Y., Castroriejo, M.P., Perez, A., Fananas, F.J. *EJOC* 62 (2007).

[5]Salvi, L., Jeon, S.-J., Fisher, E.L., Carroll, P.J., Walsh, P.J. *JACS* **129**, 16119 (2007).

N-(t-Butyl)phenylsulfinimidoyl chloride.

Dehydrogenation.   Direct introduction of a functional chain (e.g., malonic ester) to the β-position of a cycloalkanone involves treatment with LDA and the title reagent, and followed by the nucleophile.[1] Lactams are functionalized at the carbon α to the nitrogen atom, after deprotonation.[2]

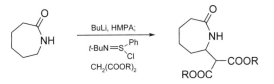

Activation of the benzylic position of 3-benzylindoles is similarly accomplished.[3]

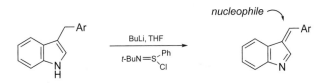

[1]Matsuo, J., Kawai, H., Ishibashi, H. *TL* **48**, 3155 (2007).
[2]Matsuo, J., Tanaki, Y., Ishibashi, H. *TL* **48**, 3233 (2007).
[3]Matsuo, J., Tanaki, Y., Ishibashi, H. *T* **64**, 5262 (2008).

C

Calcium bis(hexamethyldisilazide).

Redox reaction.   Aromatic aldehydes are converted into benzyl aroates on exposure, at room temperature, to the silylamides of alkali earth metals, including those of Ca, Sr, and Ba.[1]

[1]Crimmin, M.R., Barrett, A.G.M., Hill, M.S., Procopiou, P.A. *OL* **9**, 331 (2007).

Carbonyl(chloro)hydridobis(tricyclohexylphosphine)ruthenium.

Disiloxanes.   Vinyltriorganosilanes exchange the vinyl group for a triorganosilanol to form disiloxanes, when they are heated with the Ru complex in toluene.[1]

[1]Marciniec, B., Pawluc, P., Hreczycho, G., Macina, A., Madalska, M. *TL* **49**, 1310 (2008).

Carbonyl(chloro)hydridotris(triphenylphosphine)rhodium.

Homoallylic and homopropargylic alcohols.   Redox combination of primary alcohols and conjugated dienes[1] or 1-alken-3-ynes[2] takes place in the presence of the Rh complex, providing homoallylic alcohols and homopropargylic alcohols, respectively. In the reaction of the dienes allylmetal reagents are formed on hydrogen transfer from the alcohols.

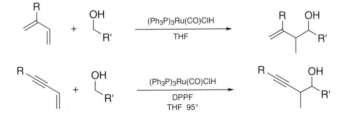

Enolization via long-range migration.   An unsaturated ketone with a remote double bond uninterrupted by a quaternary carbon or heteroatom is capable of forming a Ru-enolate by heating with the title complex in benzene. Such an enolate can be trapped by aldehydes and the resulting ruthenated aldols afford 1,3-diketones through eliminated of [Ru]-H.[3]

Fiesers' Reagents for Organic Synthesis, Volume 25. By Tse-Lok Ho
Copyright © 2010 John Wiley & Sons, Inc.

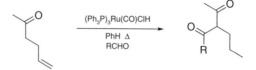

[1]Shibahara, F., Bower, J.F., Krische, M.J. *JACS* **130**, 6338 (2008).
[2]Patman, R.L., Williams, V.M., Bower, J.F., Krische, M.J. *ACIE* **47**, 5220 (2008).
[3]Fukuyama, T., Doi, T., Minamino, S., Omura, S., Ryu, I. *ACIE* **46**, 5559 (2007).

Carbonyldihydridotris(triphenylphosphine)ruthenium.

Oxidation.   The Ru complex, with Xantphos, 2 equivalents of water and crotonitrile (as hydrogen acceptor), can be used to oxidize primary alcohols. In the presence of MeOH the generation of methyl esters is realized.[1] Internal redox reaction of 1,4-butanediol under basic conditions (*t*-BuOK) leads to the formation of γ-butyrolactone.[2]

Aldoximes are oxidized to primary amides, only that reaction conditions are somewhat different: additive being TsOH · H_2O besides a phosphine ligand.[3]

Coupling reactions.   The Ru complex catalyzes replacement of the amino group of *o*-aminoaryl ketones with the carbon residue of an organoboronic ester.[4] Direct activation of a C—H bond ortho to the carbonyl group is also possible.[5]

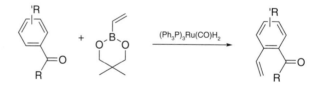

Furans and pyrroles.   2-Alkyne-1,4-diols undergo isomerization to 1,4-diones and subsequent dehydration to afford furans,[6] on heating with the Ru complex, Xantphos, and PhCOOH in toluene at 80°. The intermediates are of course convertible to pyrroles.[7]

[1]Owston, N.A., Parker, A.J., Williams, J.M.J. *CC* 624 (2008).
[2]Maytum, H.C., Tavassoli, B., Williams, J.M.J. *OL* **9**, 4387 (2007).
[3]Owston, N.A., Parker, A.J., Williams, J.M.J. *OL* **9**, 3599 (2007).
[4]Ueno, S., Chatani, N., Kakiuchi, F. *JACS* **129**, 6098 (2007).
[5]Ueno, S., Chatani, N., Kakiuchi, F. *JOC* **72**, 3600 (2007).
[6]Pridmore, S.J., Slatford, P.A., Williams, J.M.J. *TL* **48**, 5111 (2007).
[7]Pridmore, S.J., Slatford, P.A., Daniel, A., Wittlesey, M.K., Williams, J.M.J. *TL* **48**, 5115 (2007).

Cerium(IV) ammonium nitrate, CAN.

Oxidations.   Tertiary cyclopropanols undergo oxidative ring opening on exposure to CAN, an added salt provides anion to functionalize the emerging ethyl terminus.[1]

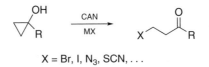

X = Br, I, N₃, SCN, ...

CAN oxidizes the OH group of glycols and 1,2-amino alcohols to initiate CC bond cleavage. The alcohol unit from a glycol monoether is released.[2] Analogously, oxidative degradation of proline and prolinol derivatives gives 2-hydroxypyrrolidines.[2] N-Arylpyroglutamic acids are further oxidized to afford succinimides (CAN–NaBrO₃ protocol).[3]

Selective conversion of a methylenedioxyarene to an o-quinone is a crucial step in a synthesis of phleblarubrone.[4] On the other hand, veratrole derivatives are cleaved to afford hexadienoic esters.[5]

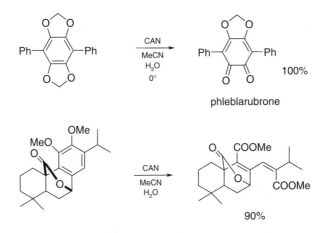

phleblarubrone

Formation of a new ring system by involving a released N-methoxymethylamine chain during arene oxidation is a surprising reward, despite the low yield.[6]

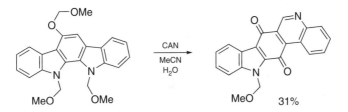

In the introduction of a sidechain to C-2 of the quinoline nucleus via 1,2-addition (extended Reissert reaction) to form an adduct with a N-(p-methoxybenzyl) group, the rearomatization step can be initiated by oxidative C—N bond cleavage with CAN.[7]

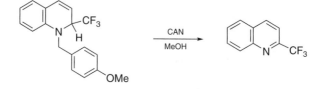

The C—N bond of *N*-arylhydrazides (with electron-deficient Ar) is reductively cleaved after oxidative activation by CAN, where MeOH serves as hydride source.[8]

Oxidative coupling.   Dialkenylsiloxanes are decomposed into 1,4-dicarbonyl compounds by oxidation with CAN.[9]

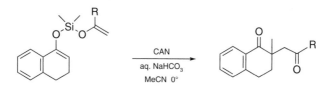

N-Benzyl-3-azabicyclo[3.1.0]hexan-1-ols, which are readily available from a Kulinkovich reaction, are converted by CAN into benzannulated 1-azabicyclo[3.3.1]-nonan-3-ones.[10] This oxidation pathway differs from that mediated by FeCl$_3$.

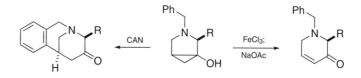

Substitution.   Acetates of Baylis–Hillman adducts of acrylic esters and ArCHO are transformed into amines via an S$_N$2' pathway, which is catalyzed by CAN.[11]

Addition reactions.   CAN serves as a catalyst for the conjugate addition of thiols and selenols to enones under solvent-free conditions.[12] Glycals are transformed into glycosides containing a nitromethyl group at C-2.[13]

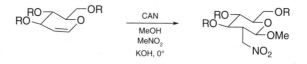

It appears that CAN acts as a Lewis acid to catalyze addition of enol ethers to *N*-arylalimines, which is followed by an intramolecular Friedel–Crafts reaction to furnish 4-alkoxy-1,2,3,4-tetrahydroquinolines.[14]

α-Nitrocinnamate esters.   Cinnamate esters are further functionalized by nitrating agent generated in situ from $NaNO_2$ and CAN.[15]

gem-Bishydroperoxides.   Various carbonyl compounds (ArCHO and ketones) are converted into the oxygen-rich compounds by aqueous H_2O_2 in the presence of catalytic CAN at room temperature.[16]

[1] Jiao, J., Nguyen, L.X., Patterson, D.R., Flowers II, R.A., *OL* **9**, 1323 (2007).
[2] Fujioka, H., Hirose, H., Ohba, Y., Murai, K., Nakahara, K., Kita, Y. *T* **63**, 625 (2007).
[3] Barman, G., Roy, M., Ray, J.K. *TL* **49**, 1405 (2008).
[4] Hayakawa, I., Watanabe, H., Kigoshi, H. *T* **64**, 5873 (2008).
[5] Marrero, J.G., San Andres, L., Luis, J.G. *SL* 1127 (2007).
[6] Sperry, J., McErlean, C.S.P., Slawin, A.M.Z., Moody, C.J. *TL* **48**, 231 (2007).
[7] Loska, R., Majcher, M., Makosza, M. *JOC* **72**, 5574 (2007).
[8] Stefane, B., Polanc, S. *SL* 1279 (2008).
[9] Clift, M.D., Taylor, C.N., Thomson, R.J. *OL* **9**, 4667 (2007).
[10] Jida, M., Guillot, R., Ollivier, J. *TL* **48**, 8765 (2007).
[11] Paira, M., Mandal, S.K., Ray, S.C. *TL* **49**, 2432 (2008).
[12] Chu, C.-M., Gao, S., Sastry, M.N.V., Kuo, C.-W., Lu, C., Liu, J.-T., Yao, C.-F. *T* **63**, 1863 (2007).
[13] Elamparuthi, E., Linker, T. *OL* **10**, 1361 (2008).
[14] Sridharan, V., Avendano, C., Menedez, J.C. *SL* 1079 (2007).
[15] Buevich, A.V., Wu, Y., Chan, T.-M., Stamford, A. *TL* **49**, 2132 (2008).
[16] Das, B., Krishnaiah, M., Veeranjaneyulu, B., Ravikanth, B. *TL* **48**, 6286 (2007).

Cerium(III) chloride.

Reduction.   Alkyl azides are reduced to primary amines with $CeCl_3 \cdot 7H_2O$ – NaI in hot MeCN.[1]

[1] Bartoli, G., Di Antonio, G., Giovannini, R., Giuli, S., Lanari, S., Paoletti, M., Marcantoni, E. *JOC* **73**, 1919 (2008).

Cesium fluoride.

Substitution.   CsF is an excellent fluoride ion source for converting alkyl mesylates to RF, especially in the presence of the imidazolium mesylate **1**.[1] The effect, due to hydrogen bonding to the tertiary alcohol to render the fluoride ion more nucleophilic but less basic (so as to minimize elinination [H-OMs]), is also manifested in a polymer-linked **1**,[2] and a combination of *t*-AmOH and a polymer-supported ionic liquid.[3]

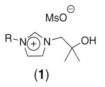

(1)

Transacylation.   Only catalytic amounts of CsF are needed for protection of hydroxy-indoles as *N*-Boc derivatives without affecting hydroxyl group[4] and transesterification of β-keto esters.[5]

Aryne generation.   The desilylative route (e.g., by CsF) is the most expedient method for generation of arynes. 1,2-Functionalization of arenes from 2-trimethylsilylaryl triflates is readily achieved as long as noninterfering co-reactants are used. Thus trapping by organoazides leads to benzotriazoles,[6] by phenyliodonium diacylmethylides leads to 3-acylbenzofurans,[7] and reaction in the presence of benzoic esters *o*-substituted with XH groups (X = O, S, NH) gives xanthones, thioxanthones, and acridones.[8] [Direct diazotization of anthranilic acid affords acridone.[9]]

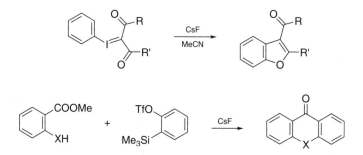

An amide anion simultaneously released by desilylation of Me_3SiNR_2 is able to add to the aryne and then trapped with an aldehyde. Deployment of this strategy realizes a synthesis of *o*-aminobenzyl alcohols.[10]

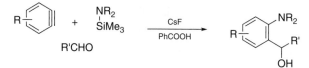

1,2-Bisphenylselenoarenes are readily prepared on generating the arynes in the presence of PhSeSePh.[11]

An electrophile role is played by an aryne also in the arylation (at the α-position) of β-amino-α,β-unsaturated ketones and esters.[12]

Most interesting is the coupling reactions at two adjacent position of an aromatic ring, for example, on basis of Pd-catalyzed reactions. Examples include the one-pot synthesis of 2-(1-alkenyl)biphenyls[13] and *o*-allylarylalkynes.[14]

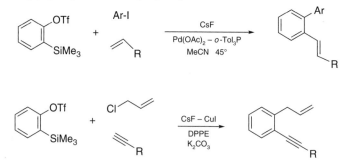

With co-presence of ArI, alkynes, TlOAc and (dba)$_2$Pd in the reaction pot in which the aryne is generated, multicomponent coupling directed toward phenanthrenes is realized.[15]

[1]Shinde, S.S., Lee, B.S., Chi, D.Y. *OL* **10**, 733 (2008).
[2]Shinde, S.S., Lee, B.S., Chi, D.Y. *TL* **49**, 42453 (2008).
[3]Kim, D.W., Jeong, H.-J., Lim, S.T., Sohn, M.-H., Chi, D.Y. *T* **64**, 4209 (2008).
[4]Inahashi, N., Matsumiya, A., Sato, T. *SL* **294** (2008).
[5]Inahashi, N., Fujiwara, T., Sato, T. *SL* **605** (2008).
[6]Shi, F., Waldo, J.P., Chen, Y., Larock, R.C. *OL* **10**, 2409 (2008).
[7]Huang, X.-C., Liu, Y.-L., Liang, Y., Pi, S.-F., Wang, F., Li, J.-H. *OL* **10**, 1525 (2008).
[8]Zhao, J., Larock, R.C. *JOC* **72**, 583 (2007).
[9]Ho, T.-L., Jou, D.-G. *JCCS(T)* **48**, 81 (2001).
[10]Yoshida, H., Morishita, T., Fukushima, H., Ohshita, J., Kunai, A. *OL* **9**, 3367 (2007).
[11]Toledo, F.T., Marques, H., Comasseto, J.V., Raminelli, C. *TL* **48**, 8125 (2007).
[12]Ramtohul, Y.K., Chartrand, A. *OL* **9**, 1029 (2007).
[13]Henderson, J.L., Edwards, A.S., Greaney, M.F. *OL* **9**, 5589 (2007).
[14]Xie, C., Liu, L., Zhang, Y., Xu, P. *OL* **10**, 2393 (2008).
[15]Liu, Z., Larork, R.C. *ACIE* **46**, 2535 (2007).

Chiral auxiliaries and catalysts.

Kinetic resolution. Various situations that dynamic kinetic resolution applies are reviewed.[1] Resolution by enantioselective benzoylation of 2-benzoylamino-1,3-propanediols is accomplished in the presence of the CuCl$_2$-complex of ligand **1A**.[2] Selective esterification catalyzed by **2** is for dynamical kinetic resolution of hemiaminals and aminals,[3] whereas tosylation of α-hydroxycarboxamides proceeds well under the influence of Cu(OTf)$_2$ – *ent-***1B**.[4]

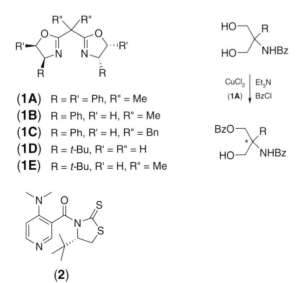

(**1A**) R = R' = Ph, R" = Me
(**1B**) R = Ph, R' = H, R" = Me
(**1C**) R = Ph, R' = H, R" = Bn
(**1D**) R = *t*-Bu, R' = R" = H
(**1E**) R = *t*-Bu, R' = H, R" = Me

(**2**)

The iridium complex **3** has found use in kinetic resolution of secondary benzylic alcohols (indanol, α-tetralol, . . .) by its mediation of enantioselective aerobic oxidation.[5,6]

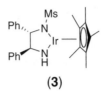

(3)

TADDOL spiroannulated to a cyclohexane proves useful for deracemization of α-benzyloxy ketones.[7]

Members belonging to one enantiomeric series of 5-substituted 2-cyclohexenones remain for being more resistant to attack by R_2Zn in the presence of a Cu(I) salt and the peptide derivative **4**.[8]

By rapid reduction of one enantiomeric series of benzylic hydroperoxides, [2.2]para-cyclophane-based diphosphine **5** kinetically resolves those active compounds.[9]

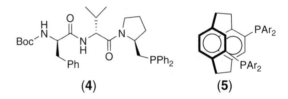

(4)   **(5)**

Oxidative ring cleavage in the presence of **6** is the basis of a kinetic resolution of *N*-acyloxazolidines.[10]

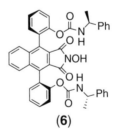

(6)

Desymmetrization.   Selective reaction of *meso*-compounds to provide desired chiral products is highly valued. By furnishing the *t*-Bu-PHOX **7** to form a proper Rh(I)-catalyst, reaction of organozinc reagents with *cis*-2,4-dimethylglutaric anhydride delivers chiral δ-keto acids.[11]

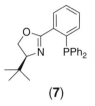

(7)

Aminolysis of *meso*-epoxides is facilitated by Sc(OTf)$_3$. In the presence of bipyridyldiol **8** chiral products are obtained.[12] *meso-N*-Acylaziridines react with Me$_3$SiN$_3$ to provide β-azido amines, and a chiral Bronsted acid (e.g., **9**) renders the ring opening asymmetrical.[13]

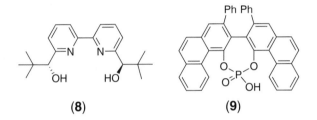

(8)   **(9)**

Reductive conversion of *meso*-cyclic imides to ω-hydroxyalkanamides is enantioselective when rendered by the hydrogenation catalyst **10**.[14]

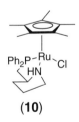

(10)

meso-1,2-Diols are desymmetrized by benzoylation in the presence of **11**[15] and tosylation in the presence of *ent-***1B**.[16] By virtue of diastereoselective selection enantioselective *N*-benzoylation of α-amino esters with the chiral reagent **12** has been achieved.[17]

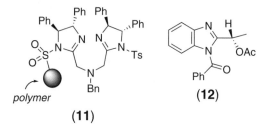

(11)   **(12)**

A synthesis of (-)-lobeline from the *cis*-diol precursor has been completed. Selective esterification followed by oxoidation and saponification are involved, the critical esterification step is mediated by **13**.[18]

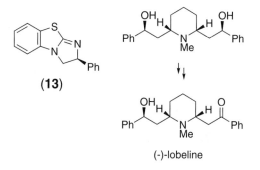

(13)

(-)-lobeline

Finding effective chiral ligands is the key to formation of semistabilized chiral lithioalkanes. Success has been demonstrated from a combination of BuLi and **1C** for benzyl trifluoromethyl sulfones[19] and that of *t*-BuLi and the sparteine surrogate **14** for *N*-Boc pyrrolidine.[20]

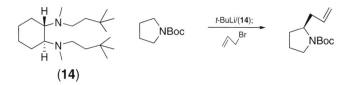

(14)

Chiral lithioamide bases of the α-phenthylamine type are known to perform asymmetric lithiation of *meso*-ketones. A fluorous analogue **15** (as LiCl complex) is now available for the purpose.[21]

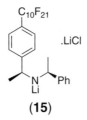

(15)

[1]Pellissier, H. *T* **64**, 1563 (2008).
[2]Hong, M.S., Kim, T.W., Jung, B., Kang, S.H. *CEJ* **14**, 3290 (2008).
[3]Yamada, S., Yamashita, K. *TL* **49**, 32 (2008).
[4]Onomura, O., Mitsuda, M., Nguyen, M.T.T., Demizu, Y. *TL* **48**, 9080 (2007).
[5]Arita, S., Koike, T., Kayaki, Y., Ikariya, T. *ACIE* **47**, 2447 (2008).
[6]Wills, M. *ACIE* **47**, 4264 (2008).
[7]Matsumoto, K., Otsuka, K., Okamoto, T., Mogi, H. *SL* 729 (2007).
[8]Soeta, T., Selim, K., Kuriyama, M., Tomioka, K. *T* **63**, 6573 (2007).
[9]Driver, T.G., Harris, J.R., Woerpel, K.A. *JACS* **129**, 3836 (2007).
[10]Nechab, M., Kumar, D.N., Philouze, C., Einhorn, C., Einhorn, J. *ACIE* **46**, 3080 (2007).
[11]Cook, M.J., Rovis, T. *JACS* **129**, 9302 (2007).
[12]Mai, E., Schneider, C. *CEJ* **13**, 2729 (2007).
[13]Rowland, E.B., Rowland, G.B., Rivera-Otero, E., Antilla, J.C. *JACS* **129**, 12084 (2007).
[14]Ito, M., Sakaguchi, A., Kobayashi, C., Ikariya, T. *JACS* **129**, 290 (2007).
[15]Arai, T., Mizukami, T., Yanagisawa, A. *OL* **9**, 1145 (2007).
[16]Demizu, Y., Matsumoto, K., Onomura, O., Matsumura, Y. *TL* **48**, 7605 (2007).
[17]Karnik, A.V., Kamath, S.S. *JOC* **72**, 7435 (2007).
[18]Birman, V.B., Jiang, H., Li, X. *OL* **9**, 3237 (2007).
[19]Nakamura, S., Hirata, N., Kita, T., Yamada, R., Nakane, D., Shibata, N., Toru, T. *ACIE* **46**, 7648 (2007).
[20]Stead, D., O'Brien, P., Sanderson, A. *OL* **10**, 1409 (2008).
[21]Matsubara, H., Maeda, L., Sugiyama, H., Ryu, I. *S* 2901 (2007).

Electrophilic substitution. Enantioselective fluorination by $(PhSO_2)_2NF$ is carried out with the aid of a Pd complex of **16A**.[1]

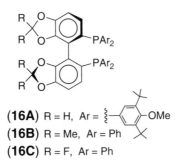

(16A) R = H, Ar = $\quad$—OMe
(16B) R = Me, Ar = Ph
(16C) R = F, Ar = Ph

Recent efforts pertaining to direct attachment of chiral sidechain to a heteroarene show that enantioselective Pictet–Spengler reaction is achievable in the presence of a multifunctional thiourea.[2] An interesting observation is that annulation of pyrroles by this method (with **17**) can give different isomers from CC bond formation at an α- or β-carbon.[3]

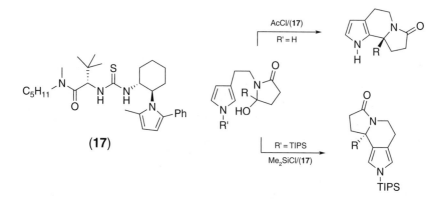

(17)

For asymmetric propargylation of furans the Ru complex **18** in which chirality instruction is furnished by a benzylic sulfide group is used.[4]

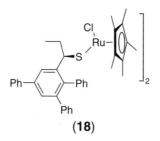

(18)

A rather general method for homologating secondary alkyl halides (e.g., bromides) is by the Ni-catalyzed reaction with organoboranes. Asymmetric induction by the C_2-symmetric diamine **19** is now realized.[5]

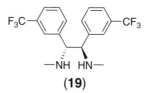

(19)

Potassium alkenyltrifluoroborates are activated by CAN for α-alkenylation of aldehydes which become chiral nucleophiles on condensation with **20A**.[6]

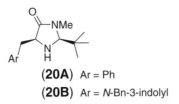

(20A) Ar = Ph
(20B) Ar = N-Bn-3-indolyl

Alkylation of tin enolates in the presence of a chiral (salen)-Cr complex shows moderate ee, due to rate differences for the reaction of two geometrically isomeric enol stannyl ethers.[7]

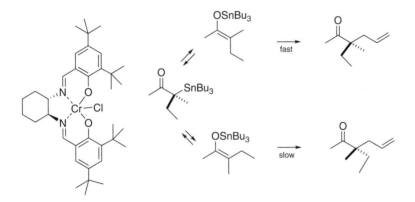

N,N′-Disubstituted thioureas exemplified by **21** help ionize the chlorine atom of a 1-chloroisochroman to generate oxocarbenium species that can be trapped in an enantio-selective sense.[8]

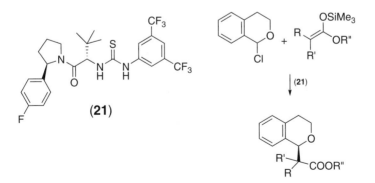

Alkylation of aldehydes with enol silyl ethers is accomplished on oxidation of the latter species with CAN. By forming chiral enamines from the aldehydes and the imidazolidinone **20** in situ the reaction furnishes optically active products.[9]

The utility of the Pd-complex of **22** is further extended to synthesis of chiral 2-alkoxy-4-pentenals from *vic*-alkoxyalkenyl allyl carbonates.[10]

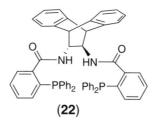

(22)

[1]Suzuki, T., Goto, T., Hamashima, Y., Sodoka, M. *JOC* **72**, 246 (2007).
[2]Raheem, I.T., Thiara, P.S., Peterson, E.A., Jacobsen, E.N. *JACS* **129**, 13404 (2007).
[3]Raheem, I.T., Thiara, P.S., Jacobsen, E.N. *OL* **10**, 1577 (2008).
[4]Matsuzawa, H., Migake, Y., Nishibayashi, Y. *ACIE* **46**, 6488 (2007).
[5]Saito, B., Fu, G.C. *JACS* **130**, 6694 (2008).
[6]Kim, H., MacMillan, D.W.C. *JACS* **130**, 398 (2008).
[7]Doyle, A.G., Jacobsen, E.N. *ACIE* **46**, 3701 (2007).
[8]Reisman, S.E., Doyle, A.G., Jacobsen, E.N. *JACS* **130**, 7198 (2008).
[9]Jang, H.-Y., Hong, J.-B., MacMillan, D.W.C. *JACS* **129**, 7004 (2007).
[10]Trost, B.M., Xu, J., Reichle, M. *JACS* **129**, 282 (2007).

Allylic substitutions. Allylic substitution is still largely dependent on Pd-catalysis because of its efficiency and mechanistic understanding. Many new chiral ligands are tested for their asymmetric induction, including as diverse as the ferrocenyl S,P-ligands **23**[1]/**24**[2] and **25**.[3]

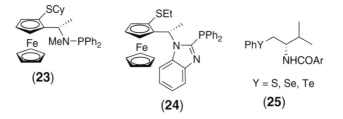

Ligands with coordination sites distributed over two different cyclopentadienyl units of the ferrocenyl nucleus are represented by **26**, for use in the Pd-catalyzed allylic substitution.[4]

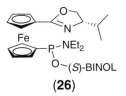

(26)

Molybdenum catalysts usually complement the Pd based species in terms of regio-chemical consequences. A particularly striking result obtained in the cinnamylation of 3-substituted oxindoles[5] indicates that, by example of difference in a 2-thienyl and a 3-methyl-2-thienyl substituents, variation in substrate structure can be significant and hence exploitable.

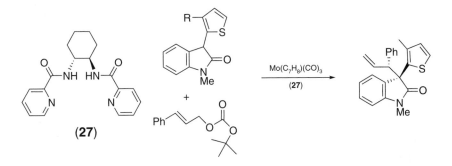

Substitution of allylic phosphates by alkenyldiisobutylalanes proceeds via the S_N2' route, $CuCl_2$ in combination with a dinuclear silver-carbene complex (**28**) are responsible for excellent asymmetric induction.[6]

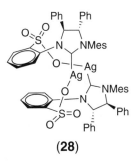

(28)

Mixed carbonates of unsymmetrical dialkenyl carbinols undergo *ipso*-substitution to afford chiral aryl ethers, with high degrees of regioselectivity and enantioselectivity when catalyzed by a Pd(0)-complex of **29**.[7]

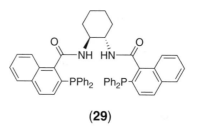

(29)

Certain *P*-chiral phosphorodiamidite ligands (e.g., **30A, 30B**) complex with iridium(I) to form catalysts for promoting reaction of cinnamyl methyl carbonate with ArOH to form chiral aryl α-vinylbenzyl ethers.[8]

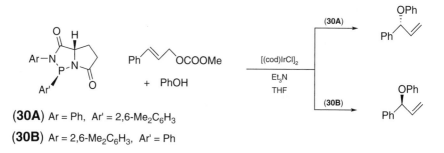

(30A) Ar = Ph, Ar' = 2,6-Me$_2$C$_6$H$_3$

(30B) Ar = 2,6-Me$_2$C$_6$H$_3$, Ar' = Ph

Branched allylic ethers are also obtained from a reaction of 1-chloro-2-alkenes with alcohols. A Ru catalyst (**31**) shows satisfactory activity.[9]

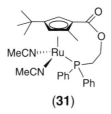

(31)

Synthesis of chiral allylic amines by substitution reaction has many choices of protocol, in terms of metal complexes and ligands. Pd catalysts having pairing with a *C$_2$*-symmetric ruthenocene (**32**) are quite novel[10] among other more conventional P,N-ligands that include **33**[11] and **34**.[12]

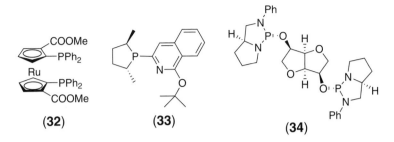

(32)   (33)   (34)

Cyclic diaminophosphine oxide **35** ligating to Pd is useful for substitution of allylic carbonates.[13] To access diamines from 2-vinylaziridines the catalyst system constituting **29** meets established standards.[14]

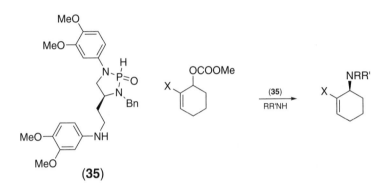

(35)

It is possible to prepare allylic boronates by a Cu-catalyzed reaction of allylic carbonates with bis(pinacolato)diboron. A chiral version of the reaction uses a QuinoxP ligand (**36**).[15]

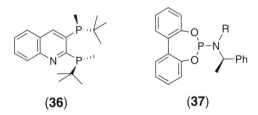

(36)   (37)

A chiral Ir(I) catalyst derived from the amino-(2,2′-biphenoxy)phosphine **37** promotes the synthesis of optically active 3-amino-1-alkenes from 2-alkenols, which are activated by (Eto)$_5$Nb.[16] 1-Vinyl-1,2,3,4-tetrahydroisoquinolines are obtained in good yields in the Pd-catalyzed process. Enantioselectivity is induced by the atropisomeric **38**.[17]

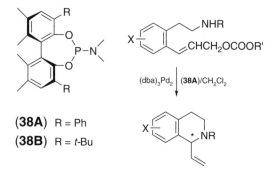

(38A) R = Ph
(38B) R = *t*-Bu

3-Butenoic esters bearing a chiral quaternary α-carbon center are prepared by Grignard reaction of α-substituted γ-chlorocrotonic esters in the presence of (4*R*, 5*R*)-diphenyl-1,3-dimesitylimidazolylidene.[18]

[1]Lam, F.L., Au-Yeung, T.T.-L., Kwong, F.Y., Zhou, Z., Wong, K.Y., Chan, A.S.C. *ACIE* **47**, 1280 (2008).

[2]Cheung, H.Y., Yu, W.-Y., Lam, F.L., Au-Yeung, T.T.-L., Zhou, Z., Chan, T.H., Chan, A.S.C. *OL* **9**, 4295 (2007).

[3]Vargas, F., Sehnem, J.A., Galetto, F.Z., Braga, A. *T* **64**, 392 (2008).

[4]Zhang, K., Peng, Q., Hou, X.-L., Wu, Y.-D. *ACIE* **47**, 1741 (2008).

[5]Trost, B.M., Zhang, Y. *JACS* **129**, 14548 (2007).

[6]Lee, Y., Akiyama, K., Gillingham, D.G., Brown, M.K., Hoveyda, A.H. *JACS* **130**, 446 (2008).

[7]Trost, B.M., Brennan, M.K. *OL* **9**, 3691 (2007).

[8]Kimura, M., Uozumi, Y. *JOC* **72**, 707 (2007).

[9]Onitsuka, K., Okuda, H., Sasai, H. *ACIE* **47**, 1454 (2008).

[10]Liu, D., Xie, F., Zhang, W. *JOC* **72**, 6992 (2007).

[11]Birkholz, M.-N., Dubrovina, N.V., Shuklov, I.A., Holz, J., Paciello, R., Waloch, C., Breit, B., Börner, A. *TA* **18**, 2055 (2007).

[12]Gavrilov, K.N., Zheglov, S.V., Vologzhanin, P.A., Maksimova, M.G., Safronov, A.S., Lyubimov, S.E., Davankov, V.A., Schäffner, B., Börner. *TL* **49**, 3120 (2008).

[13]Nemoto, T., Fukuyama, T., Yamamoto, E., Tamura, S., Fukuda, T., Matsumoto, T., Akimoto, Y., Hamada, Y. *OL* **9**, 927 (2007).

[14]Trost, B.M., Fandrick, D.R., Brodmann, T., Stiles, D.T. *ACIE* **46**, 6123 (2007).

[15]Ito, H., Ito, S., Sasaki, Y., Matsuura, K., Sawamura, M. *JACS* **129**, 14856 (2007).

[16]Yamashita, Y., Gopalarathnam, A., Hartwig, J.F. *JACS* **129**, 7508 (2007).

[17]Shi, C., Ojima, I. *T* **63**, 8563 (2007).

[18]Lee, Y., Hoveyda, A.H. *JACS* **128**, 15604 (2006).

Addition to C=O bond. Hydroxyalkylation of benzyl trifluoromethyl sulfones via lithiation can lead to chiral alcohols by adding the BOX ligand **1C** to the reaction medium.[1]

Addition of organozinc reagents to aldehydes still occupy the attention of many methodology developers, although, unfortunately, most of the works have not gone beyond certain model reactions of Et$_2$Zn and ArCHO. The addition is found to be enantioselective using a Ti complex of the fluorous TADDOL **39**.[2] Since many diamines and amino alcohols have high affinity to zinc metal it is not surprising that chiral ligands with such motifs emerge unabated. Akin in partial structure are **40**,[3] **41**,[4] **42**,[5] and **43**.[6]

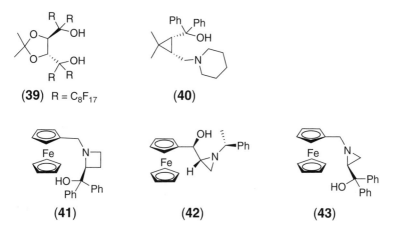

(39) R = C$_8$F$_{17}$ **(40)**

(41)   **(42)**   **(43)**

It should be noted that stoichiometric quantitites of R$_2$Zn transfer the R group to aldehydes, catalytic amounts of R$_2$Zn (and correspondingly the chiral ligands) serve as catalysts (as shown with **41**) in the addition involving organoboronic acids.[7]

The SAc group in **44** in enhancing asymmetric induction is ascribed to its strong affinity toward Zn such that the coordination sphere is more rigid. The 2-*exo*-morpholinobornane-10-thiol **45** perhaps cherishes the same advantages (in reactions involving alkenylzinc reagents).[8]

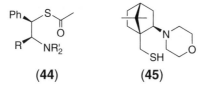

(44)   **(45)**

Ligand **46** has a secondary amino group,[9] whereas **47** performs better because it is a tridentate ligand.[10]

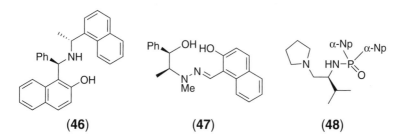

(46)   **(47)**   **(48)**

Modified diamines as ligands, including monophosphonamides (e.g., **48**[11]) and carboxamides (e.g., **49**[12]), have been scrutinized. In the use of **49** for forming complexes with the

Ni(II) ion as catalysts rather surprising and desirable results emerge. Chirality switch is observed from reactions mediated by a complex bearing one to that bearing two such ligands.

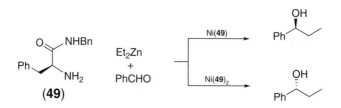

Chiral phthalides are synthesized from an *o*-iodobenzoic ester that forms a zinc compound. With a *P*-chiral diphosphinocobalt complex **50** present the addition to aldehydes follows an asymmetric course.[13]

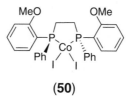

(50)

Methylmagnesium bromide modified by ZnCl$_2$ reacts with aldehydes in the presence of the Ti alkoxide derived from **51**. It leads to chiral 2-alkanols.[14] A Ti(IV) complex of the C_2-symmetric isophthalamide **52**[15] and a diastereomer of **42**[16] catalyze the addition of alkynylzinc reagents to aldehydes, whereas the polymer-linked hydroxy-imine **53** alone is used for the same purpose.[17]

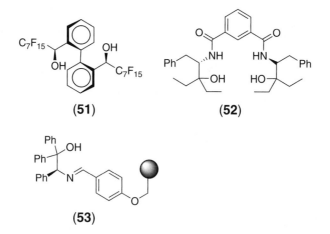

(51)   (52)

(53)

Allylic alcohols are synthesized from the reaction of aldehydes with alkenyl(ethyl)zincs that are complexed to (S)-2-diphenylhydroxymethyl-N-tritylaziridine.[18]

A piece of significant information concerning the organozinc addition is that mixed aggregates from achiral and chiral catalysts are formed and such dimers are responsible for enantiomeric reversal.[19] Another finding pertains to asymmetric amplification such that great enantiomeric enrichment of certain ligands by cooling, keeping in solution ligands of good quality. To carry out Ti-catalyzed diorganozinc addition to aldehydes in the presence of (1S,2S)-bis(triflylamino)cyclohexane, cooling a toluene solution of the ligand to −78° achieves the effect.[20]

Asymmetric addition of allylmetals to carbonyl compounds is also a well-represented reaction type. Several N-oxides (54[21], 55A[22]) are found to be effective catalysts for group transfer from allyltrichlorosilane and allylstannanes.

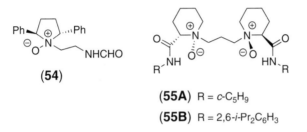

(54)

(55A) R = c-C$_5$H$_9$

(55B) R = 2,6-i-Pr$_2$C$_6$H$_3$

A Cr(III) complex of the binaphthyl that is 2,2′-disubstituted by a 7-t-butyl-8-hydroxy-quinol-2-yl group (**56**) is the source of chirality in the allenyl carbinols produced from reaction of propargylic bromides with aldehydes.[23]

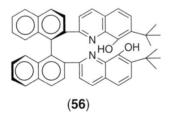

(56)

Homoallylic metallic species are generated from dienes in the presence of ent-**57B** and diastereoselective reaction with aldehydes has been observed.[24]

α-Keto esters are attacked by ArB(OH)$_2$ under catalysis by [(C$_2$H$_4$)$_2$RhCl]$_2$ and chiral tertiary benzylic alcohols are obtained when the spirodiindanyl phosphite **57A** is added to the reaction media.[25]

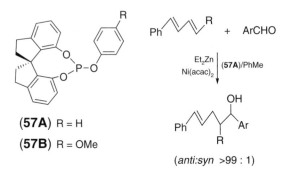

(57A) R = H
(57B) R = OMe

(*anti:syn* >99 : 1)

Addition of cyanide ion to the carbonyl group is sterically directed by a thiourea **58**, a cooperative catalyst capable of simultaneous hydrogen bonding with the oxygen atom of the acceptor molecule and guiding the cyanide ion by the protonated tertiary amine.[26]

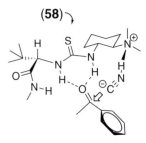

Titanium chelates of semi-salen **59**[27] and salen **60**[28] are used in asymmetric synthesis of α-cyanoalkyl ethyl carbonates from aldehydes and ethyl cyanoformate. By changing the metal atom to aluminum for complexing **60** a catalyst for elaborating α-acetoxy amides (Passerini reaction) is obtained (but enantioselectivity varies).[29]

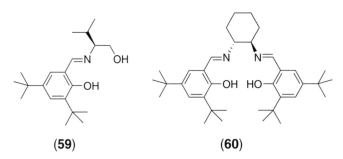

(59)   **(60)**

[1]Nakamura, S., Hirata, N., Yamada, R., Kita, T., Shibata, N., Toru, T. *CEJ* **14**, 5519 (2008).
[2]Sokeirik, Y.S., Mori, H., Omote, M., Sato, K., Tarui, A., Kumadaki, I., Ando, A. *OL* **9**, 1927 (2007).
[3]Zhong, J., Guo, H., Wang, M., Yin, M., Wang, M. *TA* **18**, 734 (2007).
[4]Wang, M.-C., Zhang, Q.-J., Zhao, W.-X., Wang, X.-D., Ding, X., Jing, T.-T., Song, M.-P. *JOC* **73**, 168 (2008).
[5]Bulut, A., Aslan, A., Izgü, E.C., Dogan, Ö. *TA* **18**, 1013 (2007).

[6]Wang, M.-C., Wang, X.-D., Ding, X., Liu, Z.-K. *T* **64**, 2559 (2008).

[7]Jin, M.-J., Sarkar, S.M., Lee, D.-H., Qiu, H. *OL* **10**, 1235 (2008).

[8]Wu, H.-L., Wu, P.-Y., Uang, B.-J. *JOC* **72**, 5935 (2007).

[9]Szatmari, I., Sillanpää, R., Fülöp, F. *TA* **19**, 612 (2008).

[10]Parrott III, R.W., Dore, D.D., Chandrashekar, S.P., Bentley, J.T., Morgan, B.S., Hitchcock, S.R. *TA* **19**, 607 (2008).

[11]Hatano, M., Miyamoto, T., Ishihara, K. *OL* **9**, 4535 (2007).

[12]Burguete, M.I., Collado, M., Escorihuela, J., Luis, S.V. *ACIE* **46**, 9002 (2007).

[13]Chang, H.-T., Jeganmohan, M., Cheng, C.-H. *CEJ* **13**, 4356 (2007).

[14]Omote, M., Tanaka, N., Tarui, A., Sato, K., Kumadaki, I., Ando, A. *TL* **48**, 2989 (2007).

[15]Hui, X.-P., Yin, C., Chen, Z.-C., Huang, L.-N., Xu, P.-F., Fan, G.-F. *T* **64**, 2553 (2008).

[16]Koyuncu, H., Dogan, O. *OL* **9**, 3477 (2007).

[17]Chen, C., Hong, L., Zhang, B., Wang, R. *TA* **19**, 191 (2008).

[18]Braga, A.L., Paixao, M.W., Westermann, B., Schneider, P.H., Wessjohann, L.A. *SL* 917 (2007).

[19]Lutz, F., Igarashi, T., Kinoshita, T., Asahina, M., Tsukiyama, K., Kawasaki, T., Soai, K. *JACS* **130**, 2956 (2008).

[20]Satyanarayana, T., Ferber, B., Kagan, H.B. *OL* **9**, 251 (2007).

[21]Simonini, V., Benaglia, M., Pignataro, L., Guizzetti, S., Celentano, G. *SL* 1061 (2008).

[22]Zheng, K., Qin, B., Liu, X., Feng, X. *JOC* **72**, 8478 (2007).

[23]Xia, G., Yamamoto, H. *JACS* **129**, 496 (2007).

[24]Yang, Y., Zhu, S.-F., Duan, H.-F., Zhou, C.-Y., Wang, L.-X., Zhou, Q.-L. *JACS* **129**, 2248 (2007).

[25]Duan, H.-F., Xie, J.-H., Qiao, X.-C., Wang, L.-X., Zhou, Q.-L. *ACIE* **47**, 4351 (2008).

[26]Zuend, S.J., Jacobsen, E.N. *JACS* **129**, 15872 (2007).

[27]Wang, W., Gou, S., Liu, X., Feng, X. *SL* 2875 (2007).

[28]Chen, S.-K., Peng, D., Zhou, H., Wang, L.-W., Chen, F.-X., Feng, X.-M. *EJOC* 639 (2007).

[29]Wang, S.-X., Wang, M.-X., Wang, D.-X., Zhu, J. *ACIE* **47**, 388 (2008).

Aldol reaction has a new list of chiral catalysts, and they include **61**,[1] **62**,[2] and **63**.[3] The presence of a hydrophobic silyl group to mask the hydroxyl residue of serine in **63** is important, L-serine itself being inactive is an indication.

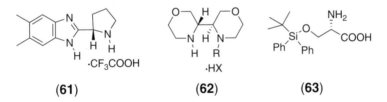

(61)   **(62)**   **(63)**

The dipeptide-derived **64**[4] is a suitable aldol reaction catalyst for handling haloacetones and α-hydroxyacetone, and the the water-compatibility of the analogous **65** underscores its utility in the reaction involving α,α′-dihydroxyacetone.[5]

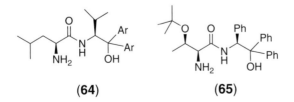

(64)   **(65)**

The chiral aldol donor *N*-azidoacetyl-4-phenylthiazolidine-2-thione forms a stable titanium enolate on treatment with $TiCl_4$ and *i*-Pr_2NEt in NMP and CH_2Cl_2 at $-78°$.[6] *syn*-Selective aldol reaction of 1,1-dimethoxy-2-alkanones is accomplished in the presence of diamine **66A**.[7] Also reported for other aldol reactions is **66B**.[8] The amide derived from phenylalanine and bispidine (**67**) promotes aldol reaction of functionalized ketone receptors.[9]

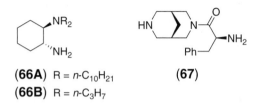

(66A) R = *n*-$C_{10}H_{21}$
(66B) R = *n*-C_3H_7

(67)

Chiral 3-hydroxyoxindoles can be synthesized from isatin by an asymmetric aldol reaction. The prolinamide **68** possesses just the right attributes of a catalyst to meet the demand.[10]

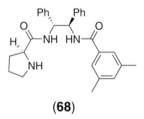

(68)

β-Hydroxyalkanitriles containing a chiral quaternary α-carbon atom are available from condensation of *N*-silyl ketene imines with aldehydes. The process is mediated by $SiCl_4$, and chiral information comes from **69**.[11]

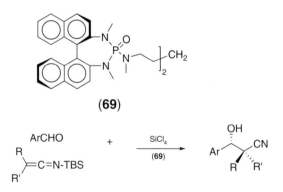

(69)

Metal chelates that find use as aldol reaction catalysts are further exemplified by the Al complex of salen **60**, which brings together α-isocyano amides and aldehyds to provide chiral 2-(α-hydroxyalkyl)-5-aminooxazoles.[12]

Copper(II) complexes of two imino nitrogen atoms belonging to chiral oxazoline and sulfoximine moieties (**70**) are able to elicit asymmetric consequences in the Mukaiyama-aldol reaction of enol silyl ethers and α-keto esters.[13]

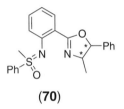

(70)

Reductive aldol reaction of 1-alken-3-ones and cinnamic esters depends on generating Rh enolates and the presence of chiral ligands turns such a process enantioselective. Effective ligands of very different structural types have been identified, and they include TADDOL phosphine **71**[14] and BOX ligand **72**.[15]

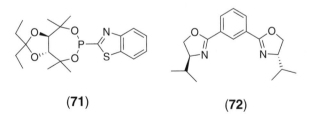

(71)   **(72)**

Copper hydride species generated in situ from hydrophenylsilane and a CuF complex initiates reductive aldol reaction by forming copper enolates (rather than enol silyl ethers). For accomplising a chiral reaction the ferrocenyl ligand **73** is added.[16]

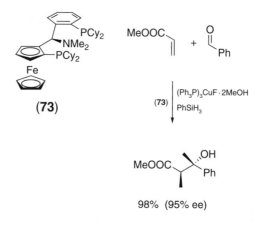

(73)

98% (95% ee)

With a Ph-BOX ligand (e.g., *ent-***1B**) to complex Cu(OTf)$_2$ for decarboxylative aldol reaction of substituted malonic acid monothioesters, *syn*-selectivity is observed.[17] This reaction operates on a different mechanism than enzyme-catalyzed decarboxylative Claisen condensation.

Intramolecular aldol reaction forms 3-hydroxycycloalkanones. Catalysis by natural α-amino acids reveals some unusual results: there is enantio-reversal in closing a 7-membered ring as compared with closure leading to 3-hydroxycyclohexanones.[18] Parenthetically, retroaldol cleavage of allyl bicyclo[3.2.0]heptan-2-on-5-yl carbonates gives 5-allyl-1,4-cycloheptanediones by asymmetric induction of **7**.[19]

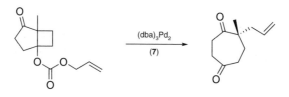

A ternary chelate of TiCl$_4$, sparteine and the *N*-acetyl derivative of the tricyclic thiazoli-dinethione **74** acts as a chiral donor in aldol reaction with aldehydes.[20] The tryptophan-derived oxazaborolidinone **75** is serviceable in completing the vinylogous Mukaiyama aldol reaction to furnish chiral products.[21]

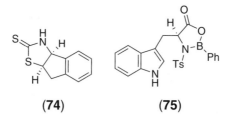

(74)   **(75)**

The asymmetric Henry reaction[22] is important because products are convertible to many other valuable bifunctional or polyfunctional compounds. Organocatalysts for the reaction containing a guanidine unit are represented by the C_2-symmetric **76**,[23] which directs the addition of nitroalkanes to α-keto esters asymmetrically.

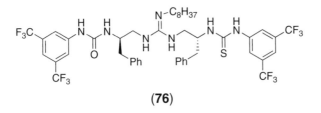

(76)

Tetraaminophosphonium salts such as **77** for catalyzing the Henry reaction have been developed.[24] 1,2-Diamines bearing chiral information to form metal complexes often can serve as catalyst for the reaction, and such is the case of N-(2-pyridylmethyl)isobornylamine (**78**) with Cu(OAc)$_2$,[25] and a Cu(II) complex of the salen **79**.[26] The bimetallic-salen complex **80A** shows catalytic activity for bringing about *anti*-selective Henry reaction.[27]

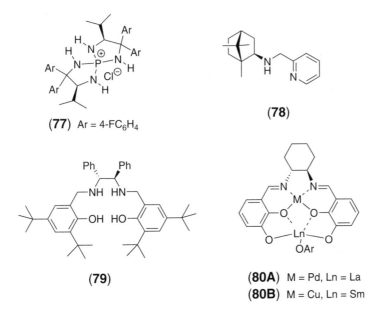

(77) Ar = 4-FC$_6$H$_4$

(78)

(79)

(80A) M = Pd, Ln = La
(80B) M = Cu, Ln = Sm

The C_2-symmetric ligands bis-N-oxide **55**[28] and cyclohexane spiroannulated to two oxazolidine units **81**,[29] in pairing with In(OTf)$_3$, and Me$_2$Zn, respectively, form active promoters for the addition of MeNO$_2$ to carbonyl compounds.

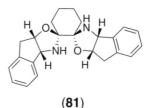

(81)

Acylsilanes are umpolung reagents that direct *C*-acylation of nitrones to give α-siloxyamino ketones. A TADDOL phosphite (**82**) engenders the reaction asymmetric.[30]

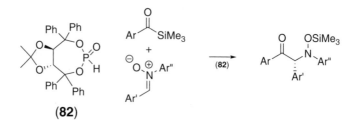

(82)

The carbonyl-ene reaction is a source of homoallylic alcohols, although the scope is somewhat limited. Synthesis of chiral tertiary α-hydroxycarboxylic esters from glyoxylic esters has been studied, and many metal catalysts of varying degree of effectiveness have been identified. The Ag-catalyzed reaction between enol silyl ethers and a glyoxylic ester is sterically controlled by the Pd-SEGPHOS complex.[31]

Other metal complexes reported for catalytic activities for the carbonyl-ene reaction are **83**,[32] **84**,[33] and Cu(OTf)$_2$-**85**.[34]

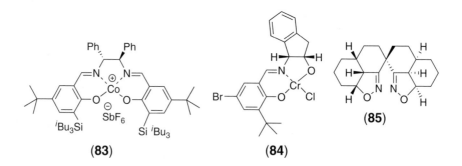

(83)   **(84)**   **(85)**

[1]Lacoste, E., Vaique, E., Berlande, M., Pianet, I., Vincent, J.-M., Landais, Y. *EJOC* 167 (2007).
[2]Kanger, T., Kriis, K., Laars, M., Kailas, T., Müürisepp, A.-M., Pehk, T., Lopp, M. *JOC* **72**, 5168 (2007).
[3]Teo, Y.-C. *TA* **18**, 1155 (2007).
[4]Xu, X.-Y., Wang, Y.-Z., Gong, L.-Z. *OL* **9**, 4247 (2007).
[5]Ramasastry, S.S.V., Albertshofer, K., Utsumi, N., Babas III, C.F. *OL* **10**, 1621 (2008).

[6]Patel, J., Clave, G., Renard, P.-Y., Franck, X. *ACIE* **47**, 4224 (2008).

[7]Luo, S., Xu, H., Chen, L., Cheng, J.-P. *OL* **10**, 1775 (2008).

[8]Luo, S., Xu, H., Li, J., Zhang, L., Cheng, J.-P. *JACS* **129**, 3074 (2007).

[9]Liu, J., Yang, Z., Wang, Z., Wang, F., Chen, X., Liu, X., Feng, X., Su, Z., Hu, C. *JACS* **130**, 5654 (2008).

[10]Chen, J.-R., Liu, X.-P., Zhu, X.-Y., Li, L., Qiao, Y.-F., Zhang, J.-M., Xiao, W.-J. *T* **63**, 10437 (2007).

[11]Denmark, S.E., Wilson, T.W., Burk, M.T., Heemstra Jr, J.R. *JACS* **129**, 14864 (2007).

[12]Wang, S.-X., Wang, M.-X., Wang, D.-X., Zhu, J. *OL* **9**, 3615 (2007).

[13]Sendelmeier, J., Hammerer, T., Bolm, C. *OL* **10**, 917 (2008).

[14]Bee, C., Han, S.B., Hassan, A., Krische, M.J. *JACS* **130**, 2746 (2008).

[15]Shiomi, T., Nishiyama, H. *OL* **9**, 1651 (2007).

[16]Deschamp, J., Chuzel, O., Hannedouche, J., Riant, O. *ACIE* **45**, 1292 (2006).

[17]Fortner, K.C., Shair, M.D. *JACS* **129**, 1032 (2007).

[18]Nagamine, T., Inomata, K., Endo, Y., Paquette, L.A. *JOC* **72**, 123 (2007).

[19]Schulz, S.R., Blechert, S. *ACIE* **46**, 3966 (2007).

[20]Osorio-Lozada, A., Olivo, H.F. *OL* **10**, 617 (2008).

[21]Simsek, S., Horzella, M., Kalesse, M. *OL* **9**, 5637 (2007).

[22]Palomo, C., Oiarbide, M., Laso, A. *EJOC* 2561 (2007).

[23]Takada, K., Takemura, N., Cho, K., Sohtome, Y., Nagasawa, K. *TL* **49**, 1623 (2008).

[24]Uraguchi, D., Sasaki, S., Ooi, T. *JACS* **129**, 12392 (2007).

[25]Blay, G., Domingo, L.R., Hernandez-Olmos, V., Pedro, J.R. *CEJ* **14**, 4725 (2008).

[26]Xiong, Y., Wang, F., Huang, X., Wen, Y., Feng, X. *CEJ* **13**, 829 (2007).

[27]Handa, S., Nagawa, K., Sohtome, Y., Matsunaga, S., Shibasaki, M. *ACIE* **47**, 3230 (2008).

[28]Qin, B., Xiao, X., Liu, X., Huang, J., Wen, Y., Feng, X. *JOC* **72**, 9323 (2007).

[29]Liu, S., Wolf, C. *OL* **10**, 1831 (2008).

[30]Garrett, M.R., Tarr, J.C., Johnson, J.S. *JACS* **129**, 12944 (2007).

[31]Mikami, K., Kawakami, Y., Akiyama, K., Aikawa, K. *JACS* **129**, 12950 (2007).

[32]Hutson, G.E., Dave, A.H., Rawal, V.H. *OL* **9**, 3869 (2007).

[33]Grachan, M.L., Tudge, M.T., Jacobsen, E.N. *ACIE* **47**, 1469 (2008).

[34]Wakita, K., Bajracharya, G.B., Arai, M.A., Takizawa, S., Suzuki, T., Sasai, H. *TA* **18**, 372 (2007).

Addition to C=N bond.   Strecker-type synthesis, the addition of Me_3SiCN to imines, on extending to *N*-phosphinoyl ketimines is enantioselective in the presence of **86**, and the optimal conditions involve the addition of 10 mol% of MCPBA.[1]

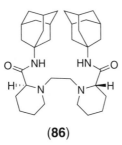

(86)

A titanium chelate of dihydroxylamine **87** helps organization of the addends so that α-amino nitriles of the (*R*)-configuration are generated.[2] Thiourea **88** with a proximal salen unit offers multiple hydrogen bonding sites for an analogous purpose of acetylcyanation of aldimines.[3]

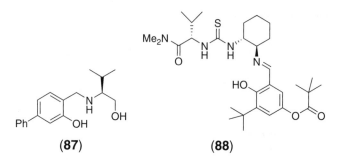

(87)   **(88)**

Organometallic reagents is subject to chiral modification therefore their addition to imines is readily rendered enantioselective. In binding R_2Zn during reaction N-(o-hydroxybenzyl)-valylphenylalanine amides **89** are versatile modifiers, considering the possibility of tuning by variation of the aryl substituent. For example, a 3,5-di-t-butyl-2-hydroxybenzyl group is particularly suitable for the addition to the N-(o-methoxyphenyl)imines of trifluoroacetylarenes.[4]

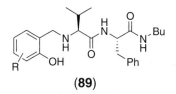

(89)

Diastereomeric benzylamines are obtained by cinnamylation of imines with alkenylsiladi-oxolane **90**. The stereochemical switch requires only a change of the N-substituent.[5] The cognate heterocycle **91** is useful catalyzing addition of ketene silyl ethers to hydrazones.[6]

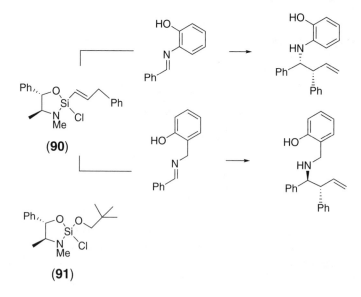

(90)

(91)

There is no denying an important role is played by the phenolic OH of the *N*-aryl group to determine the favorable transition state, although the role is less apparent in the allylation of such imines by allylstannanes, in view of a rather complicated ligand (**92**) is being employed.[7]

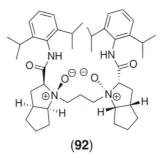

(92)

Urea **93** is a bifunctional ligand possessing a Lewis basic sulfinamide group. Its use in guiding the addition of allylindium bromide to acylhydrazones has been explored.[8] Allyl group transfer from *B*-allylpinacolatoboron to *N'*-aroylhydrazonoacetic esters is catalyzed by a zinc salt bound to the diamine **94**.[9] (Results are less than satisfactory in view of products with ee <90% being obtained.)

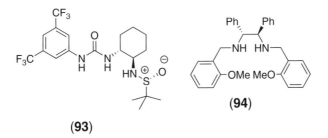

(93)   **(94)**

Chiral propargylic amines are formed by mixing aldehydes, *o*-anisidine, 1-alkynes with Me$_2$Zn and amino alcohol **95**.[10] Alkynylation of pyridinium salts is guided by CuI which is complexed to the BOX ligand **96**.[11]

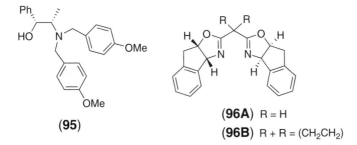

(95)

(96A) R = H
(96B) R + R = (CH$_2$CH$_2$)

The dihydroxybiaryl **97** can be used to exchange with alkenylboronate esters, bringing chirality in close proximity to the reaction site when the boronates participate in a Petasis reaction to build allylic amines.[12]

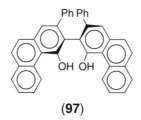

(97)

Imines are generated in situ from *N*-Boc α-sulfonylamines, therefore the adducts are useful precursors for coupling with ArB(OH)$_2$. Chiral benzhydrylamine derivatives are obtained when the Rh-catalyzed reaction is conducted in the presence of the pyrrolidinodiphosphine **98**.[13]

(98)

Synthesis of chiral allylic amines from alkynes and *N*-sulfonylaldimines involves reductive activation of the alkynes. The metal atom of iridacyclopropene intermediates also gathers the sulfonylimine as a bidentate ligand prior to bonding reorganization within the coordination sphere. The absolute stereochemical sense is governed by the chiral ligand employed (such as a member of the BIPHEP series).[14]

Articles summarizing current state of asymmetric addition to imines and highlighting Mannich reaction are available.[15,16] Special attention has also been devoted the employment of organocatalysts for the Mannich reaction.[17]

O-Silylserine **63**, the catalyst for aldol reaction, also actively promotes enantioselective Mannich reaction.[18] Excellent asymmetric induction and *anti*-selectivity are found in the Mannich reaction using 3-pyrrolidinecarboxylic acids **99**[19] and 3-triflylaminopyrrolidine **100**.[20] Since emphasis is placed on the importance of the carboxyl group of **99**, the acidic TfNH group of **100** must be similarly implicated.

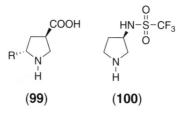

(99)   **(100)**

Considering the reactivity differences of ketones and esters as Mannich reaction donors the use of alkyl trichloromethyl ketones as surrogates of esters is a sound tactic. An asymmetric version is realized with a PYBOX ligated lanthanum aryloxide and LiOAr.[21] Coordination of the widened BOX ligand **101** to Mg furnishes a catalyst capable of inducing asymmetric cycloaddition of 3-(isothiocyanatoacetyl)-2-oxazolidinone with N-tosylaldimines, furnishing precursors of chiral α,β-diamino acids.[22]

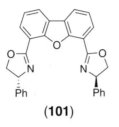

(101)

The higher reactivity of enol silyl ethers can be exploited in their reaction with imines. For addition to N-phosphonyl imines two types (SEGPHOS and DuPHOS) of ligands accommodate the variant substrates.[23]

The C_2-symmetric cyclohexane **102** that carries two thiourea groups induces the asymmetric coupling of a triphenylphosphoranylacetic ester with aldimines to give stabilized Wittig reagents containing a chirality center.[24]

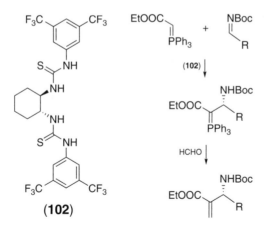

(102)

A urea (**103**) having the two nitrogen atoms as part of a β-amino alcohol and a sulfinamide, respectively, is endowed with interactive components to arrange the absolute configuration whereby nitroalkanes and aldimines react.[25] Thioureas **104**[26] and **105B**[27] are other such devices.

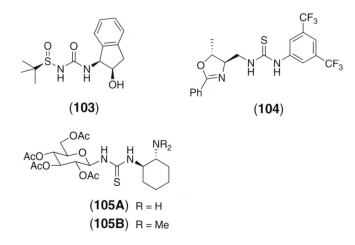

(103)   **(104)**

(105A) R = H
(105B) R = Me

A successful screening of the multitasking dinuclear zinc alkoxide **106** for catalyzing the aza-Henry reaction is no surprise.[28] On the other hand, identification of the heterobimetallic chelate **80B** is a new development.[29] The *syn : anti* product ratio of >20 : 1 and 83–98% ee in many cases vouchsafe for a general utility of the catalyst.

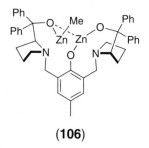

(106)

A monotriflate of tetramine (**107**) is used to engender the asymmetric addition of α-nitroalkanoic esters to imines.[30] The work seems to follow an evolving trend of partially perturbing highly efficient C_2-symmetric ligands in attempt to optimize their performance.

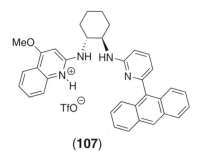

(107)

Camphor modified by replacing the 10-methyl group with a thiabicyclo[2.2.1]heptane nucleus (**108**) exerts chiral influences on the aza-Baylis–Hillman reaction.[31]

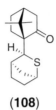

(108)

An α,β-unsaturated aldehyde adds to a nitrone to give γ-hydroxylaminoalkanoic ester when the substrates are exposed to an azolecarbene, and the reaction mixture is quenched by an alcohol. Homoenolate ion generated from the aldehyde and the carbene is the nucleophile. The use of carbene **109** engenders chiral products.[32]

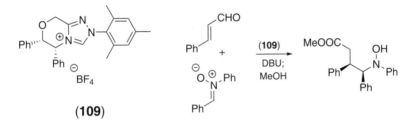

(109)

Closely related to imines are azodicarboxylic esters. Asymmetric amination is of course an important subject and a report on the utility of iridium complex **3** is on record.[33]

[1]Huang, J., Liu, X., Wen, Y., Qin, B., Feng, X. *JOC* **72**, 204 (2007).
[2]Banphavichit, V., Bhanthumravin, W., Vilaivan, T. *T* **63**, 8727 (2007).
[3]Pan, S.C., List, B. *OL* **9**, 1149 (2007).
[4]Fu, P., Snapper, M.L., Hoveyda, A.H. *JACS* **130**, 5530 (2008).
[5]Huber, J.D., Leighton, J.L. *JACS* **129**, 14552 (2007).
[6]Notte, G.T., Leighton, J.L. *JACS* **130**, 6676 (2008).
[7]Li, X., Liu, X., Fu, Y., Wang, L., Zhou, L., Feng, X. *CEJ* **14**, 4796 (2008).
[8]Tan, K.L., Jacobsen, E.N. *ACIE* **46**, 1315 (2007).
[9]Fujita, M., Nagano, T., Schneider, U., Hamada, T., Ogawa, C., Kobayashi, S. *JACS* **130**, 2914 (2008).
[10]Zani, L., Eichhorn, T., Bolm, C. *CEJ* **13**, 2587 (2007).
[11]Sun, Z., Yu, S., Ding, Z., Ma, D. *JACS* **129**, 9300 (2007).
[12]Lou, S., Schaus, S.E. *JACS* **130**, 6922 (2008).
[13]Nakagawa, H., Rech, J.C., Sindelar, R.W., Ellman, J.A. *OL* **9**, 5155 (2007).
[14]Ngai, M.-Y., Barchuk, A., Krische, M.J. *JACS* **129**, 12644 (2007).
[15]Ferraris, D. *T* **63**, 9581 (2007).
[16]Marques, M.M.B. *ACIE* **45**, 348 (2006).
[17]Ting, A., Schaus, S.E. *EJOC* 5797 (2007).
[18]Teo, Y.-C., Lau, J.-J., Wu, M.-C. *TA* **19**, 186 (2008).
[19]Zhang, H., Mitsumori, S., Utsumi, N., Imai, M., Garcia-Delgado, N., Mifsud, M., Albertshofer, K., Cheong, P.H.-Y., Houk, K.N., Tanaka, F., Barbas III, C.F. *JACS* **130**, 875 (2008).

[20]Pouliquen, M., Blanchet, J., Lasne, M.-C., Rouden, J. *OL* **10**, 1029 (2008).

[21]Marimoto, H., Lu, G., Aoyama, N., Matsunaga, S., Shibasaki, M. *JACS* **129**, 9588 (2007).

[22]Cutting, G.A., Stainforth, N.E., John, M.P., Kociok-Köhn, G., Willis, M.C. *JACS* **129**, 10632 (2007).

[23]Suto, Y., Kanai, M., Shibasaki, M. *JACS* **129**, 500 (2007).

[24]Zhang, Y., Liu, Y.-K., Kang, T.-R., Hu, Z.-K., Chen, Y.-C. *JACS* **130**, 2456 (2008).

[25]Robak, M.T., Trincado, M., Ellman, J.A. *JACS* **129**, 15110 (2007).

[26]Chang, Y., Yang, J., Dang, J., Xue, Y. *SL* 2283 (2007).

[27]Wang, C., Zhou, Z., Tang, C. *OL* **10**, 1707 (2008).

[28]Trost, B.M., Lupton, D.W. *OL* **9**, 2023 (2007).

[29]Handa, S., Gnanadesikan, V., Matsunaga, S., Shibasaki, M. *JACS* **129**, 4900 (2007).

[30]Singh, A., Johnston, J.N. *JACS* **130**, 5866 (2008).

[31]Myers, E.L., de Vries, J.G., Aggarwal, V.K. *ACIE* **46**, 1893 (2007).

[32]Phillips, E.M., Reynolds, T.E., Scheidt, K.A. *JACS* **130**, 2416 (2008).

[33]Hasegawa, Y., Watanabe, M., Gridnev, I.D., Ikariya, T. *JACS* **130**, 2158 (2008).

Conjugate additions.   Organocatalysis is enjoying great popularity, therefore a host of information has accumulated. Past years have witnessed publication of reviews on the usage of organocatalysts for conjugate additions.[1,2] While many of these catalysts are derived from (*S*)-proline and cinchona alkaloids, the tryptophan derivative **20B** has found an application in mediating transfer of alkenyl groups (from potassium alkenyltrifluoroborates) to 2-butenal.[3]

Asymmetric addition of split TBS-CN to enones is effectively performed by a gadolinium complex of **110**.[4] Traditional conjugate addition of organometallic reagents in the presence of a copper salt is subject to intervention by chiral ligands, and the discovery that simple monodentates such as **111** works well (addition of R_2Zn and R_3Al) is a revelation.[5] The congeneric O,O'-biaryl phosphoramidite **112** shows the same level of activity as expected.[6]

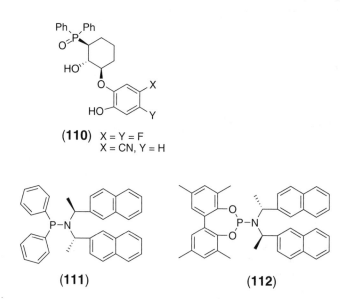

(110) X = Y = F
          X = CN, Y = H

(111)   **(112)**

A flexible biphenyl residue contributes to the effectiveness of such ligands (**113, 114**) for the Rh-catalyzed delivery of aryl groups from ArB(OH)$_2$ to enones.[7] But for the conjugate addition of a bulky alkyne to enones the pairing of Rh(I) with SEGPHOS **16A** is designed for exploiting steric advantages.[8]

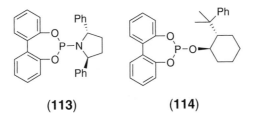

(113)   **(114)**

Copper(II) triflate forms many complexes with BOX ligands including **1E**, which negotiates the delivery of the allyl group from allyltrimethylsilane to α-methoxycarbonylated cycloalkenones while establishing a new stereocenter in the (*R*)-configuration.[9]

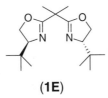

(1E)

Calcium isopropoxide complexed to the simplest Ph-BOX ligand serves as a Bronsted base and chiral catalyst for rendering glycine *t*-butyl ester into a nucleophile toward acrylic esters.[10]

Aryl transfer from ArSi(OEt)$_3$ to conjugated ketones, lactones and lactams is achieved with the aid of a palladium(II) salt supported by **116**.[11] It is a variation of the reaction involving ArB(OH)$_2$ with a similar system. The *P,P'*-dioxide of the same ligand complements CuOTf to serve as catalyst for the addition of R$_2$Zn to nitroalkenes.[12]

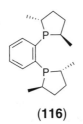

(116)

The Cu(II) complex of **117** outperforms the congener possessing a BINOL moiety in conjugate addition of organometallics to nitroalkenes, the *o,o′*-dioxydiphenylmethane unit is subject to conformational changes as determined by the bis-(α-phenethyl)amino chirality.[13]

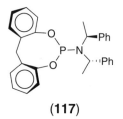

(117)

Ferrocenyldiphosphine **118A** issues chiral information on complexation to CuBr to direct 1,6-addition of Grignard reagents to α,β; γ,δ-unsaturated carbonyl compounds.[14]

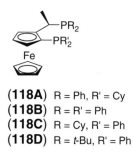

(118A) R = Ph, R' = Cy
(118B) R = R' = Ph
(118C) R = Cy, R' = Ph
(118D) R = *t*-Bu, R' = Ph

A series of 1,2-diarylethane-1,2-diamines and/or their metal complexes are effective conjugate addition catalysts involving stabilized donors. In reaction of enamides with alkylidenemalonic esters[15] a Cu(II) complex of **119** is employed, whereas the strontium complex of the bis(sulfonamide) **120** mediates the addition of malonic esters to enones.[16]

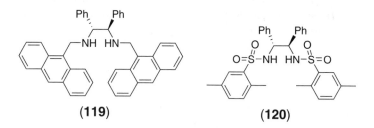

(119)   **(120)**

The differentially modified diamine **121**, having an areneulfonyl substituent at one end and a thiocarbamoyl group attached to the other nitrogen atom, tests well for catalytic activity in the addition of β-diketones to nitroalkenes.[17] Incorporation of the two amino groups into a cyclic guanidine, resulting in **122**, a new chiral catalyst for addition of *t*-butyl diphenyl-methyleneiminoacetate to acrylic esters.[18]

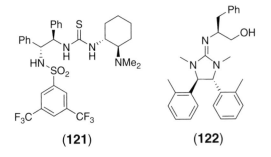

(121)   **(122)**

Thiourea **123** while bearing only one chiral carbon atom is an adequate catalyst.[19] Although much less commonly employed in the present context for calcium salts, one such appears to be able to team up with *ent*-**96A** to direct asymmetric Michael reaction involving a glycine derivative.[20]

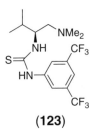

(123)

An aluminum complex of **56** is found useful to direct enantioselective addition to conjugated ketophosphonates.[21]

Unusually concentrated efforts have been spent to optimizing the conjugate addition to nitroalkenes. Useful organocatalysts for ketones donors are **124**,[22] **125**,[23] and **105A**.[24] The unsymmetrical 3,3′-dimorpholine **62** (**R** = *i*-**Pr**) is targeted for use in the case of aldehydes.[25]

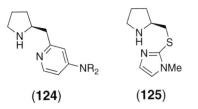

(124)   **(125)**

Owing to its Lewis acidity, zinc complex **106** lends itself to schemes for alkylation of electron-rich arenes and heteroarenes such as pyrroles by way of conjugate addition.[26] An even more valuable method is the 3-component condensation that unites indole with a nitroalkene and an aldehyde. Three contiguous stereocenters are established in a controlled manner and in an absolute sense by conducting the reaction in the presence of CuOTf, **126**, and hexafluoroisopropanol.[27]

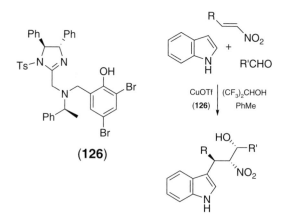

A synthesis of chiral 5-substituted 3-pyrazolidinones involves addition of hydrazines to conjugated imides, catalysis by the Mg complex of **96B**.[28]

Initiated by conjugate addition of iodide ion, which is under stereocontrol by the chiral auxiliary of an *N*-alkenyl-2-oxazolidinone, a tandem intramolecular alkylation is also enantioselective. Based on this reasoning it is possible to prepare cyclic compounds with new stereocenters of defined absolute configuration.[29]

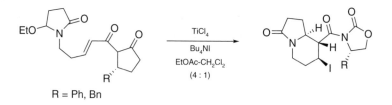

R = Ph, Bn

[1] Almasi, D., Alonso, D.A., Najera, C. *TA* **18**, 299 (2007).

[2] Tsogoeva, S.B. *EJOC* 1701 (2007).

[3] Lee, S., MacMillan, D.W.C. *JACS* **129**, 15438 (2007).

[4] Tanaka, Y., Kanai, M., Shibasaki, M. *JACS* **130**, 6072 (2008).

[5]Palais, L., Mikhel, I.S., Bournaud, C., Micouin, L., Falciola, C.A., Vuagnoux-d'Augustin, M., Rosset, S., Bernardinelli, G., Alexakis, A. *ACIE* **46**, 7462 (2007).

[6]Vagnoux-d'Augustin, M., Kehrli, S., Alexakis, A. *SL* 2057 (2007).

[7]Monti, C., Gennari, C., Piarulli, U. *CEJ* **13**, 1547 (2007).

[8]Nishimura, T., Guo, X.-X., Uchiyama, N., Katoh, T., Hayashi, T. *JACS* **130**, 1576 (2008).

[9]Shizuka, M., Snapper, M.L. *ACIE* **47**, 5049 (2008).

[10]Saito, S., Tsubogo, T., Kobayashi, S. *JACS* **129**, 5364 (2007).

[11]Gini, F., Hessen, B., Feringa, B.L., Minnaard, A.J. *CC* 710 (2007).

[12]Cote, A., Lindsay, V.N.G., Charette, A.B. *OL* **9**, 85 (2007).

[13]Wakabayashi, K., Aikawa, K., Kawauchi, S., Mikami, K. *JACS* **130**, 5012 (2008).

[14]den Hartog, T., Harutyunyan, S.R., Font, D., Minnaard, A.J., Feringa, B.L. *ACIE* **47**, 398 (2008).

[15]Berthiol, F., Matsubara, R., Kawai, N., Kobayashi, S. *ACIE* **46**, 7803 (2007).

[16]Agostinho, M., Kobayashi, S. *JACS* **130**, 2430 (2008).

[17]Wang, C.-J., Zhang, Z.-H., Dong, X.-Q., Wu, X.-J. *CC* 1431 (2008).

[18]Ryoda, A., Yajima, N., Haga, T., Kumamoto, T., Nakanishi, W., Kawahata, M., Yamaguchi, K., Ishikawa, T. *JOC* **73**, 133 (2008).

[19]Andres, J.M., Manzano, R., Pedrosa, R. *CEJ* **14**, 5116 (2008).

[20]Kobayashi, S., Tsubogo, T., Saito, S., Yamashita, Y. *OL* **10**, 807 (2008).

[21]Takenaka, N., Abell, J.P., Yamamoto, H. *JACS* **129**, 742 (2007).

[22]Ishii, T., Fujioka, S., Sekiguchi, Y., Kotsuki, H. *JACS* **126**, 9558 (2004).

[23]Xu, D.-Q., Wang, L.-P., Luo, S.-P., Wang, Y.-F., Zhang, S., Xu, Z.-Y. *EJOC* 1049 (2008).

[24]Liu, K., Cui, H.-F., Nie, J., Dong, K.-Y., Li, X.-J., Ma, J.-A. *OL* **9**, 923 (2007).

[25]Sulzer-Mosse, S., Laars, M., Kriis, K., Kanger, T., Alexakis, A. *S* 1729 (2007).

[26]Trost, B.M., Müller, C. *JACS* **130**, 2438 (2008).

[27]Arai, T., Yokoyama, N. *ACIE* **47**, 4989 (2008).

[28]Sibi, M., Soeta, T. *JACS* **129**, 4522 (2007).

[29]Koseki, Y., Fujino, K., Takeshita, A., Sato, H., Nagasaka, T. *TA* **18**, 1533 (2007).

Cycloadditions. Asymmetric Simmons–Smith reaction of allylic alcohols performed in the presence of an aluminum complex of the salen **127** has been reported.[1]

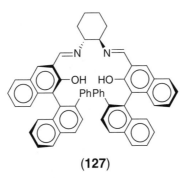

(127)

A Cu complex of the bispidine **128** catalyzes the decomposition of ethyl diazoacetate. Trapping of the carbenoid with alkenes (e.g., styrene) gives chiral cyclopropanecarboxylic esters. In the case of styrene, ethyl *cis*-2-phenylcyclopropanecarboxylate is obtained in 91% ee, although much lower value (79% ee) for the *trans*-isomer. Both products have an

(S)-configuration at C-1.[2] With a change to the Ir complex of salen **127** in which the metal is also σ-bonded to an aromatic ring, *t*-butyl *cis*-2-arylcyclopropanecarboxylates are obtained almost exclusively and ee value reaches 97–99%.[3]

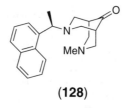

(128)

3-Substituted (2*R*,3*R*)-ethyl aziridine-2-carboxylates are synthesized from imines and ethyl diazoacetate. The catalyst system is composed from (PhO)$_3$B and (S)-VAPOL.[4]

Asymmetric cyclopropanation of electron-deficient alkenes can be carried out with a Co(II) porphyrinate in which chiral substituents are set in two disjunct *meso*-positions.[5]

The presence of **129**[6] or **130**[7] renders the Corey–Chaykovsky method for cyclopropanation of conjugated aldehydes asymmetric. Thus it is easy to access (1*S*,2*R*)-2-formylcyclopropyl ketones from enals and acymethylsulfonium ylides.[7]

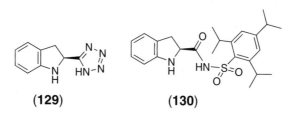

(129)   **(130)**

Another modification of the reaction entails the use of a La-Li$_3$ complex of **131**.[8]

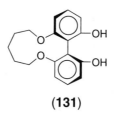

(131)

Guanidinium ylide **132** generated in situ reacts with aldehydes to give aziridine-2-carboxylic esters (*cis/trans* isomer mixtures).[9] Chiral information located four bonds away is transmitted and it guides the alignment of the reactants.

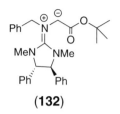

(132)

By forming adducts with azolecarbenes, cycloaddition of ketenes with imines to form β-lactams is facilitated. Azolecarbenes such as **133**[10] and **134**[11] induce chirality because in the enolates (initial adducts) the elements of asymmetry can dictate the approach of the reactants. Of particular interest is the reaction between enals and conjugated imines that leads to bicyclic structures.[11]

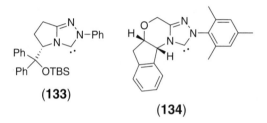

(133)

(134)

A powerful catalyst for [2+2]cycloaddition is created from the bicyclic oxazaborolidine **135** and AlBr₃. A hydrindanone frequently used in total synthesis of natural products is readily available from one such adduct in chiral form.[12]

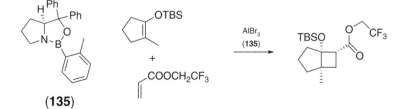

(135)

Other notable catalysts are **136**[13] and **137**,[14] the use of the latter is in conjunction with Cu(NTf₂)₂.

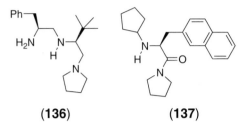

(136)   **(137)**

Molecules containing allene and alkene units that are separated by several bonds are subject to cycloisomerization. With a catalyst derived from AuCl and a SEGPHOS ligand **16**, chiral products of formal [2+2]cycloadditon are obtained.[15]

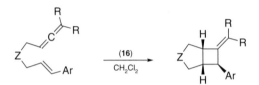

Asymmetric 1,3-dipolar cycloaddition reactions have been reviewed,[16] but ongoing research is still vigorous. The effectiveness of **138**, a phosphine modified from serine, for synthesizing cyclopentenecarboxylic esters from 2,3-alkadienoic esters and electron-deficient alkenes has been validated.[17] Phosphine-thiourea **139** is useful for directing enantioselective combination of allenes and imines.[18]

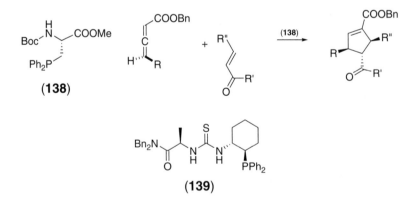

(138)

(139)

Besides organocatalysts, metal complexes show catalytic activities in various types of [3+2]cycloaddition. Thus Ru complex **140** performs a role in the reaction of nitrile oxides with enals,[19] and gold(I) benzoate complex of Cy-SEGPHOS is involved in the reaction of münchnones with alkenes.[20]

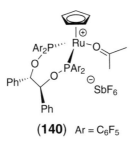

(140) Ar = C$_6$F$_5$

Azomethine ylides are typical 1,3-dipoles, their participation in [3+2]cycloaddition is greatly influenced by Cu catalysts. Asymmetric reactions are realized by the addition of chiral ferrocene ligands such as **141** (called Fesulphos)[21] and **142**.[22]

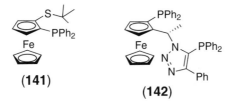

(141)   **(142)**

The ferrocenyl P,N-ligands **143A** and **143B** differ in ability to form hydrogen bonds in the transition state of 1,3-dipolar cyloaddition involving azomethine ylides and dimethyl maleate, and they give rise to adducts of opposite enantiomeric series.[23]

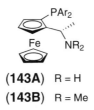

(143A) R = H
(143B) R = Me

An analogous ligand in which the amino group (of **143A/B**) is replaced by a *p*-anisylthio residue has also been scrutinized.[24]

Decomposition of an aryl diazoacetate by CuOTf in the presence of a conjugated carbonyl compound leads to a 2,3-dihydrofuran-2-carboxylate, the result of a formal [4+1]cycloaddition. To acquire a chiral product the presence of bipyridyl **144** is needed.[25]

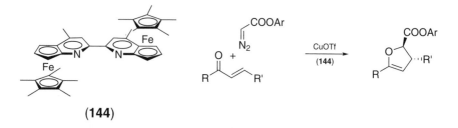

(144)

The great importance of the Diels–Alder and hetero-Diels–Alder reactions in synthesis is a strong stimulus for finding new aspects about them, especially those methodologically related, and chiral catalysts rank high in such a context. Accordingly, **145**,[26] **1D**,[27] and **146**[28] are valuable additions to the list of the metal-free entities, even **1D** is somewhat inferior due to relatively low asymmetric induction (up to 70% ee) it tenders during the reaction of anthrones and maleimides.

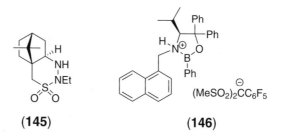

(145)   **(146)**

The versatility of **135**-AlBr$_3$ complex for catalyzing enantioselective reactions is further demonstrated by its gainful use in the Diels–Alder reaction.[29] Cu-BOX **147** is reusable and it is recovered from the reaction mixture by precipitation as a charge complex with a trinitrofluorenone.[30] The nickel complex of **101** is effective in catalyzing the hetero-Diels–Alder reaction of N-sulfonyl-1-aza-1,3-dienes with enol ethers.[31]

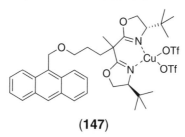

(147)

Cu-BOX ligands show desirable affinity for alkenyl 2-(N-oxidopyridyl) ketones, and they are useful asymmetric inducers in the Diels–Alder reaction.[32] The corresponding pyridyl ketones form adducts with low ee, indicating the transition state for a large portion of the reaction does not involve such a metal complex, even if it is present.

Siloxymethyl alkenyl ketones **148**, in which chirality is created by placing a methoxy substituent at an α-carbon of the silyl group, form complexes with Mg(OTf)$_2$ to become chiral dienophiles.[33] 2-(1-Methylimidazolyl) alkenyl ketones undergo enantioselective Diels–Alder reaction in water, in the presence of a DNA-based catalyst.[34]

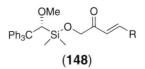

(148)

It is found that chiral dienes form better performing cationic Rh complexes than diphosphines, for use in catalyzing intramolecular Diels–Alder reaction of conjugate diene and alkyne units.[35] A cationic Ru(I) catalyst **140** operates on the basis of one-point association of the dienophile prior to establishment of the transition state for the Diels–Alder reaction. The most effective case demonstrated thus far is an intramolecular process.[36]

By ensconcing an In(III) ion in the *N*-oxide **55B** originated from two homochiral pipecolinic amides a catalyst for hetero-Diels–Alder reaction involving the Danishefsky diene and aldehydes is obtained.[37] Cr(salen) **149** appears to have similar capability.[38]

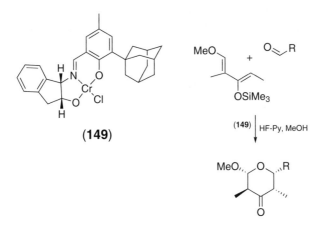

A representative of the hetero-Diels–Alder reaction of inverse electron demand is the cycloaddition of *N*-sulfonyl-1-azadienes with vinyl ethers. It is amenable to asymmetric catalysis, for example, by a nickel(II) complex of **101**.[39]

When a conjugated ketene generated in situ adds to the Er(III) complex of amino alcohol **150**, its conformation is fixed. The metal center also attracts an aldehyde to proceed with the hetero-Diels–Alder reaction.[40]

(150)

Adducts of diorganozinc reagents with ethyl 2,3-butadienoate also show diene character toward the carbonyl group. The hetero-Diels–Alder reaction is asymmetrized in the presence of Cu(OAc)$_2$ and DIFLUORPHOS.[41]

Synthesis of tropanes by a [4+3]cycloaddition becomes asymmetric when it is directed by the Rh carboxylate **151**.[42]

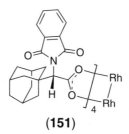

(151)

The Pauson–Khand reaction belongs to the [2+2+1]cycloaddition category. The Rh-catalyzed version is made asymmetric by ligating the metal center to **152**.[43] Substrate-control via 1,3-asymmetric induction for the establishment of a new stereocenter at C-4 of the emerging cyclopentenone system is the key to an approach to the hemigerans.[44]

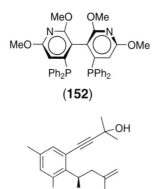

(152)

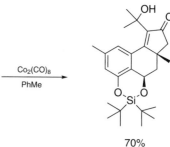

70%

Bicyclic enones originated from 1,6-enynes are also prepared using the (cod)IrPF$_6$ complex of **7**.[45]

In the elaboration of cyclopentenones the Nazarov cyclization (not a cycloaddition reaction) also offers some advantages. Metal complexes including a V(IV) chelate of salen **60**[46] and the Cu complex of *ent*-**96B**[47] have been employed as catalysts. By structural demand of the substrate to induce rearrangement following the cyclization a synthesis of spirocycles is realized.

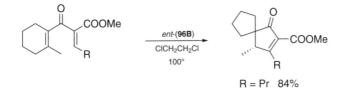

R = Pr 84%

Cross-conjugated dienones carrying a chiral auxiliary, suitable for Nazarov cyclization to provide chiral cyclopentenones, have been prepared from reaction of lithiated ethers **153/154**[48] and **155**[49] with *N*-alkenoylmorpholines.

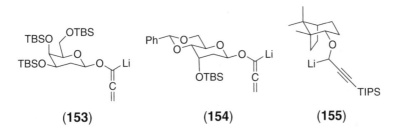

[1]Shitama, H., Katsuki, T. *ACIE* **47**, 2450 (2008).
[2]Lesme, G., Cattenati, C., Pilati, T., Sacchetti, A., Silvani, A. *TA* **18**, 659 (2007).
[3]Kanchiku, S., Suematsu, H., Matsumoto, K., Uchida, T., Katsuki, T. *ACIE* **46**, 3889 (2007).
[4]Lu, Z., Zhang, Y., Wulff, W.D. *JACS* **129**, 7185 (2007).
[5]Chen, Y., Ruppel, J.V., Zhang, X.P. *JACS* **129**, 12074 (2007).
[6]Hartikka, A., Arvidsson, P.I. *JOC* **72**, 5874 (2007).
[7]Hartikka, A., Slosarczyk, A.T., Arvidsson, P.I. *TA* **18**, 1403 (2007).
[8]Kakei, H., Sone, T., Sohtome, Y., Matsunaga, S., Shibasaki, M. *JACS* **129**, 13410 (2007).
[9]Disadee, W., Ishikawa, T. *JOC* **70**, 9399 (2005).
[10]Zhang, Y.-R., He, L., Wu, X., Shao, P.-L., Ye, S. *OL* **10**, 277 (2008).
[11]He, M., Bode, J.W. *JACS* **130**, 418 (2008).
[12]Canales, E., Corey, E.J. *JACS* **129**, 12686 (2007).
[13]Ishihara, K., Nakano, K. *JACS* **129**, 8930 (2007).
[14]Ishihara, K., Fushimi, M. *JACS* **130**, 7532 (2008).
[15]Luzung, M.R., Mauleon, P., Toste, F.D. *JACS* **129**, 12402 (2007).
[16]Pelliser, H. *T* **63**, 3235 (2007).
[17]Cowen, B.J., Miller, S.J. *JACS* **129**, 10988 (2007).

[18]Fang, Y.-Q., Jacobsen, E.N. *JACS* **130**, 5660 (2008).

[19]Brinkman, Y., Madhushaw, R.J., Jazzar, R., Bernardinelli, G., Kündig, E.P. *T* **63**, 8413 (2007).

[20]Melhado, A.D., Luparia, M., Toste, F.D. *JACS* **129**, 12638 (2007).

[21]Cabrera, S., Arrayás, R.G., Martín-Matute, B., Cossío, F.P., Carretero, J.C. *T* **63**, 6587 (2007).

[22]Fukuzawa, S., Oki, H. *OL* **10**, 1747 (2008).

[23]Zeng, W., Chen, G.-Y., Zhou, Y.-G., Li, Y.-X. *JACS* **129**, 750 (2007).

[24]Zeng, W., Zhou, Y.-G. *TL* **48**, 4619 (2007).

[25]Son, S., Fu, G.C. *JACS* **129**, 1046 (2007).

[26]He, H., Pei, B.-J., Chou, H.-H., Tian, T., Chan, W.-H., Lee, A.W.M. *OL* **10**, 2421 (2008).

[27]Akalay, D., Dürner, G., Göbel, M.W. *EJOC* 2365 (2008).

[28]Payette, J.N., Yamamoto, H. *JACS* **129**, 9536 (2007).

[29]Liu, D., Canales, E., Corey, E.J. *JACS* **129**, 1498 (2007).

[30]Chollet, G., Guillerez, M.-G., Schulz, E. *CEJ* **13**, 992 (2007).

[31]Esquivias, J., Arrayás, R.G., Carretero, J.C. *JACS* **129**, 1480 (2007).

[32]Barroso, S., Blay, G., Pedro, J.R. *OL* **9**, 1983 (2007).

[33]Campagna, M., Trzoss, M., Bienz, S. *OL* **9**, 3793 (2007).

[34]Boersma, A.J., Feringa, B.L., Roelfes, G. *OL* **9**, 3647 (2007).

[35]Shintani, R., Sannohe, Y., Tsuji, T., Hayashi, T. *ACIE* **46**, 7277 (2007).

[36]Rickerby, J., Vallet, M., Bernardinelli, G., Viton, F., Kündig, E.P. *CEJ* **13**, 3354 (2007).

[37]Yu, Z., Liu, X., Dong, Z., Xie, M., Feng, X. *ACIE* **47**, 1308 (2008).

[38]Dilger, A.K., Gopalsamuthiram, V., Burke, S.D. *JACS* **129**, 16273 (2007).

[39]Esquivias, J., Arrayás, R.G., Carretero, J.C. *JACS* **129**, 1480 (2007).

[40]Tiseni, P.S., Peters, R. *OL* **10**, 2019 (2008).

[41]Oisaki, K., Zhao, D., Kanai, M., Shibasaki, M. *JACS* **129**, 7439 (2007).

[42]Reddy, R.P., Davies, H.M. *JACS* **129**, 10312 (2007).

[43]Lee, H.W., Kwong, F.Y., Chan, A.S.C. *SL* 1553 (2008).

[44]Madu, C.E., Lovely, C.J. *OL* **9**, 4697 (2007).

[45]Lu, Z.-L., Neumann, E., Pfaltz, A. *EJOC* 4189 (2007).

[46]Walz, I., Bertogg, A., Togni, A. *EJOC* 2650 (2007).

[47]Huang, J., Frontier, A.J. *JACS* **129**, 8060 (2007).

[48]Banaag, A.R., Tius, M.A. *JACS* **129**, 5328 (2007).

[49]Dhoro, F., Kristensen, T.E., Stockman, V., Tius, M.A. *JACS* **129**, 7256 (2007).

Epoxidation and other oxidation reactions.   Regenerative pre-oxidants of the type **156A** are derived from a pyranose. They are employed in conjunction with a expendable agent, such as oxone for epoxidation of conjugated *cis*-enynes,[1] and H_2O_2 to epoxidize alkenes.[2]

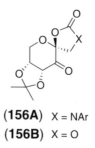

(156A) X = NAr
(156B) X = O

Among new oxidation systems based on H_2O_2 for asymmetric epoxidation of styrenes the catalytic component comprises either the Ti(IV) complex of SALAN **157**,[3] or a mixture of $FeCl_3$, 2,6-pyridinedicarboxylic acid, and **158**.[4]

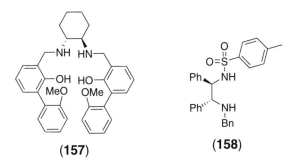

(157)   **(158)**

For enantioselective epoxidation of allylic alcohols with a hydroperoxide, a catalyst derived from vanadyl isopropoxide and a C_2-symmetric (*R,R*)-bis[*N*-hydroxy-*N*-(3,3,3-triarylpropanoyl)]-1,2-cyclohexanediamine has been developed[5]

Two chiral epoxidizing agents for enones are brucine *N*-oxide[6] and cyclo(L-Pro-L-Pro) hydroperoxide,[7] but the latter reagent produces epoxides of low ee.

cis-Dimethoxylation of *o*-hydroxystyrenes with a bimetallic Pd-Cu catalyst system under O_2 is enantioselective by the introduction of a chiral 2-(2-quinolyl)-4-isopropyloxazoline.[8]

L-Menthyl *syn*-3-amino-2-hydroxyalkanoates are acquired from menthyl acetate via enolization, Mannich reaction, and oxidation with an oxabornanesultam.[9]

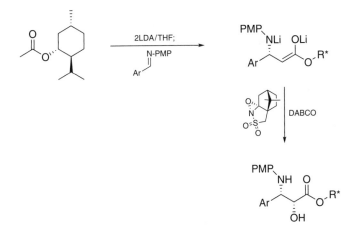

Optimization of asymmetric oxidation of sulfides catalyzed by a Fe(salan) complex has been carried out.[10] Using alkyl hydroperoxide as oxidant for sulfides, a Ti(IV) chelate of mixed tartaric esters in which one of the alkoxy residue is a ω-methylated polyethylene glycol chain provides chiral instructions.[11]

Enantioselective oxidation of hindered disulfides to monosulfoxides is accomplished in the presence of **156B**.[12]

[1]Burke, C.P., Shi, Y. *JOC* **72**, 4093 (2007).
[2]Burke, C.P., Shu, L., Shi, Y. *JOC* **72**, 6320 (2007).
[3]Shimada, Y., Kondo, S., Ohara, Y., Matsumoto, K., Katsuki, T. *SL* 2445 (2007).
[4]Gelalcha, F.G., Bitterlich, B., Anilkumar, G., Tse, M.K., Beller, M. *ACIE* **46**, 7293 (2007).
[5]Zhang, W., Yamamoto, H. *JACS* **129**, 286 (2007).
[6]Oh, K., Ryu, J. *TL* **49**, 1935 (2008).
[7]Kienle, M., Argyrakis, W., Baro, A., Laschat, S. *TL* **49**, 1971 (2008).
[8]Zhang, Y., Sigman, M.S. *JACS* **129**, 3076 (2007).
[9]Hata, S., Tomioka, K. *T* **63**, 8514 (2007).
[10]Egami, H., Katsuki, T. *SL* 1543 (2008).
[11]Gao, J., Guo, H., Liu, S., Wang, M. *TL* **48**, 8453 (2007).
[12]Khiar, N., Mallouk, S., Valdivia, V., Bougrin, K., Soufiaoui, M., Fernandez, I. *OL* **9**, 1255 (2007).

Hydrogenation and reduction of C=C bonds. For asymmetric hydrogenation of styrenes several iridium complexes are serviceable. These metal complexes (e.g., **159**,[1] **160**,[2] and **161**[3]) are prepared from [(cod)IrCl]₂ via ligand exchange. In using complex **143** the best solvent seems to be propylene carbonate; high ee of 1-methyltetralin is obtainable from hydrogenation of 1-methylenetetralin, and lesser amount of substrate is isomerized to 1-methyl-3,4-dihydronaphthalene which undergoes hydrogenation to give the methyltetralin product in the enantiomeric series. A QUINAP complex of iridium is valued for asymmetric hydrogenation of styrenes.[4]

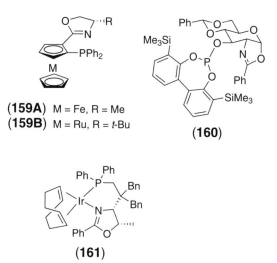

The complex **162** is an outstanding catalyst because it can be used in hydrogenation of unactivated tetrasubstituted alkenes.[5] With **163** diastereoselective hydrogenation of farnesol isomers is achieved: $(E, E) \rightarrow (R, R), (Z, E) \rightarrow (S, R), (E, Z) \rightarrow (R, S), (Z, Z) \rightarrow (S, S)$, all with ee >99%.[6]

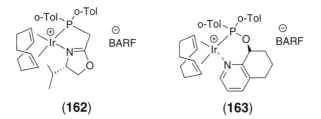

(162)   **(163)**

Little or no C-F bond hydrogenolysis occurs during hydrogenation of fluoroalkenes in the presence of catalyst **164** therefore its use on such occasions is recommended.[7]

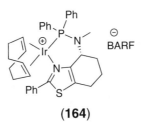

(164)

Rh(I) supported by ferrocenylphosphines **165** shows comparable activity to that of the catalyst containing **107B** (Josiphos) in hydrogenation of dimethyl itaconate,[8] therefore no advantage is gained in its use.

(165)

A more complicated ferrocenylphosphine ligand is **166** and its complex with Rh(I) catalyzes asymmetric hydrogenation of β-phthalimidomethylcinnamic esters.[9]

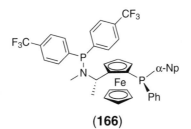

(166)

More Rh catalysts have been tested for the hydrogenation of dehydroamino ester deriva-tives. Two of them incorporate ligands **167**[10] and **168**.[11] The latter seems to have a broader substrate scope, for example in reduction of acrylic esters. Also useful for the same purpose is an iridium(I) complex of **38B**.[12] Furthermore, **169**[13] is active for hydrogenation of enol phosphinates.

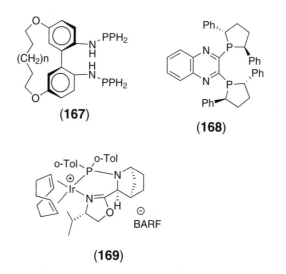

(167)

(168)

(169)

Rh complex **170**[14] and the one derived from [(cod)₂Rh]BF₄ and diphosphine **171**[15] are active in promoting reduction of α-ureido-α,β-unsaturated esters and enamides, respectively.

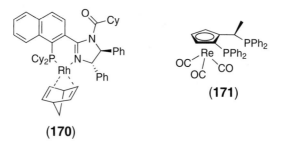

(170)

(171)

N-Boc pyrroles are subject to partial or complete reduction. With a Ru complex and 2,2′-bis(1-diphenylphosphinoethyl)-1,1′-biferrocenyl ligand to effect hydrogenation chiral pyrrolidines are synthesized.[16]

Reductive amination of α-chloroacetophenones by trichlorosilane leads to benzylic amines which are useful precursors of 2-arylaziridines. Optically active amines are obtained when the reduction is conducted in the presence of a N-formyl-N-methylvalinamide.[17] The conjugated double bond of a β-methylcinnamonitrile is reduced by PMHS, and since the reaction involves a copper hydride, the metal coordinated with a ligand such as 118C[18] forms a chiral catalyst. Similarly, reduction of conjugated esters is accomplished by PMHS with CuOAc and 118D present.[19] Reduction of alkenyl sulfones is similarly manipulatable, and the P,P'-dioxide of 116 is an alternative ligand for the copper center.[20]

Pinacolatoborylalkenes are subject to asymmetric hydrogenation using (nbd)₂RhBF₄ and 173 (Walphos 1).[21] Secondary boronates thus acquired are sources of many chiral functional molecules.

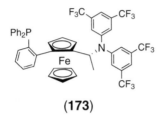

(173)

[1]Li, X., Li, Q., Wu, X., Gao, Y., Xu, D., Kong, L. *TA* **18**, 629 (2007).
[2]Dieguez, M., Mazuela, J., Pamies, O., Verendel, J.J., Andersson, P.G. *JACS* **130**, 7208 (2008).
[3]Bayardon, J., Holz, J., Schäffner, B., Andrushko, V., Verevkin, S., Preetz, A., Börner, A. *ACIE* **46**, 5971 (2007).
[4]Li, X., Kong, L., Gao, Y., Wang, X. *TL* **48**, 3915 (2007).
[5]Schrems, M.G., Neumaun, E., Pfaltz, A. *ACIE* **46**, 8274 (2007).
[6]Wang, A., Wüstenberg, B., Pfaltz, A. *ACIE* **47**, 2298 (2008).
[7]Engman, M., Diesen, J.S., Paptchikhine, A., Andersson, P.G. *JACS* **129**, 4536 (2007).
[8]Almassy, A., Barta, K., Francio, G., Sebesta, R., Leitner, W., Toma, S. *TA* **18**, 1893 (2007).
[9]Deng, J., Duan, Z.-C., Huang, J.-D., Hu, X.-P., Wang, D.-Y., Yu, S.-B., Xu, X.-F., Zheng, Z. *OL* **9**, 4825 (2007).
[10]Wei, H., Zhang, Y.J., Dai, Y., Zhang, J., Zhang, W. *TL* **49**, 4106 (2008).
[11]Fox, M.E., Jackson, M., Lennon, I.C., Klosin, J., Abboud, K.A. *JOC* **73**, 775 (2008).
[12]Giacomina, F., Meetsma, A., Panella, L., Lefort, L., de Vries, A.H.M., de Vries, J.G. *ACIE* **46**, 1497 (2007).
[13]Cheruku, P., Gohil, S., Andersson, P.G. *OL* **9**, 1659 (2007).
[14]Busacca, C.A., Lorenz, J.C., Grinberg, N., Haddad, N., Lee, H., Li, Z., Liang, M., Reeves, D., Sahe, A., Varsolona, R., Senanayake, C.H. *OL* **10**, 341 (2008).
[15]Stemmler, R.T., Bolm, C. *TL* **48**, 6189 (2007).
[16]Kuwano, R., Kashiwabara, M., Ohsumi, M., Kusano, H. *JACS* **130**, 808 (2008).
[17]Malkov, A.V., Stoncius, S., Kocovsky, P. *ACIE* **46**, 3722 (2007).
[18]Lee, D., Yang, Y., Yun, J. *S* 2233 (2007).
[19]Lipshutz, B.H., Lee, C.-T., Servesko, J.M. *OL* **9**, 4713 (2007).
[20]Desrosiers, J.-N., Charette, A.B. *ACIE* **46**, 5955 (2007).
[21]Moran, W.J., Morken, J.P. *OL* **8**, 2413 (2006).

Hydrogenation and reduction of C=O bond.   Three metals make up the major classes of homogeneous hydrogenation catalysts for the reduction of the carbonyl group. A Cp*Ir complex **3** with a bidentate ligand of monosulfonylated diphenylethanediamine is able to promote reduction of α-hydroxy ketones[1] and with the same catalyst (or enantiomer) to rapidly transfer hydrogenate aryl ketones even in the air.[2] For the latter purpose two other useful Ir complexes are that which contain the ligands **174**[3] and **175**.[4]

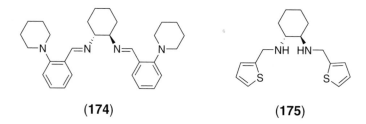

(174)   (175)

Transfer hydrogenation of aryl ketones from *i*-PrOH under basic conditions is also cata-lyzed by Rh(II) complexes. A series of 1,2-cyclohexanediamine arenesulfonamides **176**[5] as well as the hydroxamic acid **177**[6] derived from valine prove effective as chiral modifiers. Complexation of a chiral bis(*o*-[diphenylphosphino]ferrocenyl)-methanol [**178** is the (*R*,*R*)-isomer] to a Rh(I) salt provides a hydrogenation catalyst for benzenesulfonylmethyl aryl ketones.[7]

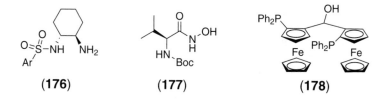

(176)   (177)   (178)

Complexation of Rh or Ru center to *N*-α-phenethyl amides of *N*-Boc-α-thioamino acids (but not the amino acid derivatives) forms highly effective catalysts for asymmetric hydrogenation of aryl ketones.[8]

Following the long tradition of ruthenium-based complexes related to homogeneous hydrogenation of ketones, further studies have shown that biaryldiphosphines **179**[9] and **16B**[10] are valuable contributors of chiral instruction in forming catalysts for hydrogenating α-keto esters. In the employment of Ru(II)-**16B** the reduction of ethyl mesitoylformate is dramatically improved by $CeCl_3 \cdot 7H_2O$, without which the conversion falls below 5%.

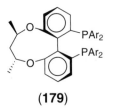

(179)

Transfer hydrogenation of aryl ketones in the presence of **159B** has been reported.[11] Catalysts catering to reduction of chloromethyl aryl ketones and α-amino ketones are **180**[12] and **181**,[13] respectively, besides their other uses. An analogue of **180** assists reduction of cyclic β-keto esters to cis-β-hydroxy esters by HCOOH–Et₃N via dynamic kinetic resolution.[14]

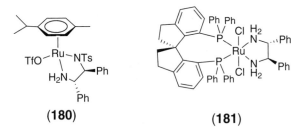

(180)   **(181)**

α-Keto esters undergo transfer hydrogenation from Hantzsch esters and the process is rendered asymmetric by a BOX-Cu(OTf)₂ complex.[15]

Diphosphine **37** forms a Pd complex to catalyze allylic substitution, and it also derives a Ru catalyst for asymmetric hydrogenation of β-keto esters.[16]

Hydrosilylation of aryl ketones is subject to sterocontrol by binding the metal catalyst with a chiral ligand. The reaction based on Fe(OAc)₂ – PMHS is modified by **116**,[17] whereas the one catalyzed by AgBF₄ and [(cod)MCl]₂ is dominated by **182**.[18] Interestingly, a change of the central metal in the complex from Rh to Ir reverses the enantioselectivity.

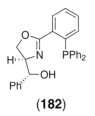

(182)

When o-substituted aryl ketones are reduced using a copper catalyst, the steric course is influenced by $(2S,4S)$-bis(diphenylphosphino)pentane.[19]

The possibility of chiral modification of $NaBH_4$ for reduction of ketones has been scrutinized. Some success is reported by adding the inexpensive cyclic borinate **183** derived from tartaric acid and $PhB(OH)_2$ to the reaction medium.[20] Atropo-enantioselective reduction of biaryl lactones occurs on exposure to $NaBH_4$ in the presence of the cobalt chelate **184**.[21]

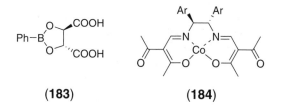

(183)   **(184)**

The (S)-diphenylprolinol-derived oxazaborolidine with an ethanediolated boron atom is a new catalyst for the asymmetric reduction of ketones with $BH_3 \cdot SMe_2$.[22] 1,3-Cyclo-alkanediones undergo CBS-reduction to provide $(3R)$-hydroxycycloalkanones. Based on this method a very short synthesis of chiral estrone methyl ether is completed.[23]

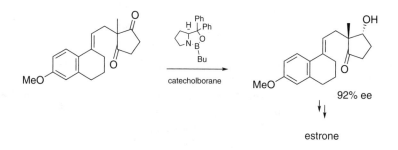

estrone

[1]Ohkuma, T., Utsumi, N., Watanabe, M., Tsutsumi, K., Arai, N., Murata, K. *OL* **9**, 2565 (2007).
[2]Wu, X., Li, X., Zanotti-Gerosa, A., Pettman, A., Liu, J., Mills, A.J., Xiao, J. *CEJ* **14**, 2209 (2008).
[3]Shen, W.-Y., Zhang, H., Zhang, H.-L., Gao, J.-X. *TA* **18**, 729 (2007).
[4]Zhang, X.-Q., Li, Y.-Y., Zhang, H., Gao, J.-X. *TA* **18**, 2049 (2007).
[5]Cortez, N.A., Aguirre, G., Parra-Hake, M., Somanathan, R. *TA* **19**, 1304 (2008).
[6]Ahlford, K., Zaitsev, A.B., Ekström, J., Adolfsson, H. *SL* 2541 (2007).
[7]Zhang, H.-L., Hou, X.-L., Dai, L.-X., Luo, Z.-B. *TA* **18**, 224 (2007).
[8]Zaitsev, A., Adolfsson, H. *OL* **8**, 5129 (2006).
[9]Sun, X., Zhou, L., Li, W., Zhang, X. *JOC* **73**, 1143 (2008).
[10]Meng, Q., Sun, Y., Ratovelomanana-Vidal, V., Genet, J.-P., Zhang, Z. *JOC* **73**, 3842 (2008).
[11]Liu, D., Xie, F., Zhao, X., Zhang, W. *T* **64**, 3561 (2008).
[12]Ohkuma, T., Tsutsumi, K., Utsumi, N., Arai, N., Noyori, R., Murata, K. *OL* **9**, 255 (2007).
[13]Liu, S., Xie, J.-H., Wang, L.-X., Zhou, Q.-L. *ACIE* **46**, 7506 (2007).

[14]Ros, A., Magriz, A., Dietrich, H., Lassaletta, J.M., Fernández, R. *T* **63**, 7532 (2007).
[15]Yang, J.W., List, B. *OL* **8**, 5653 (2006).
[16]Imamoto, T., Nishimura, M., Koide, A., Yoshida, K. *JOC* **72**, 7413 (2007).
[17]Shaikh, N.S., Enthaler, S., Junge, K., Beller, M. *ACIE* **47**, 2497 (2008).
[18]Frölander, A., Moberg, C. *OL* **9**, 1371 (2007).
[19]Shimizu, H., Igarashi, D., Kuriyama, W., Yusa, Y., Sayo, N., Saito, T. *OL* **9**, 1655 (2007).
[20]Eagon, S., Kim, J., Yan, K., Haddenham, D., Singaram, B. *TL* **48**, 9025 (2007).
[21]Ashizawa, T., Tanaka, S., Yamada, T. *OL* **10**, 2521 (2008).
[22]Stepanenko, V., De Jesus, M., Correa, W., Guzman, I., Vazquez, C., de la Cruz, W., Ortiz-Marciales, M., Barnes, C.L. *TL* **48**, 5799 (2007).
[23]Yeung, Y.-Y., Chein, R.-J., Corey, E.J. *JACS* **129**, 10346 (2007).

Hydrogenation and reduction of C=N bond. Chiral Bronsted acids possessing a bulky backbone such as VAPOL derivative **9** attract and hold imine molecules in the concave space, and this reasoning has led to successful development of a protocol for the synthesis of α-amino acid derivatives from imino precursors by transfer hydrogenation (from Hantzsch ester).[1]

The iridium(I) salt containing **185** has also been employed as catalyst in hydrogenation of imines.[2] In the hydrogenation of 2-substituted quinolines to provide chiral tetrahydro derivatives iridium complexes of SYNPHOS and DIFLUORPHOS prove to be effective catalysts.[3,4] Oximes undergo enantioselective reduction (and N—O bond cleavage) on treatment with borane and spirocyclic boronate **186**.[5]

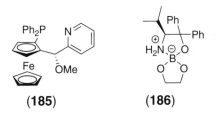

(185)   **(186)**

Reduction via hydrosilylation with trichlorosilane does not requires a metal. Asymmetric reduction is achieved in the presence of the picolinic amide of (1*R*,2*S*)-ephedrine,[6] or an S-chiral sulfinamide.[7]

[1]Li, G., Liang, Y., Antilla, J.C. *JACS* **129**, 5830 (2007).
[2]Cheemala, M.N., Knochel, P. *OL* **9**, 3089 (2007).
[3]Chan, S.H., Lam, K.H., Li, Y.-M., Xu, L., Tang, W., Lam, F.L., Lo, W.H., Yu, W.Y., Fan, Q., Chan, A.S.C. *TA* **18**, 2625 (2007).
[4]Deport, C., Buchotte, M., Abecassis, K., Tadaoka, H., Ayad, T., Ohshima, T., Genet, J.-P., Mashima, K., Ratovelomanana-Vidal, V. *SL* 2743 (2007).
[5]Huang, K., Merced, F.G., Ortiz-Marciales, M., Melender, H.J., Correa, W., De Jesus, M. *JOC* **73**, 4017 (2008).
[6]Zheng, H., Deng, J., Lin, W., Zhang, X. *TL* **48**, 7934 (2007).
[7]Pei, D., Wang, Z., Wei, S., Zhang, Y., Sun, J. *OL* **8**, 5913 (2006).

Isomerization and rearrangements. The 1,3-hydrogen shift of an enal to form ketene, when mediated by azolecarbene **109**, generates a chiral intermediate which would add onto a proximal C=O group stereoselectively. Desymmetrization thus leads to a chiral cycloalkene on decarboxylation of the cycloadduct.[1]

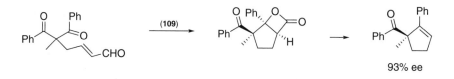

[2,3]Wittig rearrangement is rendered asymmetric by the presence of a BOX ligand (**1E**). The access to chiral aryl homoallyl carbinols in this manner is well adaptable to a synthetic route to lignans.[2]

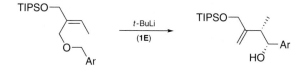

The Overman rearrangement has been extended to the synthesis of secondary allyl aryl ethers[3] and allylic thiol carbamates, using a palladized oxazolinylcobaltocene-type complex (**187**).[4] The former reaction could very well be considered as an SN2' process.

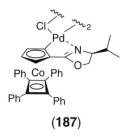

(**187**)

Ferrocenyl analogues **188**,[5] and a precatalyst version[6] combined with a silver salt, have been tapped for enantioselective aza-Claisen rearrangement. Excellent asymmetric induction ensues.

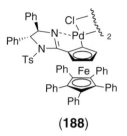

(188)

Diastereoselective control during enolization of allyl esters for Claisen rearrangement leads to predefined stereomers, and amide bases such as enantiomeric **189** are capable of generating chiral products.[7]

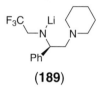

(189)

[1]Wadamoto, M., Phillips, E.M., Reynolds, T.E., Scheidt, K.A. *JACS* **129**, 10098 (2007).
[2]Hirokawa, Y., Kitamura, M., Maezaki, N. *TA* **19**, 1167 (2008).
[3]Kirsch, S.F., Overman, L.E., White, N.S. *OL* **9**, 911 (2007).
[4]Overman, L.E., Roberts, S.W., Sneddon, H.F. *OL* **10**, 1485 (2008).
[5]Fischer, D.F., Xin, Z.-q., Peters, R. *ACIE* **46**, 7704 (2007).
[6]Jautze, S., Seiler, P., Peters, R. *CEJ* **14**, 1430 (2008).
[7]Qin, Y.-c., Stivala, C.E., Zakarian, A. *ACIE* **46**, 7466 (2007).

Insertion reactions. The reaction of α-diazoalkanoic esters with an aldehyde is formally a carbene insertion into the CC bond between the formyl group and the α-carbon of the aldehyde. Taking advantage of substrate control for the reaction, esterification of the α-diazoalkanoic acid with an appropriate chiral alcohol provides the required substrate for conversion into the desired product.[1]

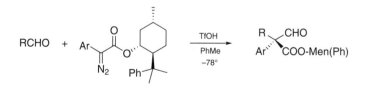

The most general method for preparation of chiral α-substituted esters is based on the insertion reaction that can be intimately influenced by a chiral additive. Ligands for Cu salts are good candidates because decomposition of diazoalkanes (including α-diazo-alkanoic esters) is known to be catalyzed by them. BOX **190** with a spirocyclic backbone is such a representative and its participation in the carbenoid insertion into the O—H bond of alcohols,[2] and the N—H bond of amines[3] is now on record. Another valuable ligand to complement Cu and Ag for N—H bond insertion is **144**.[4]

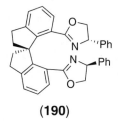

(190)

On the other hand, insertion of an amino group into C—H bonds in benzylic and allylic positions is accomplished by Rh catalysis (e.g., **191**) under oxidative conditions involving ArS(O)(NTs)NH$_2$.[5]

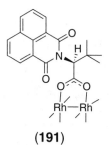

(191)

Redox condensation between two carbonyl groups by virtue of intramolecularity can be initiated by an azolecarbene. The more electrophilic formyl group that is attached to an aromatic ring undergoes umpolung by accepting the carbene/ylide then the adduct adds across a proximal carbonyl group. The use of a chiral carbene (e.g., **192**) naturally empowers enantiomerization of the reaction.[6]

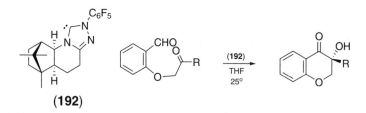

(192)

A different reaction course is followed when dicarbonyl compounds are treated with a Rh(I) salt. The acylrhodium hydride that is formed on insertion of the metal ion into the formyl C—H bond acts as reducing agent for the other carbonyl group. As expected, with Rh complexed to a chiral ligand the reaction gives optically active lactones.[7]

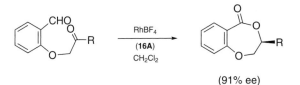

(91% ee)

[1]Hashimoto, T., Naganawa, Y., Maruoka, K. *JACS* **130**, 2434 (2008).
[2]Chen, C., Zhu, S.-F., Liu, B., Wang, L.-X., Zhon, Q.-L. *JACS* **129**, 12616 (2007).
[3]Liu, B., Zhu, S.-F., Zhang, W., Chen, C., Zhou, Q.-L. *JACS* **129**, 5834 (2007).
[4]Lee, E.C., Fu, G.-C. *JACS* **129**, 12066 (2007).
[5]Liang, C., Collet, F., Robert-Peillard, F., Müller, P., Dodd, R.H., Dauban, P. *JACS* **130**, 343 (2008).
[6]Li, Y., Feng, Z., You, S.-L. *CC* 2263 (2008).
[7]Shen, Z., Khan, H.A., Dong, V.M. *JACS* **130**, 2916 (2008).

Coupling reactions. 3-Substituted 2-methyleneindanones are obtained by a Heck reaction of *o*-alkenoylaryl triflates. The most remarkable feature of the reaction is the source of the exocyclic methylene group, it being originated from the *N*-methyl unit of the additive, 1,2,2,6,6-pentamethylpiperidine. Of course the chiral ligand **193** is the contributor of chirality.[1] The P,N-ligand **160** derived from glucosamine binds with Pd to catalyze Heck reaction of 2,3-dihydrofuran.[2]

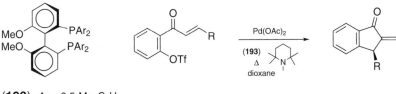

(**193**) Ar = 3,5-Me$_2$C$_6$H$_3$

A pleasing result pertains to enantioselective *o*-arylation of one of two identical aryl substituents of 2-diarylmethylpyridines.[3] Stereocontrol in the Pd insertion step is crucial and the chiral ligand (**194**) is the determining factor.

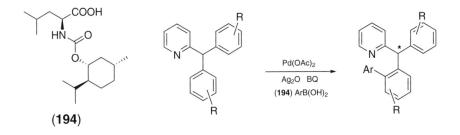

(**194**)

The organopalladium species generated from coupling reaction of ArB(OH)$_2$ with an allene is readily trapped by a properly distanced carbonyl group. Accordingly, 5,6-alkadienals are transformed into *cis*-2-(α-styryl)cyclopentanols. Adding (*S*)-SEGPHOS to complex the Pd salt has the desirable effect of asymmetric induction.[4]

Chiral 2,2-disubstituted dihydrobenzofurans in which one of the substituents is an alkenyl group can be synthesized from 2-allylphenols. A biaryl-2,2'-bisoxazoline (**195**) is the chiral-enabling ligand for the Pd-catalyst.[5]

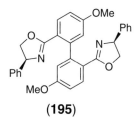

(195)

Petasis reaction on quinoline following an analogous course to the Reissert reaction is catalyzed by the thiourea **196**, water and NaHCO$_3$ are facilitating additives.[6]

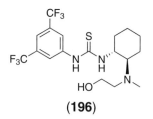

(196)

The *N,N'*—diBoc derivative of 5,6-diazabicyclo[2.2.1]hept-2-ene undergoes arylative ring opening and N—N bond cleavage on reaction with ArB(OH)$_2$ to produce *trans*-2-aryl-3-cyclopentenylhydrazines. Chiral products are obtained by using a chiral ligand-bound Rh catalyst.[7] (Hydroarylation and N—N bond cleavage occur when hetero-arylboronic acids serve as the reaction partners.)

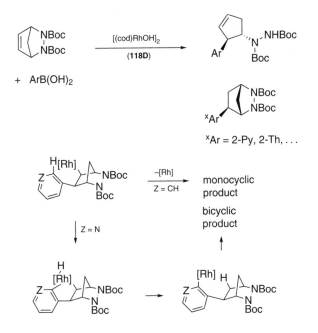

Intramolecular hydroamination of *N*-benzenesulfonyl-4-alkenylamine and analogues is catalyzed by Cu(OTf)$_2$. Oxidative cyclization involving the benzene ring is promoted by MnO$_2$ that is added.[8] When a chiral BOX ligand (*ent*-**1B**) is present amidocuprate species are formed prior to hydroamination and the transition state is regulated.

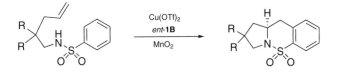

Indolizinones are formed by combining *N*-(4-pentenyl) isocyanate and 1-alkynes on Rh-catalysis. Interestingly, two different types of structures arise depending on whether the alkyne is an arylethyne or alkylethyne (using slightly different TADDOL-type ligands **197** and **198**). The absolute configuration of the angular carbon atom also differs from one series to the other, which is apparent from the chiral version of the reaction.[9]

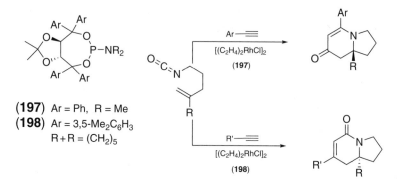

(197) Ar = Ph, R = Me
(198) Ar = 3,5-Me₂C₆H₃
R + R = (CH₂)₅

From two conjugated carbonyl compounds the cross-benzoin condensation initiated by a chiral azolecarbene (**199**) sets up a sequence of oxy-Cope rearrangement, aldol reaction and decarboxylation.[10]

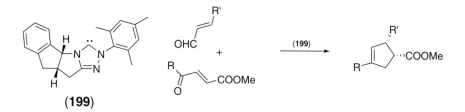

[1]Minatti, A., Zheng, X., Buchwald, S.L. *JOC* **72**, 9253 (2007).
[2]Mata, Y., Pamies, O., Dieguez, M. *CEJ* **13**, 3296 (2007).
[3]Shi, B.-F., Maugel, N., Zhang, Y.-H., Yu, J.-Q. *ACIE* **47**, 4882 (2008).
[4]Tsukamoto, H., Matsumoto, T., Kondo, Y. *OL* **10**, 1047 (2008).
[5]Wang, F., Zhang, Y.J., Yang, G., Zhang, W. *TL* **48**, 4179 (2007).
[6]Yamaoka, Y., Miyabe, H., Takemoto, Y. *JACS* **129**, 6686 (2007).
[7]Menard, F., Lautens, M. *ACIE* **47**, 2085 (2008).
[8]Zeng, W., Chemler, S.R. *JACS* **129**, 12948 (2007).
[9]Lee, E.E., Rovis, T. *OL* **10**, 1231 (2008).
[10]Chiang, P.-C., Kaeobamrung, J., Bode, J.W. *JACS* **129**, 3520 (2007).

o-Chloranil.

Thiobenzylation.   Benzyl sulfides are dehydrogenated to give benzalsulfonium salts on heating with *o*-chloranil at 80°. These can be used to react with β-keto esters. If a larger excess of *o*-chloranil is present (3 equiv.) the initial products are converted into the benzylidene derivatives.[1]

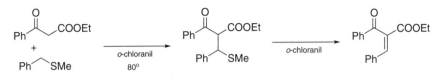

[1]Li, Z., Li, H., Guo, X., Cao, L., Yu, R., Li, H., Pan, S. *OL* **10**, 803 (2008).

1-Chlorobenzotriazole.

Sulfonyl azides.   A method for the preparation of sulfonyl azides starts from reaction of organometallic reagents with SO_2, followed by treatment with 1-chlorobenzotriazole. The sulfonyltriazoles are very reactive toward NaN_3.[1]

[1]Katritzky, A., Widyan, K., Gyanda, K. *S* 1201 (2008).

Chloro(1,5-cyclooctadiene)pentamethylcyclopentadienylruthenium(I).

Cycloaddition.   The [2+2+2]cycloaddition of a diyne with an alkyne, catalyzed by the title complex, is adaptable to a synthesis of cannabinol.[1]

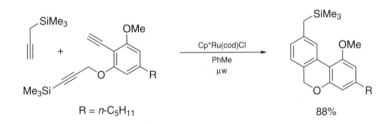

$R = n-C_5H_{11}$ 88%

Another type of useful ring closure assembles two double bonds and a triple bond.[2]

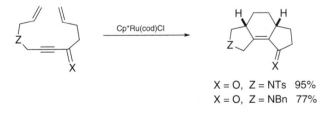

X = O, Z = NTs 95%
X = O, Z = NBn 77%

Reductive homologation and cyclopropanation are made to 1,6-enynes in their exposure to Me_3SiCHN_2 and Cp*Ru(cod)Cl.[3]

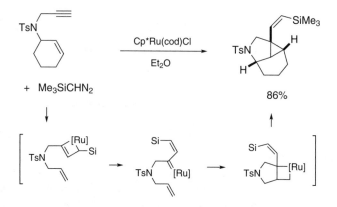

[1]Teske, J.A., Deiters, A. *OL* **10**, 2195 (2008).
[2]Tanaka, D., Sato, Y., Mori, M. *JACS* **129**, 7730 (2007).
[3]Monnier, F., Vovard-Le Bray, C., Castillo, D., Aubert, V., Derien, S., Dixneuf, P.H., Toupet, L., Ienco, A., Mealli, C. *JACS* **129**, 6037 (2007).

Chloro(cyclopentadienyl)bis(triphenylphosphine)ruthenium(I).

Cycloaddition.   7-Azabenzonorbornenes undergo [3+2]cycloaddition with alkynes, while heating with CpRuCl(PPh₃)₂ to form benzindoles.[1]

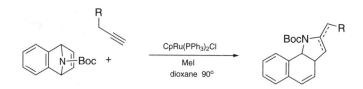

[1]Tenaglia, A., Marc, S. *JOC* **73**, 1397 (2008).

1-Chloromethyl-4-fluoro-1,4-diazoniabicyclo[2.2.2]octane bis(tetrafluoroborate), Selectfluor®.

Fluorohydroxylation.   Allenes are functionalized by treatment with Selectfluor® in aq. MeCN (1:10) at room temperature. For 1,2-alkadienes the addition yields 2-fluoro-1-alken-3-ols.[1]

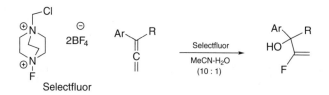

[1]Zhou, C., Li, J., Lü, B., Fu, C., Ma, S. *OL* **10**, 581 (2008).

m-Chloroperoxybenzoic acid, MCPBA.

Oxidation.   MCPBA is useful for oxidation of α-dithiolactones to give 1,2-dithietan-3-one 1-oxides.[1]

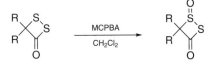

Hypervalent iodine reagents.   Co-oxidation of electron-rich iodoarenes and iodine leads to diaryliodonium species which are conveniently isolated as tosylate or triflate salts.[2] For access to an unsymmetrical diaryliondium salt the oxidation is carried out with a mixture of a ArI and Ar'B(OH)$_2$, in the presence of BF$_3$ · OEt$_2$ (to form a tetrafluoroborate salt) at room temperature.[3]

A hypervalent iodine species to initiate spirolactamization of 3-(*p*-anisyl)propanamides is created from *p*-tolyl iodide (catalytic) and MCPBA (stoichiometric) in trifluoroethanol.[4]

[1]Shigetomi, T., Okuma, K., Yokomori, Y. *TL* **49**, 36 (2008).
[2]Zhu, M., Jalalian, N., Olofsson, B. *SL* 592 (2008).
[3]Bielawski, M., Aili, D., Olofsson, B. *JOC* **73**, 4602 (2008).
[4]Dohi, T., Maruyama, A., Minamitsuji, Y., Takenaga, N., Kita, Y. *CC* 1224 (2007).

N-Chlorosuccinimide.

β-Chlorohydrins.   Alkenes are transformed into β-chlorohydrins by NBS in an aqueous solution, with thiourea as catalyst.[1] When the reaction is carried out in ROH β-chloroalkyl ethers are obtained.[2]

Chlorination.   Ketones are chlorinated (at an α-position) by NCS in MeOH, also in the presence of thiourea.[3]

[1]Bentley, P.A., Mei, Y., Du, J. *TL* **49**, 1425 (2008).
[2]Bentley, P.A., Mei, Y., Du, J. *TL* **49**, 2653 (2008).
[3]Mei, Y., Bentley, P.A., Du, J. *TL* **49**, 3802 (2008).

Chlorotris(triphenylphosphine)cobalt(I).

Coupling reactions.[1]   Highly functionalized tertiary benzylic bromides are reductively coupled by (Ph$_3$P)$_3$CoCl.

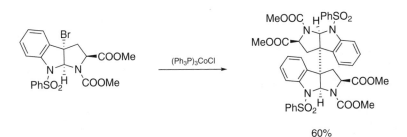

60%

[1]Movssaghi, M., Schmidt, M.A. *ACIE* **46**, 3725 (2007).

Chlorotris(triphenylphosphine)rhodium(I).

Trifluoromethylation. Zinc enolates generated from enol silyl ethers on treatment with Et_2Zn react with CF_3I in the presence of $(Ph_3P)_3RhCl$ to provide α-trifluoromethylated ketones.[1]

Reformatsky reaction. Reformatsky reagents are known to react with imines to afford β-lactams. The reaction can be applied to the synthesis of α,α-difluoro-β-lactams, and even chiral products.[2]

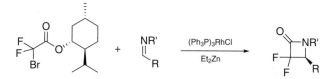

Reductive acylation. Transmetallation occurs when Et_2Zn is mixed with $(Ph_3P)_3RhCl$. The ensuing [Rh]-Et species loses ethylene rapidly and is thereby converted into a hydridorhodium compound. Enones are reduced and the Rh enolates can be acylated by $RCOCl$.[3] The reagent that is spent is Et_2Zn.

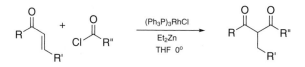

o-Hydroxyaryl ketones are formed in an atom-economical fashion by combining 1-alkene with 2-hydroxyaraldehydes in the presence of $(Ph_3P)_3RhCl$. Formation of linear products is promoted by MeCN.[4]

Reductive silylation. Enones undergo reductive silylation to afford allyl silyl ethers by Ph_2SiH_2. 1,4-Reduction ending by enolate trapping is observed by using a diarylsilane in which one of the aromatic ring is *o*-substituted by a methoxymethyl group.[5]

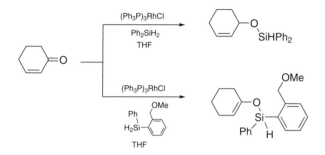

Cycloaddition.   [2+2+2]Cycloaddition of a diyne and an alkyne to form a product with a new benzene unit is efficiently catalyzed by (Ph₃P)₃RhCl. Significant applications of the reaction are found in a synthesis of bruguierol-A[6] and trypticenediols.[7]

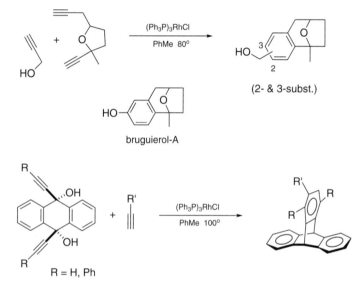

The Rh complex also catalyzes the [4+2]cycloaddition of conjugated oximes with alkynes that results in polysubstituted pyridines.[8]

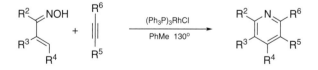

Rearrangement.   2-(3-Cyclopropenyl)pyridines undergo rearrangement to give indolizines under the influence of certain metal salts. When the cyclopropenyl group is unsymmetrically substituted the bonding reorganization can be traced. It is noted that products generated from reactions catalyzed by (Ph₃P)₃RhCl and by CuI are different.[9]

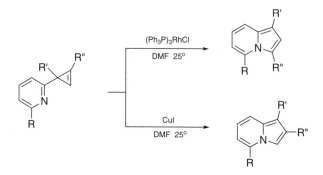

[1]Sato, K., Yuki, T., Tarui, A., Omote, M., Kumadaki, I., Ando, A. *TL* **49**, 3558 (2008).
[2]Tarui, A., Ozaki, D., Nakajima, N., Yokota, Y., Sokeirik, Y.S., Sato, K., Omote, M., Kumadake, I., Ando, A. *TL* **49**, 3839 (2008).
[3]Sato, K., Yamazoe, S., Yamamoto, R., Ohata, S., Tarui, A., Omote, M., Kumadake, I., Ando, A. *OL* **10**, 2405 (2008).
[4]Imai, M., Tanaka, M., Nagumo, S., Kawahara, N., Suemune, H. *JOC* **72**, 2543 (2007).
[5]Imao, D., Hayama, M., Ishikawa, K., Ohta, T., Ito, Y. *CL* **36**, 366 (2007).
[6]Ramana, C.V., Salian, S.R., Gonnade, R.G. *EJOC* 5483 (2007).
[7]Taylor, M.S., Swager, T.M. *OL* **9**, 3695 (2007).
[8]Parthasarathy, K., Jeganmohan, M., Cheng, C.-H. *OL* **10**, 325 (2008).
[9]Chuprakov, S., Gevorgyan, V. *OL* **9**, 4463 (2007).

Chromium – carbene complexes.

Cycloadditions. *N*-Dimethylaminoisoindoles are formed when *o*-alkynylaralaldehyde *N,N*-dimethylhydrazones are treated with a Fischer carbene complex. The isoindoles are trapped by an acetylenedicarboxylic ester in situ to generate naphthalene derivatives.[1]

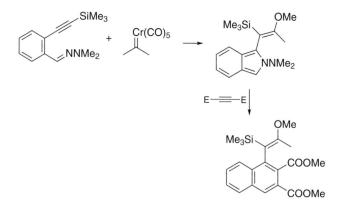

The Dötz benzannulation products involving propargyloxy derivatives are liable to elimination, leading to o-quinonemethides whch can be trapped.[2]

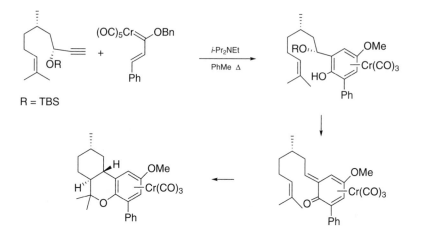

Through initiation by another [n+2]cycloaddition the usefulness of the Dötz reaction is expanded, products more varied in structural types become available.[3]

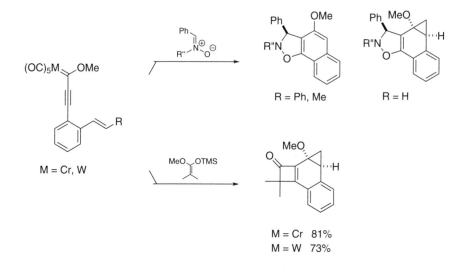

Also the intramolecular cycloaddition to form macrocycle en route to arnebinol[4] and phomactin-B$_2$[5] are remarkably efficient.

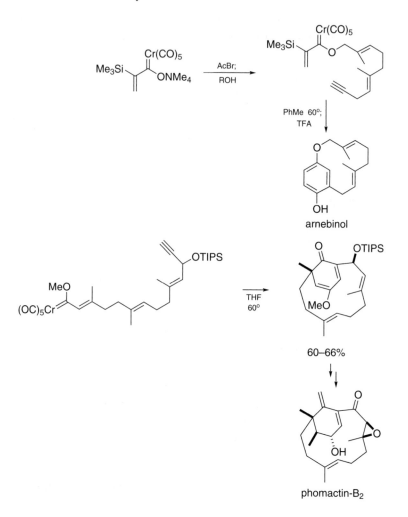

arnebinol

60–66%

phomactin-B$_2$

Conjugated Fischer carbene complexes undergo [3+2]cycloaddition to afford 2-cyclopentenones.[6] The reaction is carried out in the presence of Ni(cod)$_2$.

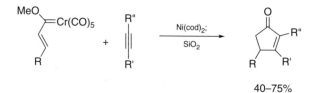

40–75%

Employing a [4+1]cycloaddition that unites a conjugated carbonyl unit with the carbenoid center of a Cr-complex paves way to a novel approach to furans.[7]

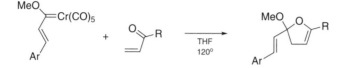

[1]Duan, S., Sinha-Mahapatra, D.K., Herndon, J.W. *OL* **10**, 1541 (2008).
[2]Korthals, K.A., Wulff, W.D. *JACS* **130**, 2898 (2008).
[3]Barluenga, J., Andina, F., Aznar, F., Valdes, C. *OL* **9**, 4143 (2007).
[4]Watanabe, M., Tanaka, K., Saikawa, Y., Nakata, M. *TL* **48**, 203 (2007).
[5]Huang, J., Wu, C., Wulff, W.D. *JACS* **129**, 13366 (2007).
[6]Barluenga, J., Barrio, P., Riesgo, L., Lopez, L.A., Tomas, M. *JACS* **129**, 14422 (2007).
[7]Barlueuga, J., Faulo, H., Lopez, S., Florez, J. *ACIE* **46**, 4136 (2007).

Chromium(II) chloride.

Addition.   Secondary and tertiary alkyl halides react with ArCHO under the influence of $CrCl_2$ – LiI and a catalytic amount of vitamin-B_{12} in DMF. It is likely the reaction proceeds via coupling of alkyl and ketyl radicals.[1]

Alkenylchromium reagents are obtained from 1,1,1-trichloroalkanes by treatment with $CrCl_2$ – LiI in THF.[2] These reagents add to aldehydes to form allylic alcohols. From 1,1,1-trichloroethanol the primary reaction products are further dehydrated to give conjugated aldehydes. α,α,α-Trichloromethylarenes afford diarylethynes.

Cyclopropanation.[3]   The double bond of acrylamides is cyclopropanated by the combination of $CrCl_2$ and $ClCH_2I$.

[1]Wessjohann, L.A., Schmidt, G., Schrekker, H.S. *T* **64**, 2134 (2008).
[2]Bejot, R., He, A., Falck, J.R., Mioskowski, C. *ACIE* **46**, 1719 (2007)
[3]Concellon, J.M., Rodriguez-Solla, H., Mejica, C., Blanco, E.G. *OL* **9**, 2981 (2007).

Cinchona alkaloid derivatives.

Desymmetrization reactions.   Cinchona alkaloids are relative abundant, morevover, the fact that the two series of quinine/cinchonidine and quinidine/cinchonine often can catalyze reactions in the opposite chirality sense makes the use of them and their derivatives very valuable in creating new stereogenic centers from prochiral substances, in one or both optical series.

Formation of chiral α-amino acids from aminomalonic acids via decarboxylative protonation is readily accomplished in the presence of thiourea **1** (ex quinidine).[1] The diastereomeric **2** (ex quinine) has been employed to generate monomethyl esters in chiral form by methanolysis *meso*-cyclic anhydrides.[2]

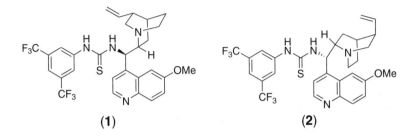

(1)   **(2)**

Electrophilic reactions.   Asymmetric bromination of alkanoyl chlorides with 1,1,3,6-tetrabromo-1,2-dihydronaphthalen-2-one is catalyzed by the quinine – derived **3**, it affords (*S*)-α-bromoalkanoic esters on alcoholysis of the products.[3] Quaternization of quinine with 9-chloromethylanthracene gives a salt that is useful for catalyzing asymmetric C-3 hydroxylation of *N*-protected oxindoles. The quaternary ammonium salt plays a dual role in that it also serves as a phase-transfer catalyst.[4]

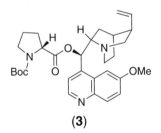

(3)

Reactions of carbonyl compounds and imines.   The salt **4** obtained from reaction of cinchonine with *m*-xylylene dibromide is shown to promote enantioselective transfer of the trifluoromethyl group from Me_3SiCF_3 to aryl ketones.[5] Also obtained from quinidine the salt containing a trifluorobenzyl group (**5**) promotes the condensation of chloromethyl phenyl sulfones with ArCHO to give benzenesulfonyl epoxides.[6]

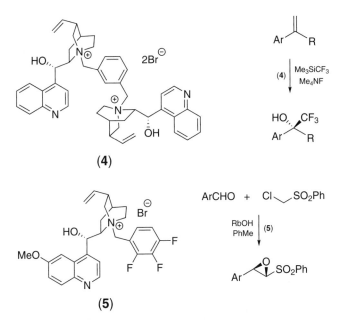

(4)

(5)

N-(α-Benzenesulfonylalkyl)carbamates are adequate surrogates of imines. Being adducts of imines, the (*S*)-1-fluoro-2-alkylamine derivatives can be prepared from the carbamates with bis(benzenesulfonyl)fluoromethane in the presence of quinidine benzochloride and CsOH, followed by twofold desulfonylation with Mg in MeOH.[7]

Strecker reaction catalyzed by (*i*-PrO)$_4$Ti whereby tosylimines accept Me$_3$SiCN enantioselectively is achieved by modifying the environment surrounding the metal center by alkoxy group exchange with 3,3'-di(β-naphthyl)-2,2'-dihydroxybiphenyl, and further complexation with cinchonine.[8]

Aza-Henry reaction is rendered asymmetric by quaternary salts of Cinchona alkaloids.[9]

Addition reactions. Changing the 9-hydroxy group of Cinchona alkaloids to a 9-epiamino group not only is synthetically expedient, such products often show excellent catalytic activities in many asymmetric reactions. Those derived from dihydrocinchona alkaloids mediate Michael reactions to good results, including addition of indole to enones,[10] and carbonyl compounds to nitroalkenes.[11] Salt **4** has also been successfully employed in the alkenylation of *t*-butyl α-aryl-α-cyanoacetate.[12]

Thiourea **2** directs entantioselective hydration of the double bond of γ-hydroxypropenyl ketones by virtue of multiple H-bonding interactions.[13]

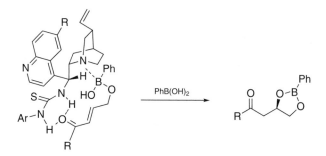

Complementary functions of quinine and quinidine derivatives are revealed again in the Michael reaction between 2-cyano-1-indanone and α-chloroacrylonitrile, with them anyone of four possible chiral diastereomers can be prepared at will.[14]

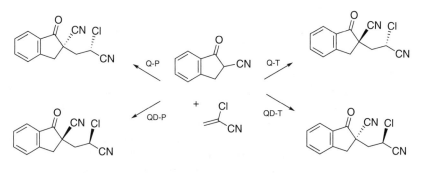

Q-P   de-O-methylquinine 9-phenanthr-9-yl ether
QD-P   de-O-methylquinidine 9-phenanthr-9-yl ether
Q-T   N-(quinin-9-yl)-N′-[3,5-bis(trifluoromethyl)phenyl]thiourea
QD-T   N-(quinidin- 9-yl)-N′-[3,5-bis(trifluoromethyl)phenyl]thiourea

A tridentate ligand is obtained from a quinine glycinate ester in which the primary amino group is further incorporated into a 3,5-di-t-butylsalicylaldimine. However, its useful scope in catalyzing enantioselective addition of Et_2Zn to ketones is quite limited.[15]

Cycloadditions. Epoxidation of 2-methyl-1,4-naphthoquinone (vitamin-K_3) by NaOCl is best catalyzed by de-O-methylquinine anthracyl-9-methochloride, according to a computer analysis.[16] Epoxidation of 2-cycloalkenones by H_2O_2 with 9-epiamino-9-deoxyquinine as catalyst shows opposite enantioselectivity as that with (R,R)-1,2-diphenylethanediamine.[17]

The ester of quinine (5) is an excellent catalyst for β-lactone synthesis from ketene and certain aldehydes.[18]

6'-Hydroxycinchonidine 9-benzoate (from quinine) also catalyzes the [3+2]cyclo-addition of α-isocyanoalkanoicesters with nitroalkenes to yield 2,3-dihydropyrrole-2-carboxylic esters.[19]

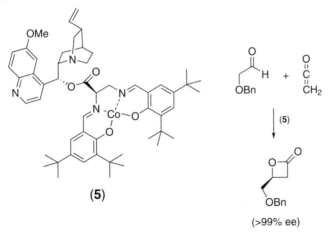

(>99% ee)

Both 9-epiamino derivatives of 9-deoxyquinine and 9-deoxyquinidine promote *exo*-selective Diels–Alder reaction of 3-hydroxy-2-pyrones with enones.[20] Actually, great latitude exists for tuning the reaction by partial structural changes of the catalysts.[21]

The hetero-Diels–Alder reaction between nascent ketenes generated from crotonyl chlorides and trichloroacetaldehyde, is effected by Sn(OTf)$_2$, and rendered asymmetric by a TMS derivative of quinidine.[22]

[1]Amere, M., Lasne, M.-C., Rouden, J. *OL* **9**, 2621 (2007).

[2]Peschiulli, A., Gun'ko, Y., Connon, S.J. *JOC* **73**, 2454 (2008).

[3]Dogo-Isonagie, C., Bekele, T., France, S., Wolfer, J., Weatherwax, A., Taggi, A.E., Paull, D.H., Dudding, T., Lectka, T. *EJOC* 1091 (2007).

[4]Sano, D., Nagata, K., Itoh, T. *OL* **10**, 1593 (2008).

[5]Mizuta, S., Shibata, N., Akiti, S., Fujimoto, H., Nakamura, S., Toru, T. *OL* **9**, 3707 (2007).

[6]Ku, J.-M., Yoo, M.-S., Park, H.-g., Jew, S.-S., Jeong, B.-S. *T* **63**, 8099 (2007).

[7]Mizuta, S., Shibata, N., Goto, Y., Furukawa, T., Nakamura, S., Toru, T. *JACS* **129**, 6395 (2007).

[8]Wang, J., Hu, X., Jiang, J., Gou, S., Huang, X., Liu, X., Feng, X. *ACIE* **46**, 8468 (2007).

[9]Gomez-Bengoa, E., Linden, A., Lopez, R., Mugica-Mendiola, I., Oiarbide, M., Palomo, C. *JACS* **130**, 7955 (2008).

[10]Bartoli, G., Bosco, M., Carlone, A., Pesciaioli, F., Sambri, L., Melchiorre, P. *OL* **9**, 1403 (2007).

[11]McCooey, S.H., Connon, S.J. *OL* **9**, 599 (2007).

[12]Bell, M., Poulsen, T.B., Jorgensen, K.A. *JOC* **72**, 3053 (2007).

[13]Li, D.R., Murugan, A., Falck, J.R. *JACS* **130**, 46 (2008).

[14]Wang, B., Wu, F., Wang, Y., Liu, X., Deng, L. *JACS* **129**, 768 (2007).

[15]Casarotto, V., Li, Z., Boucau, J., Lin, Y.-M. *TL* **48**, 5561 (2007).

[16]Berkessel, A., Guixa, M., Schmidt, F., Neudörfl, J.M., Lex, J. *CEJ* **13**, 4483 (2007).

[17]Wang, X., Reisinger, C.M., List, B. *JACS* **130**, 6070 (2008).

[18]Lin, Y.-M., Boucau, J., Li, Z., Casarotto, V., Lin, J., Nguyen, A.N., Ehrmantraut, J. *OL* **9**, 567 (2007).

[19]Guo, C., Xue, M.-X., Zhu, M.-K., Gong, L.-Z. *ACIE* **47**, 3414 (2008).

[20]Singh, R.P., Bartelson, K., Wang, Y., Su, H., Lu, X., Deng, L. *JACS* **130**, 2422 (2008).
[21]Wang, Y., Li, H., Wang, Y.-Q., Liu, Y., Foxman, B.M., Deng, L. *JACS* **129**, 6364 (2007).
[22]Tiseni, P., Peters, R. *ACIE* **46**, 5325 (2007).

Cobalt.

Coupling reactions.   Hollow nanospheres of cobalt can substitute for Pd species in Sonogashira coupling. The catalyst system still contains CuI, Ph_3P, and K_2CO_3.[1]

For conducting the Heck reaction with acrylic esters, no ligand is needed.[2]

[1]Feng, L., Liu, F., Sun, P., Bao, J. *SL* 1415 (2008).
[2]Zhou, P., Li, Y., Sun, P., Bao, J. *CC* 1418 (2007).

Cobalt–rhodium.

Carbamoylation.   Nanoparticles of the Co_2Rh_2 bimetallic species catalyze *cis* addition of $H/CONR_2$ to alkynes, where the addend groups come from R_2NH and CO.[1]

Cycloaddition.   A synthesis of bicyclic dienones by the Pauson–Khand reaction of an allene/yne is based on catalysis by Co_2Rh_2.[2] Two molecules of an allene combine with CO to form 4-alkylidene-2-cyclopentenones.[3]

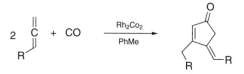

[1]Park, J.H., Kim, S.Y., Kim, S.M., Chung, Y.K. *OL* **9**, 2465 (2007).
[2]Park, J.H., Kim, S.Y., Kim, S.M., Lee, S.I., Chung, Y.K. *SL* 453 (2007).
[3]Park, J.H., Kim, E., Kim, H.-.M., Choi, S.Y., Chung, Y.K. *CC* 2388 (2008).

Cobalt(II) acetylacetonate.

Aldol reaction.   Reductive aldol reaction of conjugated amides, according to an intramolecular version as previously reported, is fully applicable to the preparation of *syn*-3-hydroxyalkanamides.[1]

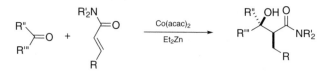

[1]Lumby, R.J.R., Joensuu, P.M., Lam, H.W. *OL* **9**, 4367 (2007).

Cobalt(II) bromide.

Biaryls.   Coupling of two ArX is mediated by Mn and catalyzed by $CoBr_2$–Ph_3P in DMF and pyridine (6:1). There is no need to prepare ArM.[1]

Cyclotrimerization.   Alkynes (e.g., PhCCH) are trimerized on exposure to $CoBr_2$ along with Zn-ZnI₂. Isomer ratio changes with respect to ligands.[2]

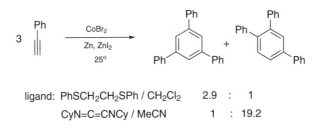

ligand: $PhSCH_2CH_2SPh$ / CH_2Cl_2 2.9   :   1

CyN=C=CNCy / MeCN   1   :   19.2

[1]Amatore, M., Gosmini, C. *ACIE* **47**, 2089 (2008).
[2]Hilt, G., Hengst, C., Hess, W. *EJOC* 2293 (2008).

Cobalt(II) chloride.

Cyclopropanation.   A bimetallic catalyst composing of $CoCl_2$, AgOAc and the ligand **1** is used for cyclopropanation of styrenes. Intramolecular cyclopropanation leads to cyclopropanolactones.[1]

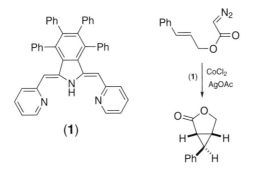

(1)

Cross-coupling.   Radicals are generated from 2-vinylsiloxy-1-iodoalkanes in the presence of an azolecarbene-complexed $CoCl_2$. Rapid transfer of the radical site to the carbon β to the silicon on ring closure precedes coupling with a Grignard reagent that is added. The method can be used to build a carbon chain containing a 1,3-diol unit.[2]

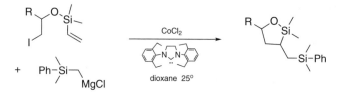

[1]Langlotz, B.K., Wakepohl, H., Gade, L.H. *ACIE* **47**, 4670 (2008).
[2]Someya, H., Ohmiya, H., Yorimitsu, H., Oshima, K. *T* **63**, 8609 (2007).

Cobalt(II) iodide/phosphine – zinc.

Cycloaddition.   3-Methylenecyclopentanols are assembled from conjugated carbonyl compounds and allenes.[1]

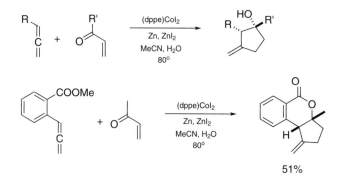

51%

Aryl sulfides.   Arylation of thiols is mediated by the title reagent mix.[2]

[1]Chang, H.-T., Jayanth, T.T., Cheng, C.-H. *JACS* **129**, 4166 (2007).
[2]Wong, Y.-C., Jayanth, T.T., Cheng, C.-H. *OL* **8**, 5613 (2006).

Copper.

Coupling reactions.   Nanosized copper is a good catalyst for Ullmann ether synthesis, using Cs_2CO_3 as base in MeCN at 50–60°.[1] Carbon-supported copper in the presence of 1,10-phenanthroline shows similar activities with the aid of microwave irradiation.[2] *N*-Arylation of *N*-heterocycles (benzimidazole, triazole, . . .) by $ArSi(OR)_3$ is mediated by Cu–$FeCl_3$ and TBAF in the air.[3]

Sonogashira coupling is conducted more conveniently and less expensively with Cu/Al_2O_3, without the need of palladium and ligand.[4] Replacement of a vinylic iodide (including 5-iodouracil) by a trifluoromethyl group can be carried out by a copper-catalyzed reaction with $(CF_3)_2Hg$ in DMA at 140°.[5]

Glaser coupling of 1-alkynes followed by [3+2]cycloaddition with organoazides affords bi-5,5′-triazolyls. Achieved in one step with Cu and $CuSO_4$ in the air, the reaction is particularly favored by adding Na_2CO_3.[6]

[1]Kidwai, M., Mishra, N.K., Bansal, V., Kumar, A., Mozumdar, S. *TL* **48**, 8883 (2007).
[2]Lipshutz, B.H., Unger, J.B., Taft, B.R. *OL* **9**, 1089 (2007).
[3]Song, R.-J., Deng, C.-L., Xie, Y.-X., Li, J.-H. *TL* **48**, 7845 (2007).
[4]Biffis, A., Scattolin, E., Ravasio, N., Zaccheria, F. *TL* **48**, 8761 (2007).
[5]Nowak, I., Robins, M.J. *JOC* **72**, 2678 (2007).
[6]Angell, Y., Burgers, K. *ACIE* **46**, 3649 (2007).

Copper(II) acetate.

Coupling reactions.   Glaser coupling of alkynyltrifluoroborate salts using $Cu(OAc)_2$ in DMSO at 60° gives conjugated diynes.[1]

Aryl sulfones are synthesized from RSO_2Na and $Ar'B(OH)_2$ under oxidative conditions (DMSO, O_2). Two similar procedures, both using $Cu(OAc)_2$ as catalyst but different *N*-heterocycle addends, have been established.[2,3] *N*-Cyclopropylation of indole is accomplished, also by following a similar recipe, with the necessary changing the coupling partner to cyclopropylboronic acid.[4] Cyclization of *N*-arylamidines to afford benzimidazoles involves activation of the *o*-C—H bond in a process realized by the action of $Cu(OAc)_2$ in DMSO (containing 2.5 equiv. HOAc) under O_2 at 100°.[5]

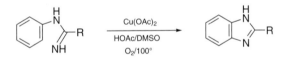

Heterocycles.   Heating a mixture of a nitroalkane and styrene with $Cu(OAc)_2$ and *N*-methylpiperidine in $CHCl_3$ at 60° leads to 5-phenylisoxazolines, no dehydrating agent is needed.[6] A method for pyridine synthesis[7] from conjugated oxime esters and alkenylboronic acids perhaps involves C—N coupling, electrocyclization and dehydrogenation in the air.

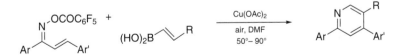

Nitroaldol reaction.   When the stereocontrolled condensation is conducted in the presence of $Cu(OAc)_2$, which is complexed to the polymer-linked diamine **1**, the catalyst is readily recovered.[8]

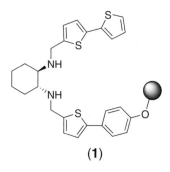

(1)

[1]Paixao, M.W., Weber, M., Braga, A.L., de Azeredo, J.B., Deobald, A.M., Stefani, H.A. *TL* **49**, 2366 (2008).
[2]Kar, A., Sayyed, I.A., Lo, W.F., Kaiser, H.M., Beller, M., Tse, M.K. *OL* **9**, 3405 (2007).
[3]Huang, F., Batey, R.A. *T* **63**, 7667 (2007).
[4]Tsuritani, T., Strotman, N.A., Yamamoto, Y., Kawasaki, M., Yasuda, N., Mase, T. *OL* **10**, 1653 (2008).
[5]Brasche, G., Buchwald, S.L. *ACIE* **47**, 1932 (2008).
[6]Cecchi, L., De Sarlo, F., Machetti, F. *SL* 2451 (2007).
[7]Liu, S., Liebeskind, L.S. *JACS* **130**, 6918 (2008).
[8]Bandini, M., Benaglia, M., Sinisi, R., Tommasi, S., Umani-Ronchi, A. *OL* **9**, 2151 (2007).

Copper(II) bis(hexafluoroacetylacetonate).

Isomerization.   2-Alkenyloxiranes give 2,5-dihydrofurans on heating with the Cu(II) complex. With the simple acetylacetonate larger amounts of unsaturated aldehydes (the other type of isomerization products) are obtained.

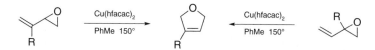

[1]Batory, L.A., McInnis, C.E., Njardarson, J.T. *JACS* **128**, 16054 (2006).

Copper(I) bromide.

Oxidative coupling.   With a mixture of CuBr – 1,10-phenanthroline and palladium(II) trifluoroacetylacetonate, and also tri(*o*-tolyl)phosphine, the coupling of aryl bromides and potassium 2-oxoalkanoates with loss of CO_2 provides aryl ketones.[1]

N-Methylamines are oxidatively activated by CuBr and NBS. They are transformed into propargylic amines on reaction with 1-alkynes that are placed in the reaction media.[2]

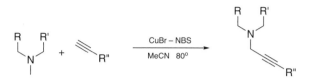

Ullmann coupling of aryl halides possessing a coordinative functionality in the ortho-position are very favorable because the arylcopper intermediates are stabilized. An oxygen atom from an acetal unit shows the beneficial effect, enabling the simplification of synthetic processes (by not having to employ the thioacetal).[3]

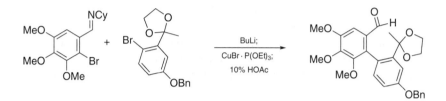

Reactions involving β-diketones. Diphenylation of β-diketones occurs when they are heated with anthranilic acid in the presence of catalytic amounts of CuBr and Cl_3CCOOH (5 mol% each) in 1,2-dichloroethane at 60°.[4]

Aldehydes are oxidized in situ by CuBr–t-BuOOH to supply O-acylating agents for β-diketones.[5]

[1]Goossen, L.J., Rudolphi, F., Oppel, C., Rodriguez, N. *ACIE* **47**, 3043 (2008).
[2]Niu, M., Yin, Z., Fu, H., Jiang, Y., Zhao, Y. *JOC* **73**, 3961 (2008).
[3]Broady, S.D., Golden, M.D., Leonard, J., Muir, J.C., Maudet, M. *TL* **48**, 4627 (2007).
[4]Yang, Y.-Y., Shou, W.-G., Wang, Y.-G. *TL* **48** 8163 (2007).
[5]Yoo, W.-J., Li, C.-J. *JOC* **71**, 6266 (2006).

Copper(II) bromide.

Dehydrogenation. Aromatization of 3,4-diaryl-2,5-dihydro derivatives of furan, thiophene, and N-arylpyrrole is accomplished in 80–91% yield by heating with $CuBr_2$ (3 equiv.) in EtOAc.[1]

Substitution. Propargylic alcohols are readily transformed into mixed ethers and sulfides when they are treated with ROH and RSH in the presence of $CuBr_2$ in $MeNO_2$ at room temperature.[2]

[1]Dang, Y., Chen, Y. *EJOC* 5661 (2007).
[2]Hui, H., Zhao, Q., Yang, M., She, D., Chen, M., Huang, G. *S* 191 (2008).

Copper(I) chloride.

Additions. Hydroboration of conjugated esters and nitriles with (bispinacolato)diboron in MeOH, promoted by CuCl in the presence of *t*-BuONa, proceeds in good yields.[1]

Under microwave irradiation an azolecarbene-complexed CuCl induces cyclization of *o*-allylaryl trichloroacetates via a free radical process. The initial adducts undergo decarboxylation and dehydrochlorination that lead to aromatization.[2,3]

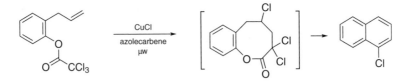

Cycloadditions. N,N'-Di-*t*-butylthiadiaziridine *S,S*-dioxide reacts with activated terminal alkenes in the presence of CuCl and Bu₃P to give five-membered heterocycles.[4] This reaction effectively performs the critical step of *vic*-diamination of the alkenes. Analogously, cyclic *N*-cyanoguanidine **1** also undergoes cycloaddition to alkenes, with Ph₃P to stabilize CuCl.[5]

1,2,3-Triazole synthesis is also catalyzed by CuCl trapped in zeolite. No ligand for the metal salt is required.[6]

Coupling reactions. Diaryl ethers are formed (Ullmann synthesis) by treatment of the reactants (ArOH and Ar'Br) with CuCl and 1-butylimidazole in toluene.[7] Diaryl sulfides can be prepared similarly, with some variation of the reaction conditions (in water, presence of 1,2-diaminocyclohexane).[8]

A three-component coupling to construct 2-(α-iminyl)pyrroles is regioselective. The reaction is suitable for the preparation of libraries of products by using different 1-alkynes.[9]

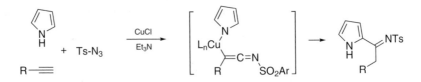

A new protocol for the condensation of aldehydes (except fomaldehyde), amines, and alkynes to give propargylic amines entails the employment of CuCl and Cu(OTf)₂ (5 mol% each) to create a cooperative catalyst system and the use of trimethylsilylalkynes.[10]

[1]Lee, J.-E., Yun, J. *ACIE* **47**, 145 (2008).
[2]Bull, J.A., Hutchings, M.G., Quayle, P. *ACIE* **46**, 1869 (2007).
[3]Bull, J.A., Hutchings, M.G., Lujan, C., Quayle, P. *TL* **49**, 1352 (2008).
[4]Zhao, B., Yuan, W., Du, H., Shi, Y. *OL* **9**, 4943 (2007).
[5]Zhao, B., Du, H., Shi, Y. *OL* **10**, 1087 (2008).
[6]Chassaing, S., Kumarraja, M., Sido, A.S.S., Pale, P., Sommer, J. *OL* **9**, 883 (2007).
[7]Schareina, T., Zapf, A., Cotte, A., Müller, N., Beller, M. *TL* **49**, 1851 (2008).
[8]Carril, M., SanMartin, R., Dominguez, E., Tellitu, I. *CEJ* **13**, 5100 (2007).
[9]Cho, S.H., Chang, S. *ACIE* **47**, 2836 (2008).
[10]Sakai, N., Uchida, N., Konakahara, T. *SL* 1515 (2008).

Copper(II) chloride.

Rearrangement.[1]   Azines are produced from [3,3]sigmatropic rearrangement of N'-Boc N'-allylhydrazones. In the presence of $CuCl_2$ and i-Pr_2NEt oxidative chlorination also occurs.

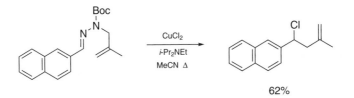

62%

Triazole synthesis.[2]   A heterogeneous catalyst for the click reaction is formed by treating $CuCl_2$ with $(s$-$BuO)_3Al$.

[1]Mundal, D.A., Lee, J.J., Thomson, R.J. *JACS* **130**, 1148 (2008).
[2]Park, I.S., Kwon, M.S., Kim, Y., Lee, J.S., Park, J. *OL* **10**, 497 (2008).

Copper(I) cyanide.

Cycloaddition.   Diazoacetic esters and alkyl lithiopropynoates form pyrazoledi-carboxylic esters in THF by a formal [3+2]cycloaddition. It is catalyzed by $CuCN \cdot 6LiCl$.[1]

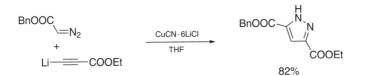

82%

[1]Qi, X., Ready, J.M. *ACIE* **46**, 3242 (2007).

Copper(II) 2-ethylhexanoate.

Cyclization. Intramolecular oxidative addition of an amidic nitrogen atom and a sp^2-hybridized carbon to a double bond is effected by the Cu(I) carboxylate.[1] This type of reaction has been accomplished by Cu(OAc)$_2$ on unsaturated sulfonamides.

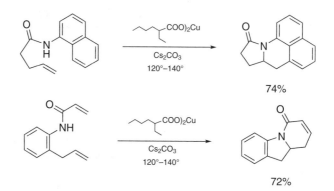

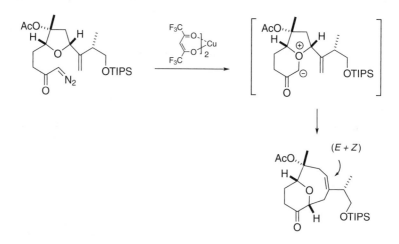

[1]Fuller, P.H., Chemler, S.R. *OL* **9**, 5477 (2007).

Copper(II) hexafluoroacetylacetonate.

Cyclization.[1] Like many other copper salts the title compound catalyzes decomposition of diazoketones. A case of heteroatom trapping followed by a [2,3]Wittig rearrangement to generate an 11-oxabicyclo[5.3.1]undecenone serves to illustrate the synthetic potentials of the process.

[1]Clark, J.S., Baxter, C.A., Dossetter, A.G., Poigny, S., Castro, J.L., Whittingham, W.G. *JOC* **73**, 1040 (2008).

Copper(I) iodide.

   Baeyer–Villiger oxidation.   The BuO-Cu(III)-NO species formed on heating CuI with Bu_4NNO_2 in *o*-xylene at 150° converts aryl isopropyl ketones and aryl trifluoromethyl ketones to butyl esters. However, the scope of this reaction is limited, ethyl and methyl ketones give low yields of the corresponding esters and phenyl and *t*-butyl ketones are not oxidized at all under such conditions.[1]

   Deoxygenation.   Amine oxides are deoxygenated by heating with CuI and *i*-Pr_2NEt in THF.[2]

   Coupling reactions.   Ether synthesis from ArX originated from Ullmann. Reaction involving mediation by CuI is improved in the presence of 3,4,7,8-tetramethyl-1,10-phenanthroline[3] or *N,N*-dimethylglycine.[4] Diaryl ethers can be synthesized by heating ArOH and Ar'X with ligand-free CuX (X = I, Br, Cl) and Cs_2CO_3 in NMP.[5] Another protocol sparing ligands is compensated by Bu_4NBr.[6]

   β-Styryl aryl ethers, sulfides, and amines are similarly prepared from β-styryl bromide. One interesting aspect of the coupling method is the employment of ethyl 2-oxocyclo-hexanecarboxylate as the ligand for CuI.[7] 9-Azajulolidine is a more general and powerful ligand for coupling reactions leading to diaryl ethers, sulfides, and amines.[8] Heteroaryl cyanides prepared from bromides and $K_4Fe(CN)_6$ is accomplished in the presence of CuI and an *N*-alkylimidazole.[9]

   A highly efficient cyclization of 2-chloro-4-sulfonamino-1-alkenes to 2-methylene-azetidines provides an avenue to β-lactams, e.g., on ozonolysis of the products.[10] Pertinent to cyclization of bromoallyl bromohomoallyl carbinols is the chemo/regioselectivity. It is actually dependent on the nature of the promoter, CuI or $Pd(OAc)_2$.[11]

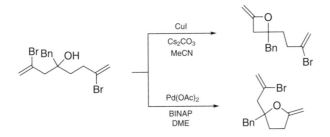

   In hetero-Ullmann coupling, i.e., arylation of phenols, thiols, and amides with ArI, a useful ligand for the CuI mediator is 1,1,1-tris(hydroxymethyl)ethane.[12] Further extension of the Ullmann reaction to the preparation of aryl cyanides from ArBr and $K_4Fe(CN)_6$ is probably quite routine.[13]

   With the additive 2-oxazolidinone in DMSO to assist CuI at 120°, *N*-arylation of amides (lactams) is readily performed.[14] A more commonly used ligand is 1,10-phenanthroline, as it is applied also to form *N*-(aryl)alkoxyamines from RNHOR'.[15] In *N*-arylation of *N*-heterocycles (indole, pyrrole, imidazole, pyrazole,...), 1,3-di(2-pyridyl)-1,3-propane-dione appears to be a useful ligand for CuI.[16]

It is fortunate that arylation of aminoalcohols at either the nitrogen or the oxygen atom can be performed at will, by changing the ligand.[17]

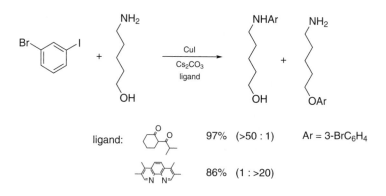

As ammonia can hardly be used in coupling with ArX, access to $ArNH_2$ needs finding a surrogate with good selectivity and CF_3CONH_2 fulfills the requirement.[18] The coupling products undergo methanolysis to deliver arylamines.

Two methods have been developed recently for pyrrole synthesis: a twofold coupling reaction of (Z,Z)-1,4-dihalo-1,3-dienes (bromides and iodides) with $BocNH_2$,[19] and stepwise coupling of alkenyl iodides with N,N′-diBoc hydrazine.[20]

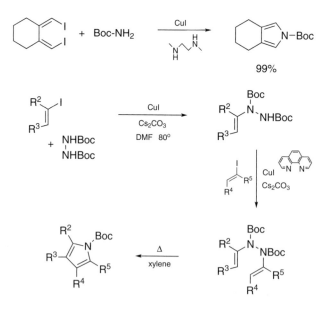

Arylation of malonic esters at room temperature occurs when the reactants are treated with CuI, picolinic acid and Cs_2CO_3.[21] And despite the general inertness of arenes, benzoxazole[22] and pentafluorobenzene[23] are found to react with ArX.

Amidine synthesis.   Alkynes, amines, and sulfonylazides (or phosphoryl azides) are combined to generate amidines. The alkynes and/or the amines can be functionalized, and their use leads to amidines containing an α-amino group or a phosphoranylalkyl group, when starting from ynamides[24] and imidophosphoranes,[25] respectively.

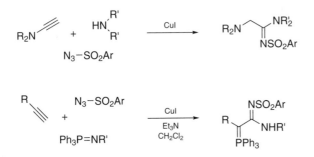

Heterocycles.   3-Aminomethylisoquinolines are obtained from *o*-ethynylaraldehydes by treatment with paraformaldehyde and amines, then *t*-BuNH₂.[26] Aminomethylation of the alkyne unit is followed by Schiff reaction and cyclization. Cyclic amidines that serve as precursors of oxindoles are assembled from *o*-ethynylarylamines and sulfonyl azides.[27]

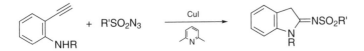

Catalyzed by CuI, *N,N'*-ditosyl-1,2-ethanediamine and *N*-tosylaminoethanol add to bromoalkynes to furnish derivatives of tetrahydropyrazine and dehydromorpholine, respectively.[28] The sulfonyl group appears to place an important role in directing the complexation of the copper atom to the triple bond.

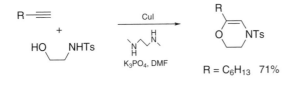

A previous observation concerning the misbehavior of sulfonyl azides in cycloaddition to alkynes prompted a study that eventually identifies the optimal condition of the reaction.[29]

The photoinitiated cyclization involving an allene and an azido group is improved by CuI, in terms of regioselectivity in the cyclization.[30] Imidocopper species are thought to undergo electrocyclization prior to demetallation.

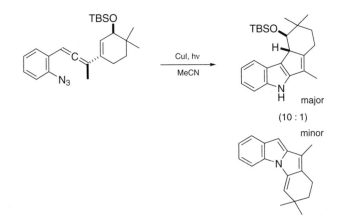

Cycloaddition of *N,N'*-di-*t*-butyldiaziridinone with conjugated dienes occurs to form 4-alkenyl-2-oxazolidinones. The less substituted double bond of the diene participates in the reaction.[31]

Addition reactions. (*Z*)-1,2-Diphosphinoalkenes are formed by a CuI-catalyzed addition of Ph_2PH to 1-phosphinoalkynes.[32] Functionalization of the triple bond of a 1-alkyne by TsN_3 (CuI, Et_3N, H_2O) leads to an *N*-tosylcarboxamide.[33]

Addition of allyltributylstannane to aldehydes is also catalyzed by CuI in DMF at room temperature.[34]

[1]Nakatani, Y., Koizumi, Y., Yamasaki, R., Saito, S. *OL* **10**, 2067 (2008).
[2]Singh, S.K., Reddy, M.S., Mangle, M., Ganesh, K.R. *T* **63**, 126 (2007).
[3]Altman, R.A., Shafir, A., Choi, A., Lichtor, P.A., Buchwald, S.L. *JOC* **73**, 284 (2008).
[4]Zhang, H., Ma, D., Cao, W. *SL* 243 (2007).
[5]Sperotto, E., de Vries, J.G., van Klink, G.P.M., van Koten, G. *TL* **48**, 7366 (2007).
[6]Chang, J.W.W., Chee, S., Mak, S., Buranaprasertsuk, P., Chavasiri, W., Chan, P.W.H. *TL* **49**, 2018 (2008).
[7]Bao, W., Liu, Y., Lv, X. *S* 1911 (2008).
[8]Wong, K.-T., Ku, S.-Y., Yen, F.-W. *TL* **48**, 5051 (2007).
[9]Schareina, T., Zapf, A., Mägerlein, W., Müller, N., Beller, M. *SL* 555 (2007).
[10]Li, H., Li, C. *OL* **8**, 5365 (2006).
[11]Fang, Y., Li, C. *JACS* **129**, 8092 (2007).
[12]Chen, Y.-J., Chen, H.-H. *OL* **8**, 5609 (2006).
[13]Schareina, T., Zapf, A., Mägerlein, W., Müller, N., Beller, M. *CEJ* **13**, 6249 (2007).

[14]Ma, H.C., Jiang, X.Z. *SL* 1335 (2008).

[15]Jones, K.L., Porzelle, A., Hall, A., Woodrow, M.D., Tomkinson, N.C.O. *OL* **10**, 797 (2008).

[16]Xi, Z., Liu, F., Zhou, Y., Chen, W. *T* **64**, 4254 (2008).

[17]Shafir, A., Lichtor, P.A., Buchwald, S.L. *JACS* **129**, 3490 (2007).

[18]Tao, C.-Z., Li, J., Fu, Y., Liu, L., Guo, Q.-X. *TL* **49**, 70 (2008).

[19]Martin, R., Larsen, C.H., Cuenca, A., Buchwald, S.L. *OL* **9**, 3379 (2007).

[20]Rivero, M.R., Buchwald, S.L. *OL* **9**, 973 (2007).

[21]Yip, S.F., Cheung, H.Y., Zhou, Z., Kwong, F.Y. *OL* **9**, 3469 (2007).

[22]Do, H.-Q., Daugulis, O. *JACS* **129**, 12404 (2007).

[23]Do, H.-Q., Daugulis, O. *JACS* **130**, 1128 (2008).

[24]Kim, J.Y., Kim, S.H., Chang, S. *TL* **49**, 1745 (2008).

[25]Cui, S.-L., Wang, J., Wang, Y.-G. *OL* **10**, 1267 (2008).

[26]Ohta, Y., Oishi, S., Fujii, N., Ohno, H. *CC* 835 (2008).

[27]Yoo, E.J., Chang, S. *OL* **10**, 1163 (2008).

[28]Fukudome, Y., Naito, H., Hata, T., Urabe, H. *JACS* **130**, 1820 (2008).

[29]Yoo, E.J., Ahlquist, M., Kim, S.H., Bae, I., Fokim, V.V., Sharpless, K.B., Chang, S. *ACIE* **46**, 1730 (2007).

[30]Feldman, K.S., Hester II, D.K., Lopez, C.S., Faza, O.N. *OL* **10**, 1665 (2008).

[31]Yuan, W., Du, H., Zhao, B., Shi, Y. *OL* **9**, 2589 (2007).

[32]Kondoh, A., Yorimitsu, H., Oshima, K. *JACS* **129**, 4099 (2007).

[33]Cho, S.H., Chang, S. *ACIE* **46**, 1897 (2007).

[34]Kalita, H.R., Borah, A.J., Phukan, P. *TL* **48**, 5047 (2007).

Copper(II) nitrate.

Oxidative cleavage. Recovery of carbonyl compounds from 2-substituted 1,3-dithianes is achieved by mixing with $Cu(NO_3)_2 \cdot 2.5H_2O$/montmorillonite-K10 in the air and irradiation with ultrasound.[1]

[1]Oksdath-Mansilla, G., Penenory, A.B. *TL* **48**, 6150 (2007).

Copper(I) oxide.

Decarboxylation. Heating with Cu_2O and 1,10-phenanthroline in quinoline and NMP causes decarboxylation of electron-deficient aroic acids.[1]

[1]Goossen, L.J., Rodriguez, N., Melzer, B., Linder, C., Deng, G., Levy, L.M. *JACS* **129**, 4824 (2007).

Copper(II) oxide.

N-Arylation. Amines are arylated with the aid of CuO nanoparticles under basic conditions (KOH, DMSO, air, 80–110°).[1] Another protocol indicates the use of $Fe(acac)_2$ as cocatalyst, as demonstrated by an effective in *N*-arylation of pyrazole.[2]

[1]Rout, L., Jammi, S., Punniyamurthy, T. *OL* **9**, 3397 (2007).

[2]Taillefer, M., Xia, N., Ouali, A. *ACIE* **46**, 934 (2007).

Copper(I) 2-thienylcarboxylate, CuTC.

Coupling reactions. Conversion of oxime ethers to *N*-substituted imines involving N—O to N—C bond exchange is empowered by CuTC. Organostannanes and organoboronic acids can supply the substituent.[1]

Diels–Alder reaction. Molecules containing a conjugate diene and a terminal alkyne units that are separated by several bonds undergo intramolecular Diels–Alder reaction, as a result of transient activation of the dienophile as an alkynylcopper species.[2]

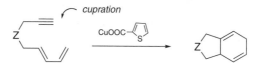

[1]Liu, S., Yu, Y., Liebeskind, L.S. *OL* **9**, 1947 (2007).
[2]Fürstner, A., Stimson, C.C. *ACIE* **46**, 8845 (2007).

Copper(I) triflate.

Diazoketone decomposition. Cyclopropanation of a proximal electron-rich double bond, for example, of an indole nucleus, is inescapable once carbenoid generation is initiated. By placing a moderately nucleophilic chain that is sterically interactable with the emerging cyclopropane, skeletal reorganization is feasible. Such transformation based on careful design is conducive to synthetic purposes.[1]

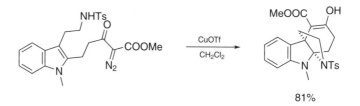

81%

Addition and cycloaddition. Two slightly different protocols are available for achieving addition of 1-alkynes to trifluoromethyl ketones:[2] use either CuOTf and *t*-BuOK with Xantphos in THF at 60°, or Cu(OTf)$_2$ and two equivalents of *t*-BuOK and 1,10-phenanthroline in toluene at 100°. Cyanoformate esters can contribute the CN group as an addend to react with organoazides in a [3+2]cycloaddition catalyzed by CuOTf.[3]

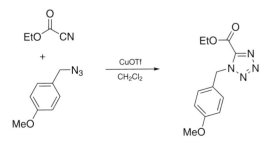

[1]Shen, L., Zhang, M., Wu, Y., Qin, Y. *ACIE* **47**, 3618 (2008).
[2]Motoki, R., Kanai, M., Shibasaki, M., *OL* **9**, 2997 (2007).
[3]Bosch, L., Vilarrasa, J. *ACIE* **46**, 3926 (2007).

Copper(II) triflate.

Isomerization.   *N*-Tosylazetidines undergo ring opening to afford the isomeric allylic amine derivatives.[1]

Substitution reactions.   Benzylic and allylic acetates are replaced on reaction with sulfonamides, no matter by what mechanism it proceeds, with the presence of Cu(OTf)$_2$ and *t*-BuOOAc.[2]

Silylene insertion into allylic ethers[3] is of interest to synthesis because it changes an electrophilic unit into a nucleophilic unit.

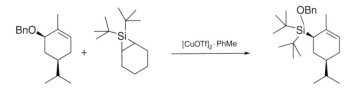

Fluorination in tandem of Nazarov cyclization succeeds in the case of alkenoylarenes.[4] That both reactions are catalyzed by Cu(OTf)$_2$ is most pleasing.

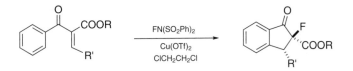

Together with *t*-BuOOAc the sulfamidation at a benzylic or allylic position in moderate yields by PhSO$_2$NHR is mediated by Cu(OTf)$_2$ – 1,10-phenanthroline.[5] Adamantane is also functionalized at C-1 by this method.

Addition and cycloaddition reactions. Henry reaction carried out in the presence of Cu(OTf)$_2$ and **1** is a demonstration of the possibility in developing reaction that is electronically and sterically tunable.[6]

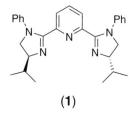

(1)

With catalysis of Cu(OTf)$_2$ reactivity of imines toward attack by RZnBr is shown to be enhanced by attaching a 2-pyridinesulfonyl group to the nitrogen atom, as in **2**.[7]

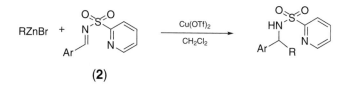

(2)

In cyclization of β-dicarbonyl compounds containing an alkyne which is extended outward to give cycloalkenones, there is reinforced activation by Cu(OTf)$_2$ and an Ag salt individually, at the active methylene group and the triple bond.[8]

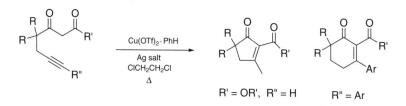

N-Tosyl carbamates form carbenoids on treatment with (py)$_4$Cu(OTf)$_2$. The reactive species are trapped by alkenes. γ,δ-Unsaturated O-tosylhydroxamic acids furnish aziridino-2-oxazolidinones.[9]

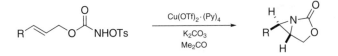

Carbenoids generated from α-diazoketones react with thiourea to give 2-aminothia-zoles.[10] Cyclodehydration follows initial trapping of the carbenoid via S-C bond formation.

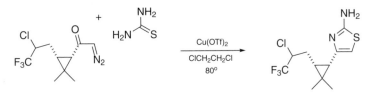

[1]Ghorai, M.K., Kumar, A., Das, K. *OL* **9**, 5441 (2007).
[2]Powell, D.A., Pelletier, G. *TL* **49**, 2495 (2008).
[3]Bourque, L.E., Cleary, P.A., Woerpel, K.A. *JACS* **129**, 12602 (2007).
[4]Nie, J., Zhu, H.-W., Cui, H.-F., Hua, M.-Q., Ma, J.-A. *OL* **9**, 3053 (2007).
[5]Pelletier, G., Powell, D.A. *OL* **8**, 6031 (2006).
[6]Ma, K., You, J. *CEJ* **13**, 1863 (2007).
[7]Esquivias, J., Arrayas, R.G., Carretero, J.C. *ACIE* **46**, 9257 (2007).
[8]Deng, C.-L., Guo, S.-M., Xie, Y.-X., Li, J.-H. *EJOC* 1457 (2007).
[9]Lebel, H., Lectard, S., Parmentier, M. *OL* **9**, 4797 (2007).
[10]Yadav, J.S., Reddy, B.V.S., Rao, Y.G., Narsaiah, A.V. *TL* **49**, 23815 (2008).

Copper(II) trifluoroacetate.

Cycloaddition. Oxaziridines are made to condense with alkenes (e.g., the more electron-rich double bond of a diene) via ring opening to give oxazolidines.[1]

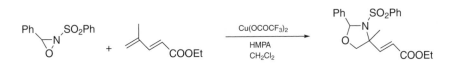

[1]Michaelis, D.J., Ischay, M.A., Yoon, T.P. *JACS* **130**, 6610 (2008).

(1,5-Cyclooctadiene)bismethallylruthenium.

Hydroamination. Amides, lactams, carbamates and ureas add to 1-alkynes to give enamide derivatives. The stereoselectivity of this anti-Markovnikov addition is sensitive to phosphine ligands that are present.[1]

[1]Goossen, L.J., Rauhaus, J.E., Deng, G. *ACIE* **44**, 4042 (2005).

(1,5-Cyclooctadiene)platinum(II) triflate.

Hydroamination.   Sulfonamides and weakly basic anilines add to alkenes with yields >95%, when catalyzed by (cod)Pt(OTf)$_2$.[1] Interestingly, (cod)RhBF$_4$ [also with DPPB ligand] induces intramolecular hydroamination in the anti-Markovnikov fashion.[2]

[1]Karshtedt, D., Bell, A.T., Tilley, T.D. *JACS* **127**, 12640 (2005).
[2]Takemiya, A., Hartwig, J.F. *JACS* **128**, 6042 (2006).

(1,3,5-Cyclooctatriene)bis(dimethyl fumarate)ruthenium.

Cross-coupling.[1]   Chain extension of *N*-vinylcarboxamides at the terminal sp^2-carbon atom with alkyl or alkenyl residue on reaction with alkenes or alkynes is catalyzed by the Ru complex.

[1]Tsujita, H., Urz, Y., Matsuki, S., Wada, K., Mitsudo, T., Kondo, T. *ACIE* **46**, 5160 (2007).

Cyclopentadienyl(η^6-naphthalene)ruthenium hexafluorophosphate.

Hydration.[1]   Anti-Markovnikov hydration of 1-alkynes to afford aldedydes is accomplished by treatment with the title complex and a 2-diphenylphosphino-6-arylpyridine.

[1]Labonne, A., Kribber, T., Hintermann, L. *OL* **8**, 5853 (2006).

(*p*-Cymene)(*N*-tosyl-1,2-diphenylethylenediamine)ruthenium.

Reduction.[1]   In the presence of the Ru complex, β-diketones are chemoselectively and enantioselectively reduced by HCOONHEt$_3$. An aliphatic ketone group is reduced in preference to an aryl ketone.

[1]Matsukawa, Y., Isobe, M., Kotsuki, H., Ichikawa, Y. *JOC* **70**, 5339 (2005).

D

Dess-Martin periodinane.

Oxidation.[1]   The title reagent is useful for oxidation of β-hydroxy-α-diazo esters at room temperature to furnish α-diazo β-keto esters.

[1]Li, P., Majireck, M.M., Korboukh, I., Weinreb, S.M. *TL* **49**, 3162 (2008).

1,4-Diazabicyclo[2.2.2]octane, DABCO.

Rearrangement.[1]   Allyl acrylates are converted into α-allylated acrylic acids on treatment with DABCO and Me₃SiCl. Rearrangement is induced by conjugate addition to generate ester enolates.

[1]Li, Y., Wang, Q., Goeke, A., Frater, G. *SL* 288 (2007).

1,8-Diazabicyclo[5.4.0]undec-7-ene, DBU.

Desilylation.   Removal of the silyl group from a silylalkyne is effected by heating with DBU at 60° in H_2O–MeCN (1 : 19).[1]

Michael reaction.   An intramolecular Michael reaction with concomitant elimination is synthetically most pleasing. The following example serves to elaborate a complex bridged ring system under mild conditions.[2]

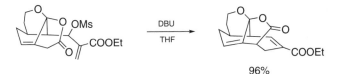

Because of steric effects epimerization can intervene in an intramolecular addition catalyzed by DBU.[3]

Fiesers' Reagents for Organic Synthesis, Volume 25. By Tse-Lok Ho
Copyright © 2010 John Wiley & Sons, Inc.

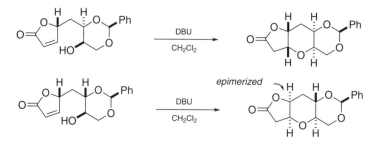

DBU also promotes conjugate addition of amines to α,β-unsaturated esters, nitriles, and ketones.[4]

Aldol reaction.   As a suitable base for catalyzing an intramolecular aldol reaction between an α-benzoyloxy ketone and an aldehyde, DBU also promotes transesterification of the product.[5]

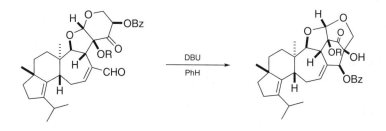

[1]Yeom, C.-E., Kim, M.J., Choi, W., Kim, B.M. *SL* 565 (2008).
[2]Prabhudas, B., Clive, D.L.J. *ACIE* **46**, 9295 (2007).
[3]Lee, H., Kim, K.W., Park, J., Kim, H., Kim, S., Kim, D., Hu, X., Yang, W., Hong, J. *ACIE* **47**, 4200 (2008).
[4]Yeom, C.-E., Kim, M.-J., Kim, B.-M. *T* **63**, 904 (2007).
[5]Watanabe, H., Nakada, M. *JACS* **130**, 1150 (2008).

Di-*t*-butyl dicarbonate.

Isothiocyanate esters.[1]   Boc₂O serves to promote elimination of H_2S from adducts of RNH_2 and CS_2. A catalytic amount of DMAP or DABCO is also added to the reaction.

Carbamates.[2]   Acylation of amines by Boc₂O is catalyzed by thiourea.

[1]Munch, H., Hansen, J.S., Pittelkow, M., Christensen, J.B., Boas, U. *TL* **49**, 3117 (2008).
[2]Khaksar, S., Heydari, A., Tajbakhsh, M., Vahdat, S.M. *TL* **49**, 3527 (2008).

Dibutyliodotin hydride.

Additive cleavage.[1]   Dibutyliodostannyl radical generated from $Bu_2Sn(I)H$ cleaves methylenecyclopropanes regioselectively. Addition of the resulting C-radical to a proximal double bond leads to a cyclic product. The alkenylstannane unit is amenable to Stille coupling.

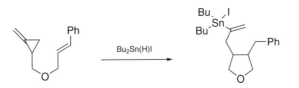

[1]Hayashi, N., Hirokawa, Y., Shibata, I., Yasuda, M., Baba, A. *JACS* **130**, 2912 (2008).

Dicarbonylhydrido-η^5-[1,3-bis(trimethylsilyl)-2-hydroxy-4,5,6,7-tetrahydroindenyl]iron.

Hydrogenation.[1]   The iron complex **1** is a highly selective hydrogenation catalyst for reducing the carbonyl group. Double bond, triple bond, halogen atoms, cyclopropane and pyridine rings are not affected.

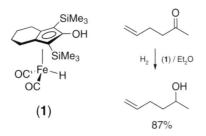

[1]Casey, C.P., Guan, H. *JACS* **129**, 5816 (2007).

Dichloramine T.

Deoximation.[1]   To the numerous procedures for recovery of carbonyl compounds from oximes is added one involving oxidation with $TsNCl_2$ in aqueous MeCN.

[1]Gupta, P.K., Manral, L., Ganesan, K. *S* 1930 (2007).

2,3-Dichloro-5,6-dicyano-1,4-benzoquinone, DDQ.

Oxidative cyclization.   An expedient method for synthesis of *cis*-2,6-disubstituted 4-pyranones involves intramolecular trapping of oxallyl cation which is generated by DDQ oxidation.[1]

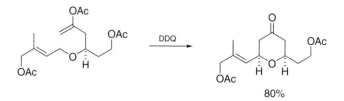

80%

Ether cleavage. An allylic *p*-methoxybenzyl ether is selectively cleaved with DDQ in a buffer solution, thereby facilitating the progress in a synthesis of multipolide-A.[2]

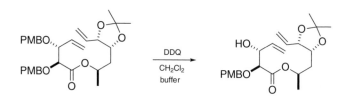

[1]Tu, W., Liu, L., Floreancig, P.E. *ACIE* **47**, 4184 (2008).
[2]Ramana, C.V., Khaladkar, T.P., Chatterjee, S., Gurjar, M.K. *JOC* **73**, 3817 (2008).

Dichloro(diethene)rhodium.

Cycloaddition. Unsaturated carbodiimides and alkynes are combined to afford bicyclic amidines, under the influence of $(C_2H_4)_2RhCl_2$ and an aminodialkoxyphosphine ligand[1].

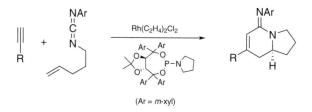

(Ar = *m*-xyl)

[1]Yu, R.T., Rovis, T. *JACS* **130**, 3262 (2008).

Dichlorobis(*p*-cymene)(triphenylphosphine)ruthenium(II).

Addition to 1-alkenes. Addition of polyhaloalkanes (e.g., CCl_4) to 1-alkenes by catalysis of Ru complexes (instead of free radical initiators) is enhanced by microwave irradiation.[1]

[1]Borguet, Y., Richel, A., Delfosse, S., Leclerc, A., Delaude, L., Demonceau, A. *TL* **48**, 6334 (2007).

Dichloro(norbornadiene)bis(triphenylphosphine)ruthenium(II).

Cycloaddition. The title complex is employed to catalyze intramolecular homo-Diels–Alder reaction.[1]

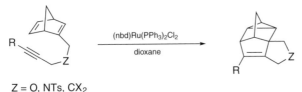

Z = O, NTs, CX$_2$

[1]Tenaglia, A., Gaillard, S. *OL* **9**, 3607 (2007).

Dichloro(pyridine-2-carboxylato)gold(III).

Cyclization. The Au(III) complex induces cyclization of silyl ethers of alkenyl/ α-oxallenyl carbinols to form easily fragmentable bicyclo[3.1.0]hexan-6-ol intermediates, which give rise to 3-acylcyclopentenes.[1] Since the substrates are prepared from conjugated carbonyl compounds the 3-step process represents a unique annulation method.

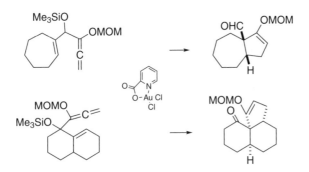

[1]Huang, X., Zhang, L. *JACS* **129**, 6398 (2007).

Dichlorotris(triphenylphosphine)ruthenium(II).

Acetals. Isomerization of allylic ethers to enol ethers by the Ru complex enables addition of alcohols to form acetals.[1]

[1]Krompiec, S., Penczek, R., Kuznik, N., Malecki, J.G., Matlengiewicz, M. *TL* **48**, 137 (2007).

Di(ethene)trispyrazolylboratoruthenium.

Hydroamination. Addition of amines to 1-alkynes afford enamines (from secondary amines) or imines (from primary amines), when the components are heated with TpRu(C$_2$H$_4$)$_2$ and Ph$_3$P in toluene at 100°.[1]

[1]Fukumoto, Y., Asai, H., Shimizu, M., Chatani, N. *JACS* **129**, 13792 (2007).

Dicobalt octacarbonyl.

Pauson–Khand reaction.   In the presence of tetramethylthiourea the Pauson–Khand reaction suceeds with enynes in which the double bond is present in a silylcyclopropene unit. It enables a synthesis of the angular triquinane sesquiterpene (-)-pentalenene.[1]

The C=N bond of a carbodiimide is shown to participate in a Pauson–Khand reaction, forming γ-imino-α,β-unsaturated-γ lactams.[2]

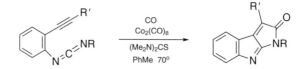

Cyclocarbonylation.   An intriguing transformation of certain epoxy enynes entails double carbonylation and cyclization.[3]

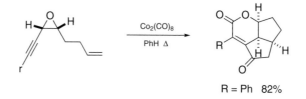

R = Ph   82%

Homologation.   Epoxides undergo ring opening and chain elongation to afford β-hudroxy esters when they are exposed to $Co_2(CO)_8$ under CO (1 atm.) in MeOH.[4]

[1]Pallerla, M.K., Fox, J.M. *OL* **9**, 5625 (2007).
[2]Aburano, D., Yoshida, T., Miyakoshi, N., Mukai, C. *JOC* **72**, 6878 (2007).
[3]Odedra, A., Lush, S.-F., Liu, R.-S. *JOC* **72**, 567 (2007).
[4]Denmark, S.E., Ahmad, M. *JOC* **72**, 9630 (2007).

Dicyclohexylboron chloride.

Hydroxyalkylation.[1]   Primary amides and aldehydes combine in the presence of Cy_2BCl and Et_3N in ether.

[1]Kiran, S., Ning, S., Williams, L.J. *TL* **48**, 7456 (2007).

Difluoro(4-trifluoromethylphenyl)bromane.

Oxidative condensation.[1]   The title reagent converts a mixture of alkynes and primary alcohols to afford enones, in the presence of $BF_3 \cdot OEt_2$.

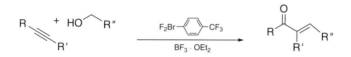

Aziridination.[2]   Ylides of the structure $[CF_3SO_2NBrC_6H_4CF_3]$ are formed by mixing the λ^3-bromane with $CF_3SO_2NH_2$ in MeCN that can cycloadd to alkenes to give *N*-triflylaziridines.

[1]Ochiai, M., Yoshimura, A., Mori, T., Nishi, Y., Hirobe, M. *JACS* **130**, 3742 (2008).
[2]Ochiai, M., Kaneaki, T., Tada, N., Miyamoto, K., Chuman, H., Shiro, M., Hayashi, S., Nakanishi, W. *JACS* **129**, 12938 (2007).

Diiodine pentoxide.

Oxidation.[1]   Benzylic alcohols are oxidized by I_2O_5 with KBr as activator in H_2O at room temperature.

[1]Liu, Z.-Q., Zhao, Y., Luo, H., Chai, L., Sheng, Q. *TL* **48**, 3017 (2007).

1,3-Diiodo-5,5-dimethylhydantoin.

Nitriles.   Primary alcohols and amines are converted to nitriles on treatment with aqueous ammonia and the title reagent.[1]

[1]Iida, S., Togo, H. *SL* 407 (2007).

Diisobutylaluminum hydride, Dibal-H.

Reduction.   At $-78°$, selective reduction of 1-alkylindole-2,3-dicarboxylic esters at the C-2 substituent (to a CHO group) by Dibal-H is observed.[1] Generally, the ester to aldehyde conversion can be performed at $0°$ with alkali metal diisobutyl(*t*-butoxy)aluminum hydride, which is formed by adding *t*-BuOM (M = Na, Li) to Dibal-H in THF.[2,3]

Rearrangement.[4]   A cyclopropyl group is liable to expand during reduction of a neighboring amidic carbonyl.

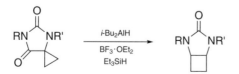

Hydroalkylation.[5]   2-Alkynoic esters form ketene Al-enolates on treatment with Dibal-H and NMO. The enolate species react with epoxides regioselectively. Further processing leads α-alkylidene-γ-butyrolactones in either the (*E*)-form or (*Z*)-form.

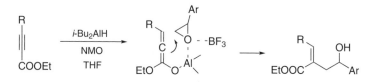

[1]Sayyed, I.A., Alex, K., Tillack, A., Schwarz, N., Spannenberg, A., Michalik, J., Beller, M. *T* **64**, 4590 (2008).
[2]Song, J.I., An, D.K. *CL* **36**, 8863 (2007).
[3]Kim, M.S., Choi, Y.M., An, D.K. *TL* **48**, 5061 (2007).
[4]Methot, J.L., Dunstan, T.A., Mampreian, D.M., Adams, B., Altman, M.D. *TL* **49**, 1155 (2008).
[5]Ramachandran, P.V., Garner, G., Pratihar, D. *OL* **9**, 4753 (2007).

N,N-Diisopropylaminoborane.

Reduction and coupling.[1]   The title reagent is generated from lithium *N,N*-diisopropylaminoborohydride on treatment with Me₃SiCl at room temperature. It reduces esters and nitriles in the presence of LiBH₄ (catalyst).

It serves as a source of boron in the preparation of arylboronic acids from ArBr by a Pd-catalyzed coupling reaction.

[1]Pasumansky, L., Haddenham, D., Clary, J.W., Fisher, G.B., Goralski, C.T., Singaram, B. *JOC* **73**, 1898 (2008).

Dilauroyl peroxide.

Radical cyclization.   Amido radicals are generated from *N*′-methyldithiocarbonyl-hydrazides by heating with dilauroyl peroxide. Setting up the functional group in juxtaposition to a double bond invites intramolecular addition that even further implications in ring formation are envisaged. As a key step in a synthesis of the amaryllidaceae alkaloid fortucine the value of such a process is demonstrated.[1]

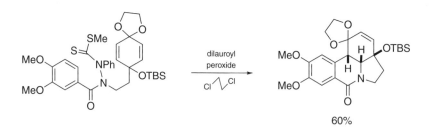

60%

Addition reactions.   3-Aryl- and 3-formylpyrroles are alkylated by heating with the xanthates (XCH$_2$SC=S)OEt and dilauroyl peroxide. The activated carbon chain CH$_2$X (X = CN, COOEt, Ac, . . .) is introduced to C-2.[2]

[1]Biechy, A., Hachisu, S., Quiclet-Sire, B., Ricard, L., Zard, S.Z. *ACIE* **47**, 1436 (2008).
[2]Guadarrama-Morales, O., Mendez, F., Miranda, L.D. *TL* **48**, 4515 (2007).

Dimanganese decacarbonyl.

Alkyl radicals.   Iodine atom abstraction from alkyl iodides occurs on irradiation with Mn$_2$(CO)$_{10}$. The carbon radicals thus generated are readily trapped by hydrazones.[1,2]

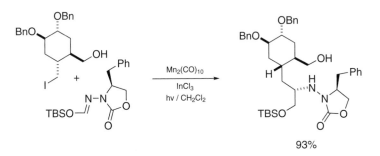

93%

[1]Korapala, C.S., Qin, J., Fristad, G.K. *OL* **9**, 4243 (2007).
[2]Fristad, G.K., Ji, A. *OL* **10**, 2311 (2008).

4-Dimethylaminopyridine, DMAP.

Esterification.   A useful protocol for esterification of alcohols by anhydrides under solvent-free conditions involves only catalytic amounts of DMAP, without any auxiliary base.[1] 6-*O*-Protected octyl β-D-glucopyranosides are selectively (>99%) acylated by treatment with an acid anhydride and DMAP in toluene maintaining at −20° to −40°.[2] Hydroxyl groups at C-2 and C-4 are untouched.

Rearrangement.[3]   Furo[2,3-*c*](3a*H*,5*H*)quinoline-2,4-diones are converted into the isomeric furo[3,4-*c*](1*H*,5*H*)quinoline-3,4-diones, on heating with DMAP. Apparently, isocyanate intermediates are formed via opening of the six-membered heterocycle.

[1]Sakakura, A., Kuwajiri, K., Ohkubo, T., Kosugi, Y., Ishihara, K. *JACS* **129**, 14775 (2007).
[2]Muramatsu, W., Kawabata, T. *TL* **48**, 5031 (2007).
[3]Kafka, S., Kosmrlj, J., Klasek, A., Pevec, A. *TL* **49**, 90 (2008).

Dimethyldioxirane.

Epoxidation.[1]   The reaction of *N*-allenylamides with dimethyldioxirane results in α-(1-oxyvinyl)iminium species, which can be trapped by pyrroles in a [4+3]cycloaddition.

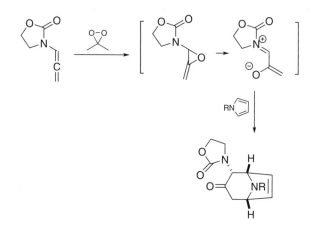

[1]Antoline, J.E., Hsung, R.P., Huang, J., Song, Z., Li, G. *OL* **9**, 1275 (2007).

Dimethylsulfide – halogen.

α-Halo-α,β-unsaturated esters.[1]   Oxidation of primary alcohols by the reagent complex in the presence of Et$_3$N and a triphenylphosphoranylacetic ester enables a Wittig reaction which is also followed by halogenation and dehydrohalogenation.

[1]Jiang, B., Dou, Y., Xu, X., Xu, M. *OL* **10**, 593 (2008).

Dimethylsulfoxonium methylide.

Methylenation.[1]   While cyclic ketones undergo Corey–Chaykovsky reaction to deliver epoxides at room temperature, excess amounts of base suppress the transformation and at high temperature the ketones are converted into 1,2-dimethylenecycloalkanes.

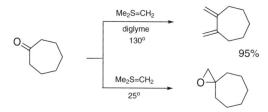

Methylene insertion.[2] *o*-Hydroxydiaryl ketones give 3-arylbenzofurans, which are apparently derived from a normal Corey–Chaykovsky reaction. Additionally, insertion of a CH_2 group between the carbonyl and one of the aryl residues also occurs.

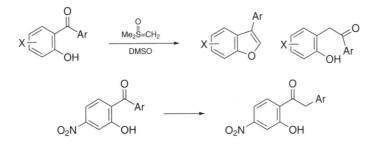

[1]Butova, E.D., Fokin, A.A., Schreiner, P.R. *JOC* **72**, 5689 (2007).
[2]Chitimalla, S.K., Chang, T.-C., Liu, T.-C., Hsieh, H.-P., Liao, C.-C. *T* **64**, 2586 (2008).

1,1-Dioxonaphtho[1,2-*b*]thiophene-2-methoxycarbonyl chloride.

Amino group protection.   Title reagent **1** is proposed to derivatize amines for their protection. Mild conditions are needed to cleave the derived carbamates.[1]

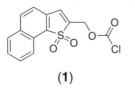

(1)

[1]Carpino, L.A., Abdel-Maksoud, A.A., Ionescu, D., Mansour, E.M.E., Zewail, M.A. *JOC* **72**, 1729 (2007).

Diphenyldiazomethane.

Benzhydryl ethers.   The title reagent protects alcohols on heating in an inert solvent.[1]

[1]Best, D., Jenkinson, S.F., Rule, S.D., Higham, R., Mercer, T.B., Newell, R.J., Weymouth-Wilson, A.C., Fleet, G.W.J., Petursson, S. *TL* **49**, 2196 (2008).

Diphenyliodonium trifluoroacetate.

Phenylation.   The reagent reacts with arylamines in refluxing DMF to give ArNHPh.[1]

[1]Carroll, M.A., Wood, R.A. *T* **63**, 11349 (2007).

S,S-Diphenyl-N-(o-nitrobenzenesulfenyl)-N'-tosylsulfodiimide.

Aziridination.   Mild thermolysis of the title reagent TsN=S(Ph$_2$)=NSAr in MeCN liberates the nitrene [ArS-N], which is intercepted by alkenes.

[1]Yoshimura, T., Fujie, T., Fujii, T. *TL* **48**, 427 (2007).

Diphenylphosphonyl azide.

Carbamoyl azides.   Azidocarbonylation of amines to form RNHCON$_3$ starts from formation of carbamate salts in the reaction with CO$_2$ (catalyzed by tetramethyl-2-phenyl-guanidine) which is followed by treatment with Ph$_2$P(O)N$_3$.[1]

[1]Garcia-Egido, E., Fernandez-Suarez, M., Munoz, L. *JOC* **73**, 2909 (2008).

Dipyridyliodonium tetrafluoroborate.

Glycosyl fluorides.[1]   *O*-Protected thioglycosides are converted to glycosyl fluorides at room temperature by the title reagent, for example, Glu(β)SPh to Glu(α)F. If the reaction medium also contains TfOH and ROH, glycosides are obtained.

[1]Huang, K.-T., Winssinger, N. *EJOC* 1887 (2007).

N,N'-Ditosylhydrazine.

Diazoacetic esters.[1]   Diazoacetic esters are prepared from reaction of bromoacetic esters and TsNHNHTs and DBU (base) in THF at 0°.

[1]Toma, T., Shimokawa, J., Fukuyama, T. *OL* **9**, 3195 (2007).

E

Erbium(III) triflate.

β-Amino alcohols. Epoxides are opened by amines in water at 60°, $Er(OTf)_3$ shows catalytic activity for this transformation.[1]

[1]Procopio, A., Gaspari, M., Nardi, M., Oliverio, M., Rosati, O. *TL* **49**, 2269 (2008).

Ethyl(carboxysulfamoyl)triethylammonium hydroxide, Burgess reagent.

Oxidative dimerization. Conversion of thiols to disulfides by the Burgess reagent is reported, despite the economic irrationality involved.[1]

Sulfilimine formation. Reaction of sulfoxides with the Burgess reagent at room temperature delivers sulfilimines.[2]

[1]Banfield, S.C., Omori, A.T., Leisch, H., Hudlicky, T. *JOC* **72**, 4989 (2007).
[2]Raghavan, S., Mustafa, S., Rathore, K. *TL* **49**, 4256 (2008).

Ethyl tribromoacetate.

Acyl bromides. Heating aldehydes with Br_3COOEt and $(PhCOO)_2$ in toluene accomplishes radical bromination.[1]

[1]Kang, D.H., Joo, T.Y., Chavasiri, W., Jang, D.O. *TL* **48**, 285 (2007).

Fiesers' Reagents for Organic Synthesis, Volume 25. By Tse-Lok Ho
Copyright © 2010 John Wiley & Sons, Inc.

F

Fluoroboric acid.

   Michael reaction. Efficient addition of thiols to conjugated carbonyl compounds proceeds in the presence of a catalyst derived from HBF$_4$ adsorbed in silica gel.[1]

[1]Sharma, G., Kumar, R., Chakraborti, A.K. *TL* **49**, 4272 (2008).

Fluorosulfuric acid – antimony(V) fluoride.

   Cyclization. Intramolecular Friedel–Crafts alkenylation of ω-aryl-1,1-difluoroalkenes can be applied to a synthesis of tetrahydro[6]helicenes.[1]

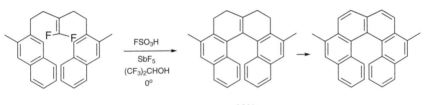

62%

[1]Ichikawa, J., Yokota, M., Kudo, T., Umezaki, S. *ACIE* **47**, 4870 (2008).

Fluorous reagents and ligands.

   Amide formation. Development of a carbodiimide reagent (**1**) containing at one end an isopropyl group and at the other end a chain ending in two polyfluorinated branches has been reported.[1]

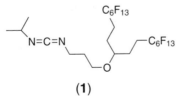

(1)

Sonogashira coupling. The method for Sonogashira coupling that employs $Pd(OSO_2C_8F_{17})_2/3\text{-}PyCH(OC_8F_{17})_2$ in a mixture of toluene and perfluorodecalin truly demonstrates the synthetic utility of fluorous compounds.[2]

Mitsunobu reagent. Pairing bis[3-(nonatrifluoro-*t*-butoxypropyl)] azodicarboxylate with Ph_3P constitutes a fluorous version of the Mitsunobu reagent.[3]

Diels–Alder reaction. Rate enhancement of the Diels–Alder reaction is noted in aqueous perfluorinated emulsions (from perfluorohexane and lithium perfluorooctane-sulfonate).[4]

Separation technique. *o*-Nitrobenzenesulfonamides that are left behind from incomplete alkylation are rapidly and quantitatively separated by treatment with a highly reactive $C_8F_{17}CH_2CH_2CH_2I$ followed by fluorous solid-phase extraction.[5]

[1]del Pozo, C., Keller, A.I., Nagashima, T., Curran, D.P. *OL* **9**, 4167 (2007).
[2]Yi, W.-B., Cai, C., Wang, X. *EJOC* **3445** (2007).
[3]Chu, Q., Henry, C., Curran, D.P. *OL* **10**, 2453 (2008).
[4]Nishimoto, K., Kim, S., Kitano, Y., Tada, M., Chiba, K. *OL* **8**, 5545 (2006).
[5]Basle, E., Jean, M., Gouault, N., Renault, J., Uriac, P. *TL* **48**, 8138 (2007).

Formaldehyde.

2-Methylene-3-butenal. The valuable dienophile is readily prepare from crotonaldehyde and aq. HCHO.[1]

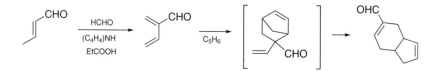

[1]Zou, Y., Wang, Q., Goeke, A. *CEJ* **14**, 5335 (2008).

Formic acid.

Double bond cleavge.[1] A surprising oxidative cleavage of 1,2,3,4-tetraaryl-2-butene-1,4-diones to afford benzils in good yields (instead of forming the furans) occurs when they are irradiated by microwaves in formic acid and a catalytic amount of conc. H_2SO_4.

[1]Rao, H.S.P., Jothilingam, S., Vasantham, K., Scheeren, H.W. *TL* **48**, 4495 (2007).

G

Gallium(III) chloride.

Diels–Alder reaction.[1] Allylsilanes and propargylsilanes condense with N-aryl-aldimines in the presence of GaCl$_3$ to provide 2,4-disubstituted tetrahydroquinolines and quinolines, respectively.

[1]Hirashita, T., Kawai, D., Araki, S. *TL* **48**, 5421 (2007).

Gallium(III) triflate.

Fluoroalkyl heterocycles.[1] Condensation of o-functionalized (OH, SH, NH$_2$) anilines with fluoroalkyl ketones leads to benzoxazolines, benzthiazolines, benzimida-zolines, . . . bearing a fluorinated carbon chain at C-2. The reaction is promoted by Ga(OTf)$_3$.

[1]Prakash, G.K.S., Mathew, T., Panja, C., Vaghoo, H., Venkataraman, K., Olah, G.A. *OL* **9**, 4627 (2007).

Gold.

O-Silyl ethers. Gold nanoparticles are found to be effective catalyst for derivatization of aldehydes by Me$_3$SiCN to afford O-trimethylsilyl cyanohydrins at room temperature.[1]

Primary alcohols are converted into silyl ethers by R$_3$SiH using nanosized gold particles supported on alumina.[2] Under the same conditions aromatic aldehydes are coupled and silylated.

Similarly supported gold particles prepared from HAuCl$_4$ and NaOH on CeO$_2$ or ZrO$_2$ have found use in the three-component condensation of aldehydes, amines and alkynes.[3]

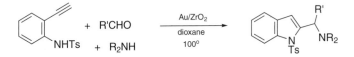

Hydrogenation. Gold-on-titanium dioxide is a special catalyst with which nitro-alkenes are converted into saturated oximes.[4] Thus only the sidechain is affected when o,β-dinitrostyrene is subjected to the hydrogenation conditions in its presence. Conventional hydrogenation (Pd/C, Pt/C) of the same compound leads to indole and diamine products.

Fiesers' Reagents for Organic Synthesis, Volume 25. By Tse-Lok Ho
Copyright © 2010 John Wiley & Sons, Inc.

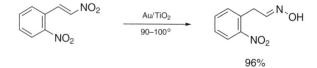

96%

[1]Cho, W.K., Lee, J.K., Kang, S.M., Chi, Y.S., Lee, H.-S., Choi, I.S. *CEJ* **13**, 6351 (2007).
[2]Raffa, P., Evangelisti, C., Vitulli, G., Salvadori, P. *TL* **49**, 3221 (2008).
[3]Zhang, X., Corma, A. *ACIE* **47**, 4358 (2008).
[4]Corma, A., Serna, P., Garcia, H. *JACS* **129**, 6358 (2007).

Gold(III) bromide.

Glycosylation. By taking advantage of the affinity of gold ions for the triple bond, 1,2-ortho esters of sugars containing a propargyloxy unit have been designated as latent glycosyl donors. Glycosylation using $AuBr_3$ gives 2-benzoyloxy glycosides.[1]

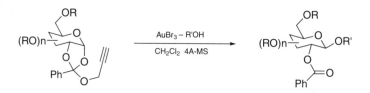

[1]Sureshkumar, G., Hotha, S. *TL* **48**, 6564 (2007).

Gold(I) chloride.

Spiroacetals. 1,1-Addition of hydroxyl groups to a triple bond in creating spiroacetal systems has been realized in the presence of AuCl.[1]

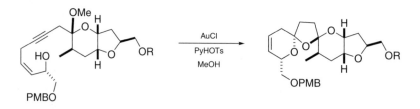

Isomerization. *o*-alkynylaroic acids undergo Au-catalyzed cyclization.[2] There is a possibility of forming either the phthalide or the isocoumarin system.[2]

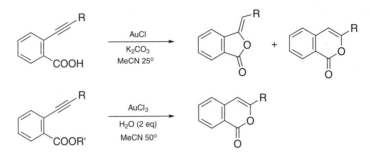

From the *S*-silyl derivatives of *o*-alkynylarylthiols, cycloisomerization leads to 3-silyl-benzothiophenes.[3] The gold salt is found to be highly effective in promoting Meyer–Schuster rearrangement.[4]

Cycloelimination.   On exposure to AuCl, esters of 1,4,5-alkatrien-3-ols are subject to cycloisomerization and elimination to provide benzene derivatives.[5]

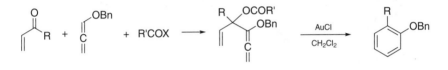

[1]Li, Y., Zhou, F., Forsyth, C.J. *ACIE* **46**, 279 (2007).
[2]Marchal, E., Uriac, P., Legouin, B., Toupet, L., van de Weghe, P. *T* **63**, 9979 (2007).
[3]Nakamura, I., Sato, T., Terada, M., Yamamoto, Y. *OL* **9**, 4081 (2007).
[4]Lopez, S.S., Engel, D.A., Dudley, G.B. *SL* 949 (2007).
[5]Huang, X., Zhang, L. *OL* **9**, 4627 (2007).

Gold(I) chloride – tertiary phosphine.

Cycloisomerization.   *N*-Allenyl-*N*-arylcarbamates cyclize to afford 1,4-dihydroquino-line derivatives, when catalyzed by AuCl and 2-(di-*t*-butylphosphino)biphenyl.[1]

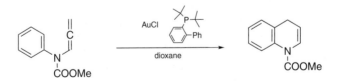

Sonogashira coupling.   Use (Ph$_3$P)AuCl instead of Pd catalyst, the Sonogashira coupling is effected without a copper salt.[2]

[1]Watanabe, T., Oishi, S., Fujii, N., Ohno, H. *OL* **9**, 4821 (2007).
[2]Gonzalez-Arellano, C., Abad, A., Corma, A., Garcia, H., Iglesias, M., Sanchez, F. *ACIE* **46**, 1536 (2007).

Gold(I) chloride – 1,3-bis(2,6-diisopropylphenyl)imidazol-2-ylidene/silver salts.

1,3-Rearrangement.   3-Acetoxy-1-alkenes undergo isomerization to give the 1-acetoxy isomers on heating with the Au(I) complex in 1,2-chichloroethane.[1]

Addition.   Addition of ROH to allenes results in the generation of allylic ethers. The Au(I)-catalyzed reaction is regioselective, the products retain the most stable double bond.[2]

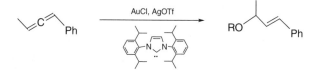

Hydrofluorination of alkynylarenes delivers (Z)-β-fluorostyrenes on treatment with 3HF Et₃N and catalytic amounts of the Au complex.[3]

Quinoline synthesis.   2-Acylarylamines and alkynes condense under the influence of the Au(I) complex.[4]

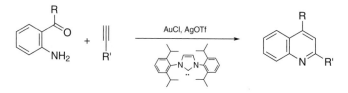

Cycloaddition.   Cyclopropanation of the double bond(s) in dienynes is observed, with the *sp*-hybridized carbon atom(s) behaving like carbene(s). The transformation is highly efficient while giving intriguing polycyclic isomers.[5]

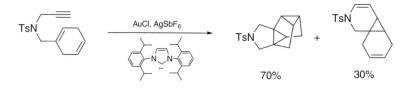

70%   30%

Oxidative transformations.   Addition of Ph₂SO to the reaction has the effect of converting one of the *sp*-hybridized carbon atoms of an enyne into a carbonyl group while modification of the molecular skeleton takes place.[6]

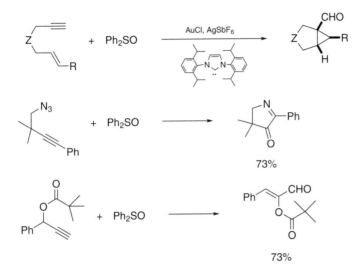

73%

73%

[1]Marion, N., Gealageas, R., Nolan, S.P. *OL* **9**, 2653 (2007).
[2]Zhang, Z., Widenhoefer, R.A. *OL* **10**, 2079 (2008).
[3]Akana, J.A., Bhattacharyya, K.X., Müller, P., Sadighi, J.P. *JACS* **129**, 7736 (2007).
[4]Liu, X.-Y., Ding, P., Huang, J.-S., Che, C.-M. *OL* **9**, 2645 (2007).
[5]Kim, S.M., Park, J.H., Choi, S.Y., Chung, Y.K. *ACIE* **46**, 6172 (2007).
[6]Witham, C.A., Mauleon, P., Shapiro, N.D., Sherry, B.D., Toste, F.D. *JACS* **129**, 5838 (2007).

Gold(I) chloride – tertiary phosphine/silver salts.

Vinyl ethers and esters.[1] From readily available ROCH=CH$_2$ and through an exchange reaction catalyzed by (Ph$_3$P)AuCl–AgOAc a great variety of vinyloxy compounds are prepared.

Substitution reactions. Alkyl 2-alkynylbenzoates are activated by the Au(I) salt toward formation of isocoumarin, thereby weakening the O—C$_{(alk)}$ bond of the esters. Attack of nucleophiles results in the cleavage of the esters.[2,3] Particularly noteworthy is the formation of tetralins by way of an intramolecular reaction involving an aromatic ring (a C-nucleophile).[3]

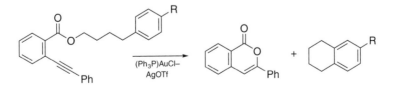

Cyclizations. Intramolecular hydroamination to form 5- and 6-membered hetero-cycles[4] occurs on unactivated C≡C bonds by heating the ammonium salts with AuCl,

ArPCy$_2$, and AgOTf in PhMe at 80°. By a similar process *O*-propargyl hydroxylamines give isoxazolines.[5] *N*-Boc derivatives of propargylic amines lose isobutene to furnish 5-alkylidene-2-oxazolidinones.[6]

Twofold addition involving heteroatom and *C*-nucleophiles are exemplified in the formation of 2,3-benzo-9-oxabicyclo[3.3.1]nonanes[7] and the tricyclic 2,2'-bipyrrolyl derivatives.[8]

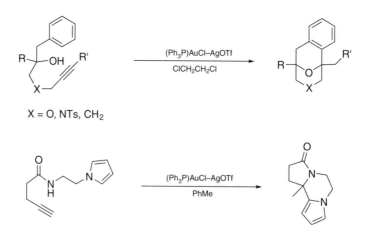

Molecules containing allene and propargylic alcohol subunits are activated toward cyclization. It is found that Au(I) and Pt(IV) salts promote different modes of reaction. Allene activation takes precedence with (Ph$_3$P)AuSbF$_6$.[9]

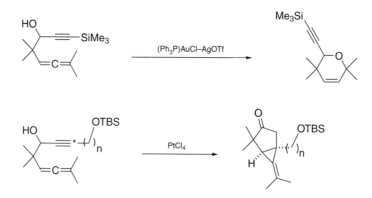

1-Alkynes can be used as external nucleophiles that participate in the second stage of the reaction. Pyrrolo[1,2-*a*]quinolines are generated.[10]

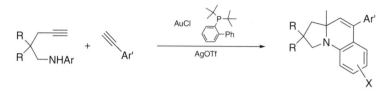

The synthesis of 2-alkoxypyridines from 1-alkoxy-3-alken-1-ynes involves trapping by nitriles and cyclization of the intermediates.[11]

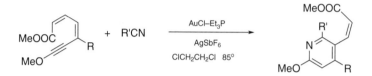

Ring closure is observed when 6-heptynals are exposed to the gold(I) salt.[12] Silyl ketene amides and carbamates cyclize accordingly.[13]

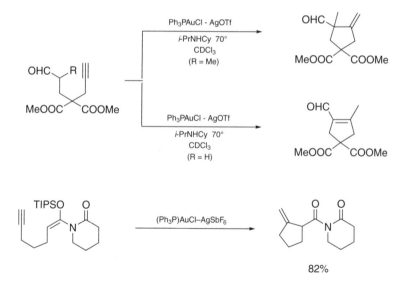

Cyclopentenes are formed from 1,5-enynes in moderate yields.[14] Conversion of 1,2,4-alkatrienes into cyclopentadienes can be effected with the gold complex[15] or PtCl$_2$ (loc. cit.).

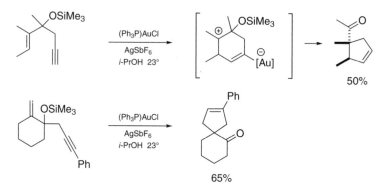

Participation of nucleophiles in the cyclization of enynes gives rise to functionalized products. Thus, substrates containing an allylic carbonate unit are transformed into derivatives of 4-cycloalkene-1,2-diols.[16] Without an internally participatory group an enyne can incorporate an amine.[17]

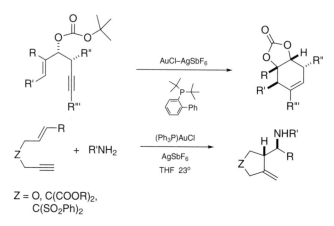

Z = O, C(COOR)₂, C(SO₂Ph)₂

In some situations, bicyclization of polyenynes leads to hydrocarbon products[18] and arenes containing an allenyl group in a sidechain are subject to cycloisomerization.[19]

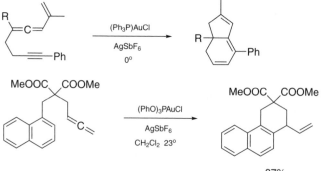

Formation of a benzene ring is readily achieved from enynols and their esters when the oxygen functionality is propargylic.[20,21]

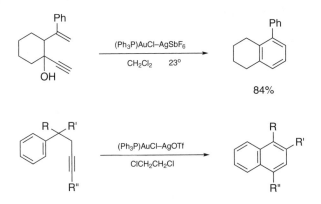

84%

Of synthetic interest is the assemblage of conjugated enynes and propargylic esters in the presence of AuCl and aromatization of the resulting cyclopropanes by (ArO)$_3$PAuX.[22]

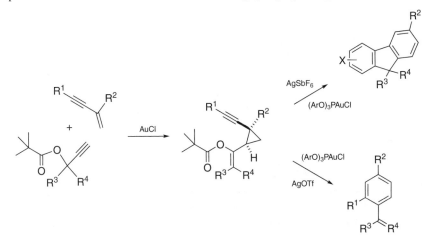

Highly substituted α-pyrones are formed when propargyl propynoates are treated with (Ph$_3$P)AuSbF$_6$.[23] The reaction proceeds via sequential Au-activation to induce [3,3]-sigmatropic rearrangement and at the conjugated triple bond for the ensuing cyclization.

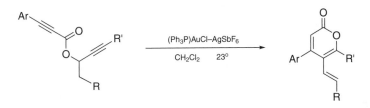

While rearrangement of propargylic esters to allenyl esters is facile, denouement of such intermediates is highly dependent on the presence of other multiple CC bonds in juxtaposition.[24,25] In any event, the generation of polycyclic compounds in one synthetic operation deserves serious consideration of the method for exploitation in the construction of significant and complex target molecules.

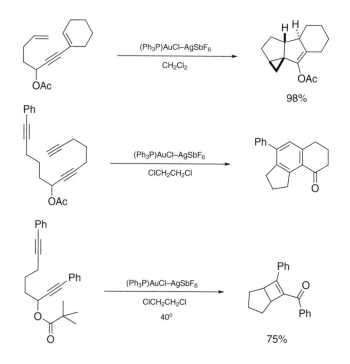

5-Alkoxyalk-6-en-1-ynes undergo cyclization to give cyclohepta-1,4-diene derivatives.[26] The analogous siloxy enynes provide cyclohept-4-enones as major products. However, a change of the phosphine ligand to a more electron-rich version diverts the reaction pathway to the formation of products of a different type.[27]

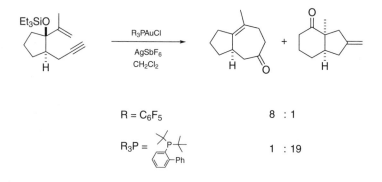

N-Allyl β-keto amides are found to cyclize in the presence of the Au(I) complex. Somewhat better results are obtained with that containing the biphenyldi-*t*-butylphosphine ligand.[28]

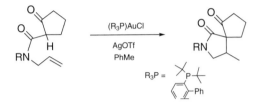

Cyclodehydration of diols in which one of the hydroxyl groups is allylic shows another reaction pattern. When 2-alkenyl-6-alkyltetrahydropyrans are produced it favors the *cis*-isomers.[29]

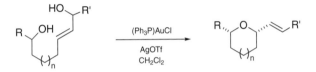

The Au(I) catalyst system transforms 2,3-benzo-7-oxabicyclo[2.2.1]hept-2-ene into a tetralin-1,4-dication equivalent, as shown by its reaction with allylsilanes.[30]

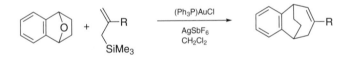

α,β;γ,δ-Dienals generate allyl cations extended by an auroxy substituent. Trapping in situ by allylsilanes and allylic alcohols provides structurally diversified products.[31]

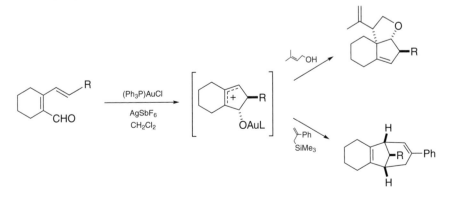

Through oxygen atom transfer acycarbenoids of gold are formed from homopropargyl phenyl sulfoxides. Cyclization ensues.[32]

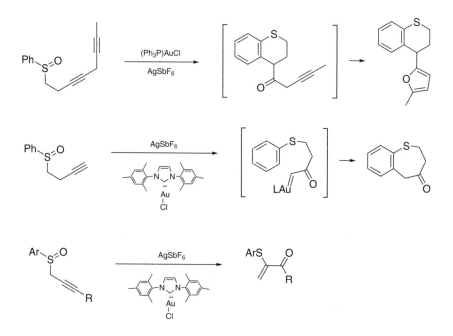

Rearrangements.   Isomerization of propargylic alcohols to 1,3-transposed conjugated carbonyl compounds is catalyzed by (Ph₃P)AuOTf and MoO₂(acac)₂.[33] Allenyl oxomolybdates are likely involved as intermediates.

Rearrangement intervenes in the hydration of α-silylated propargylic carboxylates.[34]

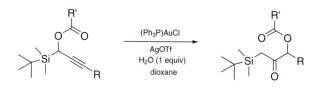

Alkynes substituted at both propargylic positions, one with an ester and the other an epoxy ring are converted into enones, attendant by transposition of both unsaturated and oxygenation sites.[35]

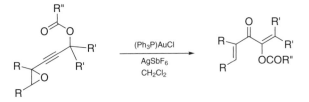

The ester unit of 1-(γ-pivaloxypropargyl)cyclobutanols undergoes 1,3-shift to afford an allene that is activatable for cyclization.[36]

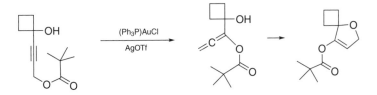

[1]Nakamura, A., Tokunaga, M. *TL* **49**, 3729 (2008).

[2]Li, Y., Yang, Y., Yu, B. *TL* **49**, 3604 (2008).

[3]Asao, N., Aikawa, H., Tago, S., Umetsu, K. *OL* **9**, 4299 (2007).

[4]Bender, C.F., Widenhoefer, R.A. *CC* 2741 (2008).

[5]Yeom, H.-S., Lee, E.-S., Shin, S. *SL* 2292 (2007).

[6]Lee, E.-S., Yeom, H.-S., Hwang, J.-H., Shin, S. *EJOC* 3503 (2007).

[7]Barluenga, J., Fernandez, A., Satrustegui, A., Dieguez, A., Rodriguez, F., Fananas, F.J. *CEJ* **14**, 4153 (2008).

[8]Yang, T., Campbell, L., Dixon, D.J. *JACS* **129**, 12070 (2007).

[9]Zriba, R., Gandon, V., Aubert, C., Fensterbank, L., Malacria, M. *CEJ* **14**, 1482 (2008).

[10]Liu, X.-Y., Che, C.-M. *ACIE* **47**, 3805 (2008).

[11]Barluenga, J., Fernandez-Rodriguez, M.A., Garcia-Garcia, P., Aguilar, E. *JACS* **130**, 2764 (2008).

[12]Binder, J.T., Crone, B., Haug, T.T., Menz, H., Kirsch, S.F. *OL* **10**, 1025 (2008).

[13]Minnihan, E.C., Colletti, S.L., Toste, F.D., Shen, H.C. *JOC* **72**, 6287 (2007).

[14]Kirsch, S.F., Binder, J.T., Crone, B., Duschek, A., Haug, T.T., Liebert, C., Menz, H. *ACIE* **46**, 2310 (2007).

[15]Lee, J.H., Toste, F.D. *ACIE* **46**, 912 (2007).

[16]Lim, C., Kang, J.-E., Lee, J.-E., Shin, S. *OL* **9**, 3539 (2007).

[17]Leseurre, L., Toullec, P.Y., Genet, J.-P., Michelet, V. *OL* **9**, 4049 (2007).

[18]Lin, G.-Y., Yang, C.-Y., Liu, R.-S. *JOC* **72**, 6753 (2007).

[19]Tarselli, M.A., Gagne, M.R. *JOC* **73**, 2439 (2008).

[20]Grise, C.M., Rodrigue, E.M., Barriault, L. *T* **64**, 797 (2008).

[21]Dudnik, A.S., Schwier, T., Gevorgyan, V. *OL* **10**, 1465 (2008).

[22]Gorin, D.J., Watson, I.D.G., Toste, F.D. *JACS* **130**, 3736 (2008).

[23]Luo, T., Schreiber, S.L. *ACIE* **46**, 8250 (2007).

[24]Lemiere, G., Gandon, V., Cariou, K., Fukuyama, T., Dhimane, A.-L., Fensterbank, L., Malacria, M. *OL* **9**, 2207 (2007).

[25]Oh, C.H., Kim, A. *SL* 777 (2008).

[26]Bae, H.J., Baskar, B., An, S.E., Cheong, J.Y., Thangadurai, D.T., Hwang, I.-C., Rhee, Y.H. *ACIE* **47**, 2263 (2008).

[27]Baskar, B., Bae, H.J., An, S.E., Cheong, J.Y., Rhee, Y.H., Duschek, A., Kirsch, S.F. *OL* **10**, 2605 (2008).
[28]Zhou, C.-Y., Che, C.-M. *JACS* **129**, 5828 (2007).
[29]Aponick, A., Li, C.-Y., Biannic, B. *OL* **10**, 669 (2008).
[30]Hsu, Y.-C., Datta, S., Ting, C.-M., Liu, R.-S. *OL* **10**, 521 (2008).
[31]Lin, C.-C., Teng, T.-M., Odedra, A., Liu, R.-S. *JACS* **129**, 3798 (2007).
[32]Shapiro, N.D., Toste, F.D. *JACS* **129**, 4160 (2007).
[33]Egi, M., Yamaguchi, Y., Fujiwara, N., Akai, S. *OL* **10**, 1867 (2008).
[34]Sakaguchi, K., Okada, T., Shimada, T., Ohfune, Y. *TL* **49**, 25 (2008).
[35]Cordonnier, M.-C., Blanc, A., Pale, P. *OL* **10**, 1569 (2008).
[36]Yeom, H.S., Yoon, S.J., Shin, S. *TL* **48**, 4817 (2007).

Gold(I) chloride – tertiary phosphine/silver hexafluoroantimonate-acetonitrile complex.

Enol esters.[1]   Propargylic esters react with nucleophiles with attendant 1,2- or 1,3-migration of the ester subunit, depending on the substitution pattern of the propargylic site.

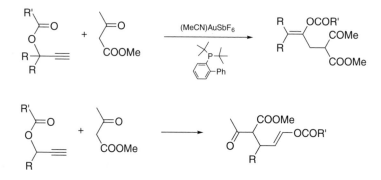

Cyclization.   Allylic triorganostannyl and acetoxy groups at the two termini of a chain are simultaneously detached in the presence of $R_3PAu(MeCN)SbF_6$. In the process the remainder skeleton forms a ring.[2]

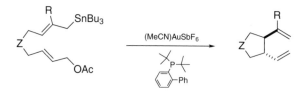

t-Butyl *N*-alkynylcarbamates cyclize to give imidazolones with loss of the *t*-butyl group.[3]

[1]Amijs, C.H.M., Lopez-Carrillo, V., Echavarren, A.M. *OL* **9**, 4021 (2007).
[2]Porcel, S., Lopez-Carrillo, V., Garicia-Yebra, C., Echavarren, A.M. *ACIE* **47**, 1883 (2008).
[3]Istrate, F.M., Buzas, A.K., Jurberg, I.D., Odabachian, Y., Gagosz, F. *OL* **10**, 925 (2008).

Gold(III) chloride.

O-Trimethylsilyl cyanohydrins. Derivatization of ketones and aldehydes is catalyzed by $AuCl_3$ at room temeperature.[1]

Insertion by nitrene. Formation of ArNHNs from arenes and PhI=NNs is mediated by $AuCl_3$. A secondary benzylic C—H bond is also reactive (e.g., 1,3,5-triisopropylbenzene gives two kinds of nitrene insertion products, and the benzylic amine derivative is predominant in a 3:2 ratio to the arylamine isomer.)[2]

Cyclization. Intramolecular addition of a hydroxy group to an allene unit results in cyclic ethers. Methoxymethyl ethers are also reactive but different regioselectivity has been noted.[3]

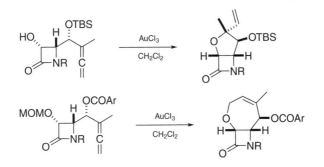

1-Bromoalka-1,2-dien-4-ones afford β-bromofurans. With these substrates $AuCl_3$ as well as AuCl are serviceable catalysts. However, there exists a remarkable ligand effect pertaining the employment of $(Ph_3P)AuX$, with $X = BF_4$ vs. $X = OTf$.[4]

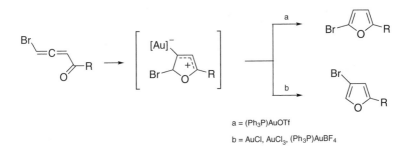

a = $(Ph_3P)AuOTf$

b = AuCl, $AuCl_3$, $(Ph_3P)AuBF_4$

Substitution. The geminal functional groups of 1-arenesulfonylcyclopropanols are both replaced in an $AuCl_3$-catalyzed reaction with amines and 1-alkynes in water.[5] It constitutes a new access to the special kind of propargylic amines. Direct conversion of allylic alcohols to the corresponding amines is also accomplished on treatment with $AuCl_3$ in MeCN at room temperature.[6]

Quinoline synthesis.   2,4-Disubstituted quinolines are synthesized in one operation from arylamines, aldehydes, and 1-alkynes. A mixture of $AuCl_3$ and CuBr is used to promote the condensation. The effectiveness of $AuCl_3$ to transform *N*-propargylarylamines to quinolines at room temperature has been independently verified.[7]

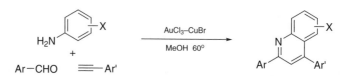

Benzyl ethers.   Addition of ROH to styrenes to provide secondary benzyl ethers by Au(III) salts alone is not practical because Au(III) ion is readily reduced to the catalytically inactive Au(0) species. The problem is solved by using an $AuCl_3$–$CuCl_2$ combination.[8]

[1]Cho, W.K., Kang, S.M., Medda, A.K., Lee, J.K., Choi, I.S., Lee, H.-S. *S* 507 (2008).
[2]Li, Z., Capretto, D.A., Rahaman, R.O., He, C. *JACS* **129**, 12058 (2007).
[3]Alcaide, B., Almendros, P., del Campo, T.M. *ACIE* **46**, 6684 (2007).
[4]Xia, Y., Dudnik, A.S., Gevorgyan, V., Li, Y. *JACS* **130**, 6940 (2008).
[5]Liu, J., An, Y., Jiang, H.-Y., Chen, Z. *TL* **49**, 490 (2008).
[6]Guo, S., Song, F., Liu, Y. *SL* 964 (2007).
[7]Xiao, F., Chen, Y., Liu, Y., Wang, J. *T* **64**, 2755 (2008).
[8]Zhang, X., Corma, A. *CC* 3080 (2007).

Gold(III) chloride – silver triflate.

Cyclization.   Allenylmalonic esters undergo cyclization in HOAc, leading to dihydro-α-pyrones.[1] Exposure of 1-aroxy-2,3-epoxypropanes to $AuCl_3$–AgOTf in dichloroethane leads to chroman-3-ols. A critical ligand has been identified.[2]

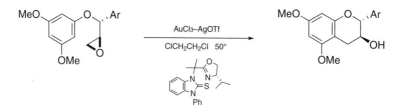

o-Alkynylarylamines and 1-alkynes are combined to generate *N*-(2-alkenyl)indoles.[3]

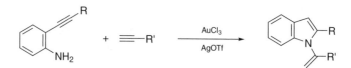

In a single step alkynones are transformed into cyclic conjugated ketones.[4] This reaction does not go through hydration.

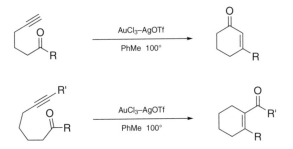

[1]Piera, J., Krumlinde, P., Strübing, D., Bäckvall, J.-E. *OL* **9**, 2235 (2007).
[2]Liu, Y., Li, X., Lin, G., Xiang, Z., Xiang, J., Zhao, M., Chen, J., Yang, Z. *JOC* **73**, 4625 (2008).
[3]Zhang, Y., Donahue, J.P., Li, C.-J. *OL* **9**, 627 (2007).
[4]Jin, T., Yamamoto, Y. *OL* **9**, 5259 (2007).

Gold(I) cyanide.

Isoflavanones.   With AuCN–Bu$_3$P to catalyze the combination of salicyladehydes and ethynylarenes, a redox transformation that proceeds via hydroauration of the alkynes by acylaurium hydrides eventually results in the formation of isoflavanones.[1]

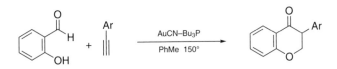

[1]Skouta, R., Li, C.-J. *ACIE* **46**, 1117 (2007).

Gold(III) oxide.

Cycloisomerization.   4-Alkynoic acids cyclize to give γ-alkylidene-γ-butyrolactones under the influence of Au$_2$O$_3$.[1]

[1]Toullec, P.Y., Genin, E., Antoniotti, S., Genet, J.-P., Michelet, V. *SL* 707 (2008).

Gold(I) triflimide – azolecarbene.

Rearrangement.   2-Pivaloxy-1,3-dienes are formed by treatment of the corresponding propargylic esters with the Au(I) complex.[1]

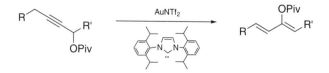

Oxygen atom transfer is observed in the reaction of homopropargyl sulfoxides.[2] Formation of the 1,3-dicarbonyl unit from homopropargyl sulfoxides that contain a distal propargylic OH also engenders a group migration.

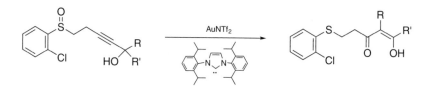

For cycloisomerization of α-alkynyl-β-keto esters the use of Tf_2NAu in conjunction with very bulky tris[(triarylsilyl)ethynyl]phosphine ligands, remarkable rate enhancements are observed.[3] The effect is attributable to the cavity environment created by the ligand to keep the nucleophilic center and the Au-activated triple bond of the substrate close.

Cycloaddition.   1-Alkynyl-1-cyclopropyl ketones generate cyclic 1,3-dipolar species and their cycloaddition with vinyl ethers is followed by ring size regulation of the cyclo-adducts (expansion of the 3-membered ring and contraction of the 6-membered ring) and demetallation.[4]

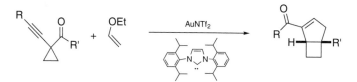

Equally interesting is trapping by carbonyl compounds, imines, and indoles, leading to polycycles containing a furan ring.[5]

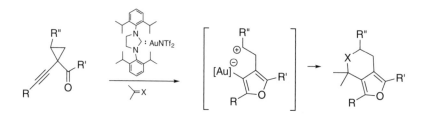

[1]Li, G., Zhang, G., Zhang, L. *JACS* **130**, 3740 (2008).
[2]Li, G., Zhang, L. *ACIE* **46**, 5156 (2007).
[3]Ochida, A., Ito, H., Sawamura, M. *JACS* **128**, 16486 (2006).
[4]Li, G., Huang, X., Zhang, L. *JACS* **130**, 6944 (2008).
[5]Zhang, G., Huang, X., Li, G., Zhang, L. *JACS* **130**, 1814 (2008).

Gold(I) triflimide – triarylphosphine complex.

Rearrangement. Rearrangement of propargylic esters as promoted by (Ph₃P)AuNTf₂ affords α-iodinated enones in the presence of NIS and a solvent system of acetone and water (800 : 1) at 0°.[1]

Esters of allenyl carbinols give 2-acyloxy-1,3-dienes on treatment with AuNTf₂, which is complexed to (2′,4′,6′-triisopropyl-2-biphenyl)dicyclohexylphosphine.[2]

An aza-Claisen rearrangement is implicated in the transformation of *N*-geranyl-*N*-(pent-2-en-4-yn-1-yl)-*p*-toluenesulfonamide into an *N*-tosylpyrrole.[3]

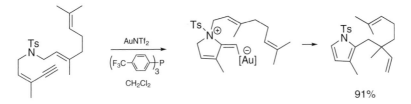

3-Allylbenzosiloles are readily prepared from [(*o*-alkynyl)aryl]allylsilanes.[4]

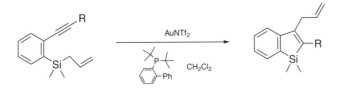

[1]Yu, M., Zhang, G., Zhang, L. *OL* **9**, 2147 (2007).
[2]Buzas, A.K., Istrate, F.M., Gagosz, F. *OL* **9**, 985 (2007).
[3]Istrate, F.M., Gagosz, F. *OL* **9**, 3181 (2007).
[4]Matsuda, T., Kadowaki, S., Yamaguchi, Y., Murakami, M. *CC* 2744 (2008).

Graphite.

Substitution. Alkylation of alcohols and arenes by alkyl halides (including benzyl halides) is easily performed on heating (116–130°) the components with graphite, either neat or in PhCl.[1]

[1]Sereda, G.A., Rajpara, V.B., Slaba, R.L. *T* **63**, 8351 (2007).

Grignard reagents.

X/magnesium exchange. The general method for preparing Grignard reagents by the exchange method using *i*-PrMgCl as applied to 3-substituted 1,2,5-tribromobenzenes in THF at −40° is dependent on the nature of the substituent R.[1] Preference for exchange of the 1-Br atom when R = H, Me, OMe; and of the 2-Br atom when R = F, Cl, CF$_3$, CN.

1-Chlorocyclopropyl phenyl sulfoxide undergoes exchange reaction to afford 1-chloro-cyclopropylmagnesium chloride (a magnesium carbenoid) that reacts with *N*-lithioaryl-amines to give *o*-cyclopropylarylamines.[2]

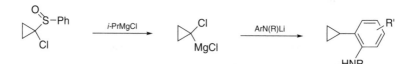

Selective insertion of magnesium carbenoid to a cyclic C—H bond instead of one at the α-position of an ester group is perhaps quite unexpected.[3]

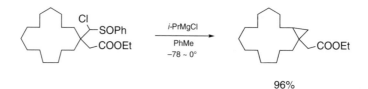

96%

Br/Mg exchange converts 1,2-dibromocyclopentene into the β-bromoalkenyl-magnesium chloride (LiCl complex), which reacts normally with carbonyl compounds. It is possible to peform a copper-mediated coupling at the β-carbon site while retaining the C—MgCl unit.[4]

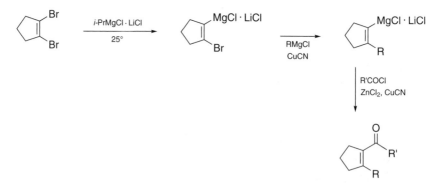

Addition reactions. RMgCl adds to pyridine *N*-oxides at C-2 to generate dienal oximes.[5] When the crude products are heated with Ac$_2$O, homologated pyridines result.[6]

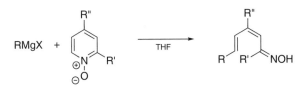

Neighboring group-direction determines the Grignard reaction of 2-methoxy-*N*-methyl-succinimide. On the other hand, a bulky TBSO group exerts its regiochemical influence.[7]

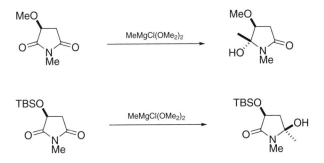

At low temperatures and over very short reaction time Grignard reaction yields ketone from esters of polyfluorinated carboxylic acids. The ketones are susceptible to reduction by RR′CHOMgX at a higher temperature.[8]

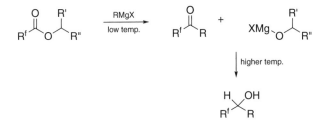

Diaryl ketones are formed in the Grignard reaction of ArCHO with Ar′MgX LiCl, when PhCHO is added during workup.[9] A redox process (Mg-Oppenauer oxidation of the halomagnesium diarylmethoxides) is involved.

o-Aminoaryl chloromethyl ketones furnish 2-substituted indoles on reaction with RMgX (or RLi), as a consequence of 1,2-aryl migration after the addition step.[10]

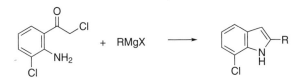

Conjugate addition of the R group from RMgX to α,β-unsaturated carboxylic acids and amides is observed when mixed with three equivalents of MeLi in THF. The methyl group of MeLi does not compete.[11]

Alkynylmagnesium bromides add to organoazides to form 4-(1,2,3-triazolyl)magnesium bromides. 1,4,5-Trisubstituted 1,2,3-triazoles are obtained after transmetallation (ZnCl₂) and Negishi coupling.[12]

Substitution.   (S)-Mesitylenesulfinimines are obtained from **1**, through a double displacement sequence and reaction with aldehydes. The first step is the Grignard reaction with MesMgBr.[13]

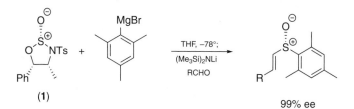

(1)

99% ee

Desulfurization and methylenation of lithiated sultams by ICH₂MgCl releases homoallylic amines.[14]

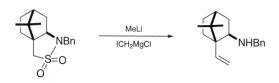

Reaction of α-chloroalkyl *t*-butanesulfinylimines with RMgX affords cyclopropylamine derivatives. But different products are obtained from the reaction involving allylmagnesium chloride.[15]

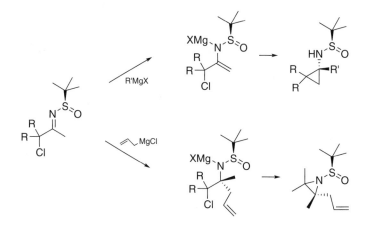

Tricyclopropylbismuth and dicyclopropylbismuth chloride are obtained by reaction of cyclopropylmagnesium bromide with $BiCl_3$ according to the required stoichiometry. These organobismuth compounds are useful for *N*-cyclopropylation of lactams such as phenanthridinone.[16]

N-Bromomagnesium enamines are formed on treatment of cyclohexanone imines of *N,N*-diethylethanamine with mesitylmagnesium bromide. These highly nucleophilic species react with even secondary alkyl fluorides.[17]

[1]Menzel, K., Mills, P.M., Frantz, D.E., Nelson, T.D., Kress, M.H. *TL* **49**, 415 (2008).
[2]Yamada, Y., Miura, M., Satoh, T. *TL* **49**, 169 (2008).
[3]Ogata, S., Saitoh, H., Wakasugi, D., Satoh, T. *T* **64**, 5711 (2008).
[4]Despotopoulou, C., Bauer, R.C., Krasovskiy, A., Mayer, P., Stryker, J.M., Knochel, P. *CEJ* **14**, 2499 (2008).
[5]Andersson, H., Wang, X., Björklund, M., Olsson, R., Almqvist, F. *TL* **48**, 6941 (2007).
[6]Andersson, H., Almqvist, F., Olsson, R. *OL* **9**, 1335 (2007).
[7]Ye, J.-L., Huang, P.-Q., Lu, X. *JOC* **72**, 35 (2007).
[8]Yamazaki, T., Terajima, T., Kawasaki-Takasuka, T. *T* **64**, 2419 (2008).
[9]Kloetzing, R.J., Krasovskiy, A., Knochel, P. *CEJ* **13**, 215 (2007).
[10]Pei, T., Chen, C.-Y., Dormer, P.G., Davies, I.W. *ACIE* **47**, 4231 (2008).
[11]Kikuchi, M., Niikura, S., Chiba, N., Terauchi, N., Asaoka, M. *CL* **36**, 736 (2007).
[12]Akao, A., Tsuritani, T., Kii, S., Sato, K., Nonoyama, N., Mase, T., Yasuda, N. *SL* 31 (2007).
[13]Sasraku-Neequaye, L., MacPherson, D., Stockman, R.A. *TL* **49**, 1129 (2008).
[14]Rogachev, V.O., Merten, S., Seiser, T., Kataerva, O., Matz, P. *TL* **49**, 133 (2008).
[15]Denolf, B., Mangelinckx, S., Törnroos, K.W., De Kimpe, N. *OL* **9**, 187 (2007).
[16]Gagnon, A., St-Onge, M., Little, K., Duplessis, M., Barabe, F. *JACS* **129**, 44 (2007).
[17]Hatakeyama, T., Ito, S., Yamane, H., Nakamura, M., Nakamura, E. *T* **63**, 8440 (2007).

Grignard reagents/cerium(III) chloride.

Addition to carbonyl group. The decrease in basicity of the $RMgX-CeCl_3$ system helps maintain the normal nucleophilic addition of the organometallic reagents, while suppressing enolization of carbonyl substrates.[1]

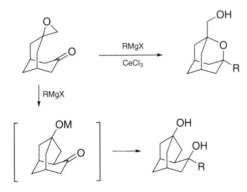

[1]Mlinaric-Majerski, K., Kragol, G., Ramljak, T.S. *SL* 405 (2008).

Grignard reagents/chromium(II) salts.

Addition. Grignard reagents such as PhMgBr add to alkynes to give alkenylmagnesium bromides when CrCl₂ and *t*-BuCOOH are present as catalysts. The adducts have a *cis*-configuration.[1]

[1]Murakami, K., Ohmiya, H., Yorimitsu, H., Oshima, K. *OL* **9**, 1569 (2007).

Grignard reagents/cobalt(II) salts.

Couplings. In the presence of CoCl₂ and 1,3-diarylimidazolium chloride, *N*-aryl-*N*-(2-iodoethyl)-*p*-toluenesulfonamie reacts with some RMgCl to afford 3-substituted *N*-toylpyrrolidines.[1]

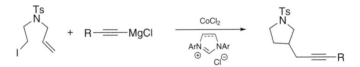

[1]Someya, H., Ohmiya, H., Yorimitsu, H., Oshima, K. *OL* **9**, 1565 (2007).

Grignard reagents/copper salts.

Substitution. *N,N*-Diorganohydroxylamine benzoates react with RMgX in the presence of CuCl₂ to provide RNR'R''.[1] Alkylation of cyclopentadienylmagnesium bromide with tertiary alkyl halides occurs when Cu(OTf)₂ is present.[2]

The S_N2' substitution of allylic picolinates is subject to chelation control.[3]

A method for synthesizing tetrahydropyridines involves Cu-catalyzed ring opening of *N*-tosylaziridnes with 2-(1,3-dioxan-2-yl)magnesium bromide.[4] 2-Alkylideneaziridines are attacked by RMgX at C-3 to generate iminomagnesium halides that can be alkylated.[5]

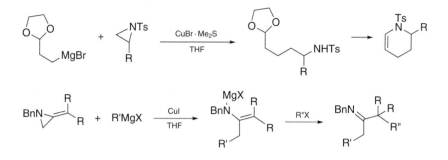

Ring-fused β-lactones open to deliver *trans*-2-organocycloalkanecarboxylic acids.[6]

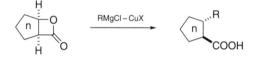

Addition reactions. As an alternative to the Baylis–Hillman approach (Z)-2-hydroxy-alkyl-2-alkenoic esters are assembled from propynoic esters via a Cu(I)-catalyzed Grignard reaction and trapping with RCHO.[7]

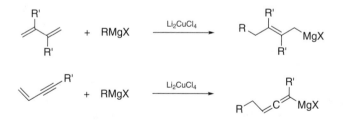

Allylic and allenyl Grignard reagents are available by conjugate carbomagnesiation of 1,3-dienes and 1,3-enynes, respectively.[8]

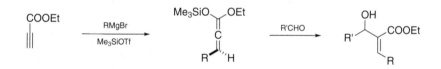

Attack of Grignard reagents on *N*-allyl-*N*-alkynyl-*N*-arenesulfonamides (CuBr · SMe₂ being present) prompts a [3,3]sigmatropic rearrangement, producing α-allyl nitriles.[9]

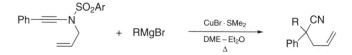

Coupling reactions. Various Grignard reagents couple with RX (reactivity profile: X = Cl < F < OMs < OTs < Br) in the presence of $CuCl_2$ and a minute amount of 1-phenylpropyne. Coupling reaction with unsymmetrical dichloroalkanes selectively replaces the primary chloride.[10] Esters are formed by coupling of RMgX with $ClCOOR'$.[11]

[1]Campbell, M.J., Johnson, J.S. *OL* **9**, 1521 (2007).
[2]Sai, M., Someya, H., Yorimitsu, H., Oshima, K. *OL* **10**, 2545 (2008).
[3]Kiyotsuka, Y., Acharya, H.P., Hyodo, T., Kobayashi, Y. *OL* **10**, 1719 (2008).
[4]Pattenden, L.C., Adams, H., Smith, S.A., Harrity, J.P.A. *T* **64**, 2951 (2008).
[5]Montagne, C., Shiers, J.J., Shipman, M. *TL* **47**, 9207 (2006).
[6]Zhang, W., Matla, A.S., Romo, D. *OL* **9**, 2111 (2007).
[7]Mueller, A.J., Jennings, M.P. *OL* **10**, 1649 (2008).
[8]Todo, H., Terao, J., Watanabe, H., Kuniyasu, H., Kambe, N. *CC* 1332 (2008).
[9]Yasui, H., Yorimitsu, H., Oshima, K. *CL* **36**, 32 (2007).
[10]Terao, J., Todo, H., Begum, S.A., Kuniyasu, H., Kambe, N. *ACIE* **46**, 2086 (2007).
[11]Bottalico, D., Fiandanese, V., Marchese, G., Punzi, A. *SL* 974 (2007).

Grignard reagents/iron salts.

Isomerization. Secondary alkylmagnesium halides are transformed into the primary isomers upon treatment with $FeCl_3$, CuBr and Bu_3P in THF at $-25°$.[1]

Coupling reactions. Heteroatom-directed arylation of 2-arylpyridines with ArMgBr and (tmeda)$ZnCl_2$ also requires $Fe(acac)_3$, 1,10-phenanthroline, and an electron acceptor (e.g., 1,2-dichloro-2-methylpropane).[2] A redox cycle of the iron species is set up during the reaction. It is further shown that TMEDA and hexamethylenetetramine have cooperative effect on $Fe(acac)_3$.[3]

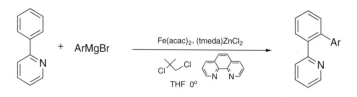

Demetallative dimerization occurs when ArMgX are exposed to $FeCl_3$ in dry air at room temperature. Conjugated dienes are similarly prepared from alkenylmagnesium halides. (For oxidative homocoupling of alkynyl and benzyl Grignard reagents, $MnCl_2$ 2LiCl is used as the catalyst.)[4]

Sulfonyl chlorides are totally defunctionalized (with loss of SO_2 also) during coupling with Grignard reagents.[5] Alkenylmagnesium bromides and alkyl halides (bromides and iodides) are coupled in the presence of $FeCl_3$ and TMEDA,[6] whereas for the formation of ArMgBr and RX a complex derived from $FeCl_3$ and **1** has been identified.[7]

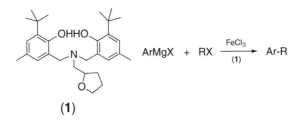

(1)

Addition reactions. *cis*-Addition of ArMgBr to alkynes leads to styrylmagnesium bromides. Regioselective generation of stilbene derivatives from arylalkynes is observed.[8] 2,3-Alkadienoic esters undergo conjugate addition with RMgX–Fe(acac)₃.[9]

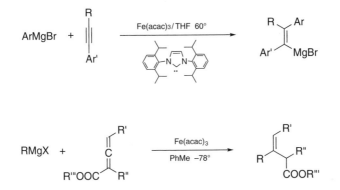

Elimination.[10] Conjugated enynes and styrenes are formed from β-hydroxy sulfides that are derived from 2-alkynyl- and 1-aryl-1,3-dithiolanes. The reaction is considered as an alternative method to McMurry coupling.

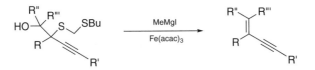

[1]Shirakawa, E., Ikeda, D., Yamaguchi, S., Hayashi, T. *CC* 1214 (2008).
[2]Norinder, J., Matsumoto, A., Yoshikai, N., Nakamura, E. *JACS* **130**, 5858 (2008).
[3]Cahiez, G., Habiak, V., Duplais, C., Moyeux, A. *ACIE* **46**, 4364 (2007).

[4]Cahiez, G., Moyeux, A., Buendia, J., Duplais, C. *JACS* **129**, 13788 (2007).
[5]Volla, C.M.R., Vogel, P. *ACIE* **47**, 1305 (2008).
[6]Guerinot, A., Reymond, S., Cossy, J. *ACIE* **46**, 6521 (2007).
[7]Chowdhury, R.R., Crane, A.K., Fowler, C., Kwong, P., Kozak, C.M. *CC* 94 (2008).
[8]Yamagami, T., Shintani, R., Shirakawa, E., Hayashi, T. *OL* **9**, 1045 (2007).
[9]Lu, Z., Chai, G., Ma, S. *JACS* **129**, 14546 (2007).
[10]Huang, L.-F., Chen, C.-W., Luh, T.-Y. *OL* **9**, 3663 (2007).

Grignard reagents/manganese salts.

Coupling. Grignard reagents couple with heteroaryl chlorides proceeds in the presence of MnCl$_2$.[1]

[1]Rueping, M., Ieawsuwan, W. *SL* 247 (2007).

Grignard reagents/nickel complexes.

Substitution. Certain alkoxy group of an alkyl naphthyl ether is susceptible to replacement on reaction with MeMgI under the influence of Ni(II).[1] Interestingly, in the presence of (dppf)NiCl$_2$ the sidechain methoxy group of 6-methoxy-2-naphthylmethyl methyl ether shows a higher reactivity.[2]

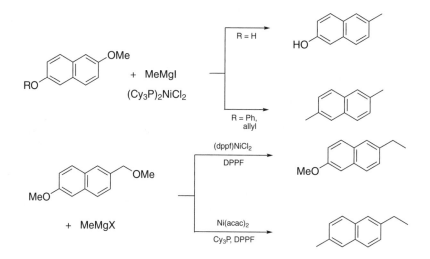

Kumada coupling. A report of biaryl synthesis from ArMgBr and Ar′Cl highlights the use of a Ni carbenoid (**1**).[3] Both bis(η^3-allyl)nickel and palladium complexes are also useful catalysts for the cross-coupling.[4]

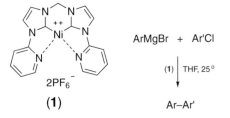

ArMgBr + Ar'Cl

(1) | THF, 25°

(1)

Ar–Ar'

[1]Guan, B.-T., Xiang, S.-K., Wu, T., Sun, Z.-P., Wang, B.-Q., Zhao, K.-Q., Shi, Z.-J. *CC* 1437 (2008).
[2]Guan, B.-T., Xiang, S.-K., Wang, B.-Q., Sun, Z.-P., Wang, Y., Zhao, K.-Q., Shi, Z.-J. *JACS* **130**, 3268 (2008).
[3]Xi, Z., Liu, B., Chen, W. *JOC* **73**, 3954 (2008).
[4]Terao, J., Naitoh, Y., Kuniyasu, H., Kambe, N. *CC* 825 (2007).

Grignard reagents/palladium complexes.

Coupling.   In polychloroarenes the chlorine atom ortho to a protic group (if such is present) is selectively replaced by R of RMgX.[1] One possible explanation of the phenomenon assigns the importance of Mg coordination to facilitate the oxidative addition of Pd to the C—Cl bond.

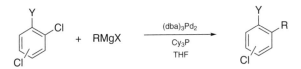

Coupling ascribing to more clearcut directing effect is the selective reaction with the ortho-Br of an alkali metal salt of 2,5-dibromobenzoic acid in the presence of $(dba)_3Pd_2$.[2] Kumada coupling at low temperature is accomplished by using $(dba)_2Pd$ with the 2'-dimethylaminobiphenyl(dicyclohexyl)phosphine ligand.[3]

Biaryl synthesis through Kumada coupling has employed a recyclable catalyst in which $PdCl_2$ is anchored in a mesoporous silica with an appended bipyridyl unit.[4] It also has addressed the steric hindrance issue. Choice of ligands appears to be important. One report describes the use of a $PdCl_2$ complex of both 1,3-bis(2,6-diisopropylphenyl)-imidazol-2-ylidene and 3-chloropyridine,[5] another indicates the effectiveness of $(t\text{-}Bu_2POH)_2PdCl_2$.[6]

[1]Ishikawa, S., Manabe, K. *OL* **9**, 5593 (2007).
[2]Houpis, I.N., van Hoeck, J.-P., Tilstam, U. *SL* 2179 (2007).
[3]Martin, R., Buchwald, S.L. *JACS* **129**, 3844 (2007).
[4]Tsai, F.-Y., Lin, B.-N., Chen, M.-J., Mou, C.-Y., Liu, S.-T. *T* **63**, 4304 (2007).
[5]Organ, M.G., Abdel-Hadi, M., Avola, S., Hadei, N., Nasielski, J., O'Brien, C.J., Valente, C. *CEJ* **13**, 150 (2007).
[6]Wolf, C., Xu, H. *JOC* **73**, 162 (2008).

Grignard reagents/silver salts.

Coupling. In the presence of catalytic amounts of $AgNO_3$ the coupling of $ArCH_2MgBr$ with alkyl halides (reactivity: tertiary > secondary) in ether takes place at room temperature.[1] Zero-valent Ag entity is produced to donate a single electron to RX, generating an alkyl radical that is to combine with Ag(0); AgX that also emerges reacts with $ArCH_2MgBr$ and the first catalytic cycle is completed when the two different organosilver species couple.

[1]Someya, H., Ohmiya, H., Yorimitsm, H., Oshima, K. *OL* **10**, 969 (2008).

Grignard reagents/titanium(IV) compounds.

Kulinkovich reaction. Optimal conditions for the preparation of 1,2-disubstituted cyclopropanols from methyl esters in ether or THF involve 1 equivalent of $(i\text{-}PrO)_4Ti$, 1.5 equivalent of MeMgX and 1.5 equiv. of RCH_2CH_2MgX.[1]

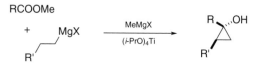

Intramolecular vinyl transfer from a vinyl ether to an ester has been reported.[2]

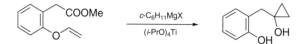

Reaction involving 3-butenylmagnesium bromide shows substrate-dependence. 1,2-Dicarbanion character and 1,4-dicarbanion character are manifested toward ester and nitrile groups, respectively.[3]

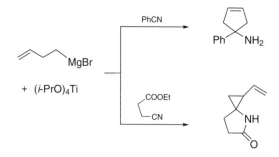

Reductive coupling. Cross coupling of allylic alcohols with alkynes leads to 1,4-dienes using $(i\text{-}PrO)_3TiCl$ and $c\text{-}C_5H_9MgCl$.[4] Allenyl carbinols react without loss of the hydroxyl group.[5]

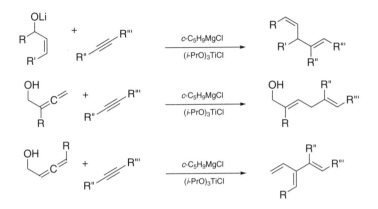

Cross-coupling of alkenes and alkynes proceeds via titanacyclopentenes, therefore protonolysis afford alkenes in which the two substituents originally attached to the *sp*-carbon atoms become *cis*-related.[6]

Imines and homallylic alcohols also combine stereoselectively to give 1,5-amino alcohols, which are valuable precursors of piperidines.[7]

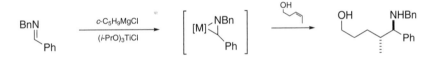

Workup by trapping with CO_2 diverts the product formation to conjugated γ-lactams.[8]

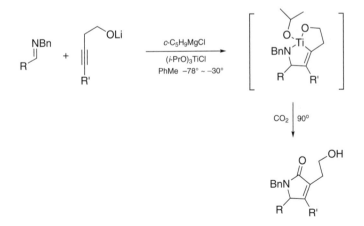

[1]Kulinkovich, O.G., Kananovich, D.G. *EJOC* 2121 (2007).
[2]Garnier, J.-M., Lecornué, F., Charnay-Pouget, F., Ollivier, J. *SL* 2827 (2007).
[3]Bertus, P., Menant, C., Tanguy, C., Szymoniak, J. *OL* **10**, 777 (2008).

[4]Kolundzic, F., Micalizio, G.C. *JACS* **129**, 15112 (2007).
[5]Shimp, H.L., Hare, A., McLaughlin, M., Micalizio, G.C. *T* **64**, 3437 (2008).
[6]Reichard, H.A., Mecalizio, G.C. *ACIE* **46**, 1440 (2007).
[7]Takahashi, M., Micalizio, G.C. *JACS* **129**, 7514 (2007).
[8]McLaughlin, M., Takahashi, M., Micalizio, G.C. *ACIE* **46**, 3912 (2007).

Grignard reagents/zirconium compounds.

Addition to ketones.   Participation of the cyclic ether moiety in Cp_2ZrCl_2-catalyzed carbomagnesiation of terminal alkenes, e.g., (**1**), has been observed.[1] Grignard reagents resulting from the transformation contain rather special structures.

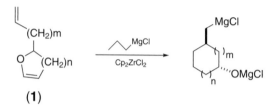

(**1**)

[1]Barluenga, J., Alvarez-Rodrigo, L., Rodriguez, F., Fananas, F.J. *OL* **9**, 3081 (2007).

H

Hafnium(IV) chloride.

Michael reaction.[1] Catalyzed by HfCl$_4$, indoles and pyrroles undergo Michael reactions with enones at C-3, and C-2/C-5, respectively. The reaction of pyrazole and imidazole takes place at an N-atom.

Sakurai reaction + Friedel–Crafts alkylation.[2] Two Lewis-catalyzed reactions to generate 3,4,4-triaryl-1-butenes can be performed in sequence in one pot using HfCl$_4$. For example, addition of an allylsilane to ArCHO is followed by alkylation of anisole or phenol.

[1]Kawatsura, M., Aburatani, S., Uenishi, J. *T* **63**, 4172 (2007).
[2]Sano, Y., Nakata, K., Otoyama, T., Umeda, S., Shiina, I. *CL* **36**, 40 (2007).

Hexabutylditin.

Condensation. Generation of stabilized free radicals from dithiocarbonate esters via C—S bond cleavage is promoted by Bu$_3$SnSnBu$_3$. By providing an alkene and a trapping agent, homologation of a carbon chain while performing functionalization, for example, carboazidation,[1] is realized.

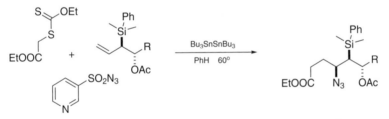

By the same principle a synthesis of piperidones is achieved in one step.[2]

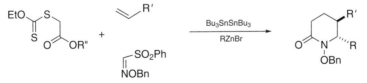

[1]Chabaud, L., Landais, Y., Renaud, P., Robert, F., Castet, F., Lucarini, M., Schenk, K. *CEJ* **14**, 2744 (2008).
[2]Godineau, E., Landais, Y. *JACS* **129**, 12662 (2007).

Fiesers' Reagents for Organic Synthesis, Volume 25. By Tse-Lok Ho
Copyright © 2010 John Wiley & Sons, Inc.

Hexafluoroacetone.

β-Hydroxycarboxamides.   Selective amidation of β-hydroxyalkanoic acids is easily performed via formation and aminolysis of 2,2-bis(trifluoromethyl)-1,3-dioxan-4-ones.[1] The heterocycles are obtained from condensation of the hydroxy acids with hexafluoroacetone in the presence of *N,N*'-diisopropylcarbodiimide.

[1]Spengler, J., Ruiz-Rodriguez, J., Yraola, F., Royo, M., Winter, M., Burger, K., Albericio, F. *JOC* **73**, 2311 (2008).

Hexakis[hydrido(triphenylphosphine)copper].

Ring expansion.   2-Allenyl-1,1-cyclopropanedicarboxylic esters are transformed into 3-methylenecyclopentenes on treatment with [(Ph₃P)CuH]₆ at room temperature in toluene containing a small amount of water.[1]

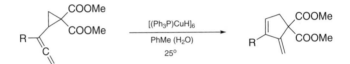

[1]Hiroi, K., Kato, F., Oguchi, T., Saito, S., Sone, T. *TL* **49**, 3567 (2008).

Hexamethylenetetramine.

Transesterification.   The title amine actively catalyzes transestericiation of β-keto esters.[1]

[1]Ribeiro, R.S., de Souza, R.O.M.A., Vasconcellos, M.L.A.A., Oliveira, B.L., Ferreira, L.C., Aguiar, L.C.S. *S* 61 (2007).

Hydrazine hydrate.

Hydrodechlorination.[1]   2,6-Dichloropurine is selectively transformed into 2-chloropurine by treatment with hydrazine and then heating with NaOH under essentially the Wolff–Kishner reduction conditions.

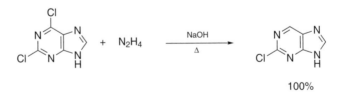

100%

[1]Uciti-Broceta, A., de las Infantas, M.J.P., Gallo, M.A., Espinosa, A. *CEJ* **13**, 1754 (2007).

Hydrogen fluoride.

Fragmentation. A moisture stable homoallylic fluorosilane which shows electrophile and pronucleophile properties is available from a 3-hydroxysilolane. The fragmentation is induced by HF.[1]

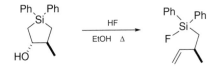

[1]Sen, S., Purushotham, M., Qi, Y., Sieburth, S.M. *OL* **9**, 4963 (2007).

Hydrogen fluoride – antimony(V) fluoride.

Cyclization. Piperidines fluorinated at C-3 and C-4 are accessible from diallylamine derivatives. Hydride and methyl shifts can intervene prior to capture of the carbocations by fluoride ion.[1]

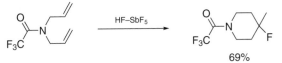

69%

1,1,1-Trifluoroalkanes.[2] These substances can be obtained by treatment of 1,1-dichloroalkenes with $HF-SbF_5$.

[1]Vardelle, E., Gamba-Sanchez, D., Matin-Mingot, A., Jouannetaud, M.-P., Thibaudeau, S., Marrot, J. *CC* 1473 (2008).
[2]Cantet, A.-C., Gesson, J.-P., Renoux, B., Jouannetaud, M.-P. *TL* **48**, 5255 (2007).

Hydrogen peroxide.

Oxidation. The oxidation of arylamines to give nitrosoarenes by H_2O_2 is catalyzed by PhSeSePh at room temperature.[1] The unstable products are trapped with conjugated dienes.

A synthesis of allenyl carbinols (in variable yields) involves treatment of 2-phenylselenyl-1,3-dienes with H_2O_2 and Et_3N at room temperature.[2]

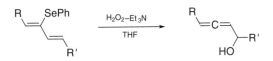

N-Oxidation of electron-deficient pyridines is achieved by H_2O_2 in MeCN at 0° if it is activated by Tf_2O (Na_2CO_3 to neutralize the acid).[3]

Epoxidation. Conjugated carbonyl compounds are epoxidized with stoichiometric H_2O_2 and NaOH while employing tetrabutylammonium peroxydisulfate as catalyst.[4]

[1]Zhao, D., Johansson, M., Bäckvall, J.-E. *EJOC* 4431 (2007).
[2]Redon, S., Berkaoui, A.-L.B., Pannecoucke, X., Outurquin, F. *T* **63**, 3707 (2007).
[3]Zhu, X., Kreutter, K.D., Hu, H., Player, M.R., Gaul, M.D. *TL* **49**, 832 (2008).
[4]Yang, S.G., Hwang, J.P., Park, M.Y., Lee, K., Kim, Y.H. *T* **63**, 5184 (2007).

Hydrogen peroxide, acidic.

Oxidation. A protocol for oxidation of sulfides to sulfoxides with H_2O_2 in EtOH includes Tf_2O.[1] The stepwise conversion of arylamines to nitrosoarenes and thence nitroarenes by H_2O_2 using heteropolyacids as catalyst has been reported.[2]

Iodination. Arenes are iodinated with electrophilic species generated in situ from NH_4I with H_2O_2 in HOAc.[3]

[1]Khodaei, M.M., Bahrami, K., Kairimi, A. *S* 1682 (2008).
[2]Tundo, P., Romanelli, G.P., Vazquez, P.G., Loris, A., Arico, F. *SL* 967 (2008).
[3]Narender, N., Reddy, K.S.K., Mohan, K.V.V.K., Kulkarni, S.J. *TL* **48**, 6124 (2007).

Hydrogen peroxide – metal catalysts.

Oxidation. For promoting oxidation of sulfides to sulfoxide by H_2O_2 metal catalysts now include $HAuCl_4 \cdot 4H_2O$[1] and a titanium complex[2] derived from tris(3-*t*-butyl-2-hydroxybenzyl)amine. In the latter report good turnover while employing only 0.01 – 1% of the catalyst has been observe.

The molybdenum-alkyne complex $CpMo(CO)_3(CCPh)$ is converted into **1** to exert its catalytic activity during oxidation of arylamines to nitrosoarenes.[3]

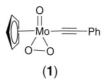

(1)

Alcohols are oxidized (to aldehydes and ketones) by H_2O_2 with catalytic quantitites of $RuCl_3$ and 3-iodobenzoic acid.[4]

Sulfoxidation is catalyzed by a Fe(III)-corrolazine, and surprisingly the active oxidant is a ferric hydroperoxide species.[5] Complete oxidation of thiols to sulfonic acids occurs on treatment with H_2O_2 and methyltrioxorhenium in MeCN at room temperature.[6]

To generate bromine in situ a mixture of HBr and KBr is oxidized by H_2O_2 in the presence of a catalytic amount of NH_4VO_3.[7]

Epoxidation. To convert H_2O_2 into an epoxidizing agent for alkenes a mixture of $FeCl_3 \cdot 6H_2O$, 2,6-pyridinedicarboxylic acid, and pyrrolidine in *t*-AmOH is added.[8]

A Pt salt, [(dppe)Pt(C_6F_5)(H_2O)]OTf, is able to catalyze selective epoxidation of a monosubstituted (terminal) alkene with H_2O_2, without affecting an internal double bond.[9]

As catalysts for epoxidation, methyltrioxorhenium is modified by converting into the more stable **2**.[10]

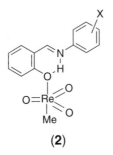

(2)

[1]Yuan, Y., Bian, Y. *TL* **48**, 8518 (2007).
[2]Mba, M., Prins, L.J., Licini, G. *OL* **9**, 21 (2007).
[3]Biradar, A.V., Kotbagi, T.V., Dongare, M.K., Umbarkar, S.B. *TL* **49**, 3616 (2008).
[4]Yusubov, M.S., Gilmkhanova, M.P., Zhdankin, V.V., Kirschning, A. *SL* 563 (2007).
[5]Kerber, W.D., Ramdhanie, B., Goldberg, D.P. *ACIE* **46**, 3718 (2007).
[6]Ballistreri, F.P., Tomaselli, G.A., Toscano, R.M. *TL* **49**, 3291 (2008).
[7]Moriuchi, T., Yamaguchi, M., Kikushima, K., Hirao, T. *TL* **48**, 2667 (2007).
[8]Anilkumar, G., Bitterlich, B., Gelalcha, F.G., Tse, M.K., Beller, M. *CC* 289 (2007).
[9]Colladon, M., Scarso, A., Sgarbossa, P., Michelin, R.A., Strukul, G. *JACS* **129**, 7680 (2007).
[10]Zhou, M.-D., Zhao, J., Li, J., Yue, S., Bao, C.-N., Mink, J., Zang, S.-L., Kühn, F.E. *CEJ* **13**, 158 (2007).

Hydrosilanes.

Reduction. Intramolecular delivery of a hydride ion to an incipient carbocation can occur with a hydrosilane.[1]

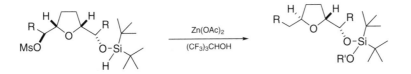

Using the siloxane [$Me_2Si(H)$]$_2$O under catalysis by (*i*-PrO)$_4$Ti, phosphine oxides are readily deoxygenated.[2]

The diiodorhodium acetate complexed to two benzimidazolylcarbene units (**1**) is a useful catalyst for hydrosilylation of ketones,[3] although a simpler system involves (EtO)$_2$SiMeH and Fe(OAc)$_2$.[4]

A method for reductive amination of aldehydes calls for using **2** (prepared from [(cod)IrCl]$_2$) as catalyst.[5] Catalyzed by FeSO$_4 \cdot$ 7H$_2$O and using (Me$_3$Si)$_2$NH and Cbz-Cl to conjugate with aldehydes or acetals, access to protected amines is allowed.[6] N-Alkylacetamides are formed when RCHO and MeCN are heated (microwave) with t-BuSiHMe$_2$ and CF$_3$CH$_2$OH.[7]

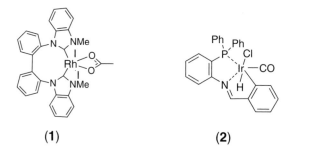

(1)   **(2)**

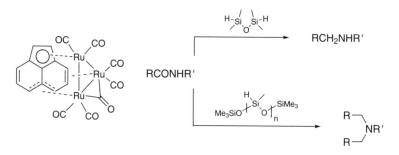

Reduction of secondary amides by Ru-catalysis to provide simple secondary amines or tertiary amines is dependent on the hydrosilane used.[8]

A highly unusual reduction whereby methyl esters are converted to methyl ethers by Et$_3$SiH at room temperature, is catalyzed by BF$_2$OTf $\cdot$ OEt$_2$ which is generated in situ from BF$_3 \cdot$ OEt$_2$ and Me$_3$SiOTf.[9] But perhaps because of the high reactivity of the catalyst toward other functional groups, the method is limited to use on relatively simple esters.

Reductive cleavage of 2-tosyl-2-aza-7-oxabicyclo[2.2.1]heptan-5-ones by Et$_3$SiH leads to either piperidine derivatives or tetrahydrofuran-3-ones, depending on the Lewis acid catalyst. Cleavage of the C(1)—O bond is favored by the presence of TiCl$_4$, whereas C(1)—N bond cleavage facilitated by SnCl$_4$ or BF$_3 \cdot$ OEt$_2$.[10]

Hydrosilylation.   The PtO_2-catalyzed reaction of diarylethynes with $Me_2Si(OEt)H$, followed by treatment with Bu_4NF, leads to (Z)-stilbenes in >95% stereoselectivity.[11]

Cycloaddition is observed when conjugated diynes and R_2SiH_2 are treated with $[Cp^*Ru(MeCN)_3]PF_6$ at room temperature.[12]

Aldol reaction.[13]   Trichlorosiloxyalkenes are generated in a debrominative silylation process from α-bromo ketones. Condensation with ArCHO leads to β-aryl enones. In the presence of the Lewis base Ph_3PO the reaction is optimized.

Ether synthesis.[14]   Carbonyl compounds undergo reductive etherification with ROH in a reaction with Et_3SiH, which is catalyzed by $FeCl_3$.

Addition.[15]   Markovnikov hydrocyanation of alkenes is accomplished employing TsCN and $PhSiH_3$ and a (salen)Co complex in ethanol at room temperature.

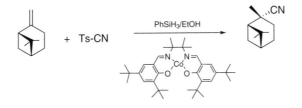

[1]Donohoe, T.J., Williams, O., Churchill, G.H. *ACIE* **47**, 2869 (2008).
[2]Berthod, M., Favre-Reguillon, A., Mohamad, J., Mignani, G., Docherty, G., Lemaire, M. *SL* 1545 (2007).
[3]Chen, T., Liu, X.-G., Shi, M. *T* **63**, 4874 (2007).
[4]Nishiyama, H., Furuta, A. *CC* 760 (2007).
[5]Lai, R.-Y., Lee, C.-I., Liu, S.-T. *T* **64**, 1213 (2008).
[6]Yang, B.-L., Tian, S.-K. *EJOC* 4646 (2007).
[7]Lehmann, F., Scobie, M. *S* 1679 (2008).
[8]Hanada, S., Ishida, T., Motoyama, Y., Nagashima, H. *JOC* **72**, 7551 (2007).
[9]Morra, N.A., Pagenkopf, B.L. *S* 511 (2008).
[10]Muthusamy, S., Krishnamurthi, J., Suresh, E. *OL* **8**, 5101 (2006).
[11]Giraud, A., Provot, O., Hamze, A., Brion, J.-D., Alami, M. *TL* **49**, 1107 (2008).
[12]Matsuda, T., Kadowaki, S., Murakami, M. *CC* 2627 (2007).
[13]Smith, J.M., Greaney, M.F. *TL* **48**, 8687 (2007).
[14]Iwanami, K., Yano, K., Oriyama, T. *CL* **36**, 38 (2007).
[15]Gaspar, B., Carreira, E.M. *ACIE* **46**, 4519 (2007).

Hydroxylamine diphenylphosphinate.

Aziridination.[1]   Under basic conditions the title reagent delivers an NH group to conjugated ketones.

[1]Armstrong, A., Baxter, C.A., Lamont, S.G., Pape, A.R., Wincewicz, R. *OL* **9**, 351 (2007).

(2-Hydroxy-5-methoxyphenyl)diphenylmethanol.

Acetalization.[1]   The title reagent condenses with carbonyl compounds to form 1,3-dioxanes which are photolabile, therefore such derivatives are of special synthetic value.

[1]Wang, P., Hu, H., Wang, Y. *OL* **9**, 1533 (2007).

Hydroxy(tosyloxy)iodobenzene.

Cyclization.   The bridging of piperazine derivative at an α-position of a nitrogen atom to an indole nucleus requires dual activation of two types of C—H bonds. Achieving the transformation with PhI(OH)OTs [Koser reagent] in one step, albeit in low yields, is quite gratifying.[1]

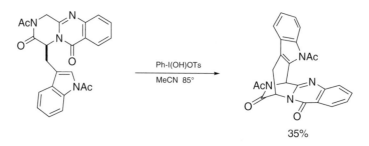

35%

Ring contraction.   Illustrated by the transformation of 1-aryl-1,2-dihydronaphthalenes to indanes[2] the Koser reagent has the same ability as Tl(NO₃)₃ but of course its use avoids the issue of toxicity.

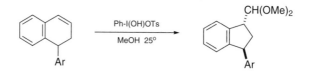

[1]Walker, S.J., Hart, D.J. *TL* **48**, 6214 (2007).
[2]Silva, L.F. Jr, Siqueira, F.A., Pedrozo, E.C., Vieira, F.Y.M., Doriguetto, A.C. *OL* **9**, 1433 (2007).

Hypofluorous acid – acetonitrile.

Oxidation.   The exposure of α-amino esters to HOF-MeCN results in the formation of the corresponding α-nitroalkanoic esters.[1] Rapid conversion of thiols and disulfides to sulfinic acids or sulfonic acids in >90% yields occurs under similar conditions.[2]

Aldehydes are transformed into the corresponding nitriles via treatment of their *N,N*-dimethylhydrazones with HOF-MeCN.[3]

[1]Harel, T., Rozen, S. *JOC* **72**, 6500 (2007).
[2]Shefer, N., Carmeli, M., Rozen, S. *TL* **48**, 8178 (2007).
[3]Carmeli, M., Shefer, N., Rozen, S. *TL* **47**, 8969 (2006).

I

Imidazole-1-sulfonyl azide hydrochloride.

Diazo group transfer.[1]   The title reagent, prepared by adding imidazole to NaN_3 and SO_2Cl_2 in MeCN, is shelf-stable. It is useful in transfer a diazo group to amine and active methylene compounds.

[1]Goddard-Borger, E.D., Stick, R.V. *OL* **9**, 3797 (2007).

Indium.

Tosylation.   Alcohols and amines are tosylated by TsCl.[1] However, the role played by indium metal in the protocol is questionable.

Substitution.   Indium-mediated reaction of iodoalkynes with glycals[2] or glycosyl acetates[3] gives *C*-glycosides.

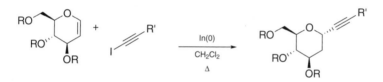

Indoles.   Indium combined with hydriodic acid under phase transfer conditions has been used to convert 2-alkynylnitroarenes into indoles.[4]

Barbier reaction.   Indium is activated by AgI or CuI/I_2 to promote reaction of unactivated alkyl halides with aldehydes in water.[5]

Formation of allylindium species for reaction with sulfinimines in water is benefited by a halide salt (of Na, Li, K, NH_4), with regard to producing good yields of the adducts and diastereoselectivity.[6]

Of special interest is the reaction of the amphoteric 2-formylaziridines, which exist in dimeric form and expose one diastereoface to allylindium reagents. Accordingly, no protective group (on the nitrogen atom) is necessary prior to the reaction.[7]

Fiesers' Reagents for Organic Synthesis, Volume 25. By Tse-Lok Ho
Copyright © 2010 John Wiley & Sons, Inc.

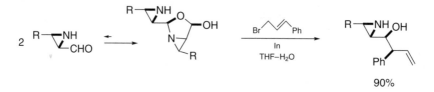

$$2 \quad \text{R—NH} \atop \text{CHO}$$

90%

Addition and substitution. Addition of allylindium reagents to allenyl carbinols affords linear products.[8] A net 1,3-butadien-2-ylation at the β-carbon of β-acetoxy-β-lactams results when they are treated with indium and 1,4-dibromo-2-butyne.[9]

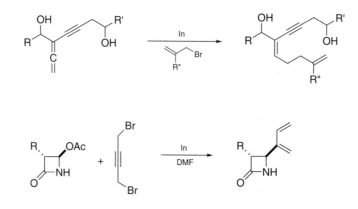

A tertiary propargylic hydroxy group is subject to replacement (with transposition) on reaction with allylindium reagents, forming allenes.[10]

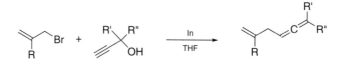

A branched enyne unit is constituted from propargyl bromide to join up to RCHO in the presence of In–InBr$_3$.[11]

A method for *N*-alkylation of α-amino esters by RCHO and R′I employing In/Ag and InCl$_3$ in aqueous MeOH has been developed.[12] It involves formation of organoindium reagents and imines, and their mutual reaction.

A π-allylpalladium moiety is changed to the nucleophilic π-allylindium counterpart by interaction with In–InCl$_3$.[13] For an ensuing coupling reaction roles played by two partner reactants are reversed.

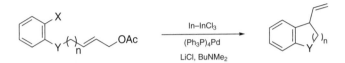

Allylindium species are cleaved to C-centered radicals therefore they are alternatives to allyltins for ring closure. UV irradiation is beneficial to the chemical step.[14]

Cleavage of 2,2,2-trichloroethyl esters.[15]   Carboxylic acids are liberated by treatment with indium from trichloroethyl esters via chlorine atom abstraction, reduction and fragmentation. However, those esters of arylacetic acids and 3-arylpropanoic acids behave differently, due to the tendency for the dichloromethyl radical to abstract a benzylic hydrogen.

[1]Kim, J.-G., Jang, D.O. *SL* 2501 (2007).
[2]Lubin-Germain, N., Hallonet, A., Huguenot, F., Palmier, S., Uziel, J., Auge, J. *OL* **9**, 3679 (2007).
[3]Lubin-Germain, N., Baltaze, J.-P., Coste, A., Hallonet, A., Laureano, H., Legrave, G., Uziel, J., Auge, J. *OL* **10**, 725 (2008).
[4]Kim, J.S., Han, J.H., Lee, J.J., Jun, Y.M., Lee, B.M., Kim, B.H. *TL* **49**, 3733 (2008).
[5]Shen, Z.-L., Yeo, Y.-L., Loh, T.-P. *JOC* **73**, 3922 (2008).
[6]Sun, X.-W., Liu, M., Xu, M.-H., Lin, G.-Q. *OL* **10**, 1259 (2008).
[7]Hili, R., Yudin, A.K. *ACIE* **47**, 4188 (2008).
[8]Kim, S., Lee, P.H. *EJOC* 2262 (2008).
[9]Lee, K., Lee, P.H. *CEJ* **14**, 8877 (2007).
[10]Lee, K., Lee, P.H. *OL* **10**, 2441 (2008).
[11]Huang, J.-M., Luo, H.-C., Chen, Z.-X., Yang, G.-C. *EJOC* 295 (2008).
[12]Shen, Z.-L., Cheong, H.-L., Loh, T.-P. *CEJ* **14**, 1875 (2008).
[13]Seomoon, D., Lee, K., Kim, H., Lee, P.H. *CEJ* **13**, 5197 (2007).
[14]Hirashita, T., Hayashi, A., Tsuji, M., Tanaka, J., Araki, S. *T* **64**, 2642 (2008).
[15]Mineno, T., Kansui, H., Kunieda, T. *TL* **48**, 5027 (2007).

Indium(III) acetate – phenylsilane.

Reduction.   In EtOH catalytic amounts of In(OAc)$_3$ and 2,6-lutidine in dry air initiate the reduction of RX (X = Br, I) to hydrocarbon products by PhSiH$_3$.[1]

Radical addition.   Under essentially identical conditions as above, alkyl iodides generate free radicals which are trapped by alkenes (e.g., acrylic esters).[2]

[1]Miura, K., Tomita, M., Yamada, Y., Hosomi, A. *JOC* **72**, 787 (2007).
[2]Miura, K., Tomita, M., Ichikawa, J., Hosomi, A. *OL* **10**, 133 (2008).

Indium(III) bromide.

3,4-Dibromotetrahydropyrans.[1]   Prins cyclization involving RCHO and homoallylic alcohols bearing a bromine atom at the far end of the double bond, as induced by $InBr_3$ and terminated by Me_3SiBr, is stereoselective.

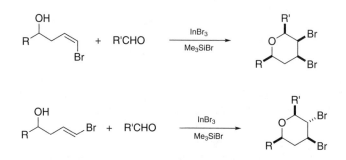

Carbocyclization.   Opening of an epoxide induced by $InBr_3$ can lead to polycarbo-cyclic products due to participation of double bond(s) and aromatic ring.[2]

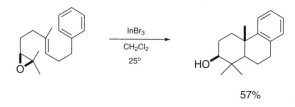

Friedel–Crafts alkenylation involving the triple bond of an *o*-isothiocyanatoarylalkyne and initiated by $InBr_3$ [or $In(OTf)_3$] leads to a 4-substituted quinoline-2-thione.[3] Best results are obtained from alkynes with a *t*-butyl group at the other end of the triple bond (which is lost during the reaction).

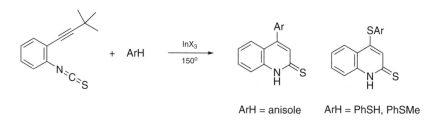

ArH = anisole   ArH = PhSH, PhSMe

Alkylation + allylation.[4]   A carbon chain is constructed from an enol silyl ether (including ketene silyl acetal, . . .) with an alkylating agent and an allylsilane with catalysis of $InBr_3$ in CH_2Cl_2 at room temperature.

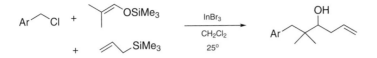

Addition reactions. With InBr$_3$ as catalyst sulfonamides are induced to add to unactivated alkenes in refluxing toluene.[5] Carbamates and arylamines do not react in the same way.

A formal hydroamination is the ring closure of N-tosyl-o-alkynylaryl amines to furnish indoles. Cyclization with sulfonyl group migration to the aryl nucleus is limited to substrates that contain an o-methoxy substituent to the nitrogen atom. It should be noted that similar but simpler N-mesylarylamines (without the methoxy group) undergo cyclization to form 3-sulfonylindoles when AuBr$_3$ instead of InBr$_3$ is used as catalyst.[6]

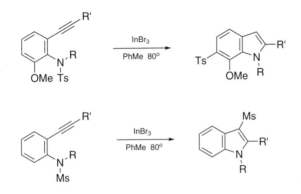

In an addition of 3-methyl-1-butyn-3-ol to aldehydes in the presence of (S)-BINOL-InBr$_3$ and an amine, acceleration by ligand is significant.[7]

[1]Liu, F., Loh, T.-P. *OL* **9**, 2063 (2007).
[2]Zhao, J.-F., Zhao, Y.-J., Loh, T.-P. *CC* 1353 (2008).
[3]Otani, T., Kunimatsu, S., Nihei, H., Abe, Y., Saito, T. *OL* **9**, 5513 (2007).
[4]Nishimoto, Y., Yasuda, M., Baba, A. *OL* **9**, 4931 (2007).
[5]Huang, J.-M., Wong, C.-M., Xu, F.-X., Loh, T.-P. *TL* **48**, 3375 (2007).
[6]Nakamura, I., Yamagishi, U., Song, D., Konta, S., Yamamoto, Y. *ACIE* **46**, 2284 (2007).
[7]Harada, S., Takita, R., Ohshima, T., Matsunaga, S., Shibasaki, M. *CC* 948 (2007).

Indium(III) chloride.

Friedel–Crafts acylation. Carboxylic acids are converted into RCOCl with Me$_2$Si(H)Cl and InCl$_3$.[1] In the presence of an activated arene (e.g., an aryl ether) an aryl ketone is formed.[1] A dual role is played by InCl$_3$.

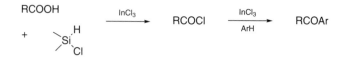

Addition to C=O. Triethoxysilane and allyltriethoxysilane react with ArCHO to give benzyl ethyl ethers. The reaction is accomplished with a mixture of $InCl_3$ and Me_3SiCl.[2]

2,3-Dimethyl-1,3-butadiene is converted into an allylating agent by a mixture of $InCl_3$ and Bu_3SnH, which forms $HInCl_2$ in situ. The regioselectivity for reaction with ketones is temperature dependent, γ-adducts are formed at room temperature, α-adducts in refluxing THF.[3]

Glucal undergoes isomerization and dehydration to give 2-furylethanediol by treatment with $InCl_3$ in an ionic liquid.[4]

Substitution reactions. Secondary alcohols are homologated to furnish *N*-tosyl-methylcarboxamides on treatment with $TsCH_2NC$ in the presence of $InCl_3$.[5]

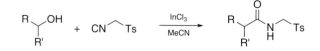

Desilylalkylation of allylsilanes by a trimethylsilyl ether involves formation of an activating complex $[Me_3SiI/InCl_3]$ for $ROSiMe_3$. Iodine is needed in addition to $InCl_3$ in this reaction.[6]

Homoallylic amine derivatives are obtained from reaction of α-aminoalkyl *p*-tolyl sulfones and allylsilanes.[7]

The intermolecular Friedel–Crafts alkylation with benzylic and allylic halides occurs at room temperature in CH_2Cl_2.[8] Diarylmethanes are also obtained in moderate yields from arenes and benzyl alcohols on heating the mixtures with $InCl_3 \cdot 4H_2O$ and acetylacetone at 120°.[9] 1-Vinyltetralin and 4-vinyltetrahydroisoquinoline derivatives are readily formed by an intramolecular allylation.[10]

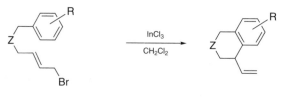

Z = NTs, C(COOEt)₂

Reaction of 1,3-diones with propargylic alcohols leads to 3-acylfurans.[11] Using other Lewis acids such as FeCl₃ the reaction stops short of cyclization. Alkylation by alkenes (styrene, norbornene, cyclopentadiene, dihydropyran, ...) affords moderate yields of the adducts.[12] Again, InCl₃ appears to show a special catalytic activity, as AlCl₃, TiCl₄, MnCl₂, BiCl₃ are ineffective.

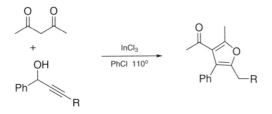

For the Pd(0)-catalyzed cross-coupling of heteroarylmetals with ArI the use of indium derivatives (prepared from the corresponding lithio compounds) is a success.[13]

[1]Babu, S.A., Yasuda, M., Baba, A. *OL* **9**, 405 (2007).
[2]Yang, M.-S., Xu, L.-W., Qiu, H.-Y., Lai, G.-Q., Jiang, J.-X. *TL* **49**, 253 (2008).
[3]Hayashi, N., Honda, H., Shibata, I., Yasuda, M., Baba, A. *SL* 1407 (2008).
[4]Teijeira, M., Fall, Y., Santamarta, F., Tojo, E. *TL* **48**, 7926 (2007).
[5]Krishna, P.R., Sekhar, E.R., Prapurna, Y.L. *TL* **48**, 9048 (2007).
[6]Saito, T., Nishimoto, Y., Yasuda, M., Baba, A. *JOC* **72**, 8588 (2007).
[7]Das, B., Damodar, K., Saritha, D., Chowdhury, N., Krishnaiah, M. *TL* **48**, 7930 (2007).
[8]Kaneko, M., Hayashi, R., Cook, G.R. *TL* **48**, 7085 (2007).
[9]Sun, H.-B., Li, B., Chen, S., Li, J., Hua, R. *T* **63**, 10185 (2007).
[10]Hayashi, R., Cook, G.R. *OL* **9**, 1311 (2007).
[11]Feng, X., Tan, Z., Chen, D., Shen, Y., Guo, C.-C., Xiang, J., Zhu, C. *TL* **49**, 4110 (2008).
[12]Yuan, Y., Shi, Z. *SL* 3219 (2007).
[13]Font-Sanchis, E., Cespedes-Guirao, F.J., Sastre-Santos, A., Fernandez-Lazaro, F. *JOC* **72**, 3589 (2007).

Indium(III) chloride – aluminum.

Reduction. Both anthraquinones and anthrones are reduced to anthracenes by aluminum at room temperature, with InCl₃ present in catalytic amounts.[1]

Pinacol coupling. In aqueous media the coupling of ArCOR is achieved by Al–InCl₃.[2]

[1]Wang, C., Wan, J., Zheng, Z., Pan, Y. *T* **63**, 5071 (2007).
[2]Wang, C., Pan, Y., Wu, A. *T* **63**, 429 (2007).

Indium(I) iodide.

Allylation. Allyl(pinacolato)boron and analogues transfer the allyl group to ketones[1] in the presence of InI. The reaction is highly chemoselective therefore many functional groups are tolerated.

The reaction with *N*-acylhydrazones is characterized by high regioselectivity and diastereoselectivity.[2]

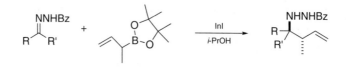

[1]Schneider, U., Kobayashi, S. *ACIE* **46**, 5909 (2007).
[2]Kobayashi, S., Konishi, H., Schneider, U. *CC* 2313 (2008).

Indium(III) triflate.

Acetalization. As catalyst for acetalization of carbonyl compounds that are acid-sensitive, In(OTf)$_3$ offers another choice.[1,2]

β-Amino alcohols. *meso*-Epoxides undergo aminolysis in the presence of In(OTf)$_3$. The reaction is rendered enantioselective by adding the bipyridyl ligand **1**.[3]

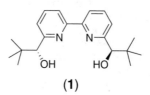

(1)

α-Methylene-γ-lactones. The regioselectivity for allylation of aldehydes with α-(pinacolatoboryl)methyl-α,β-unsaturated esters is dependent on the acidity of the catalyst.[4]

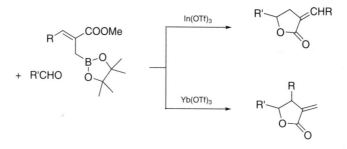

Alkenylation and allylation. 1,3-Dicarbonyl compounds are alkenylated by alkynes under solvent-free conditions.[5]

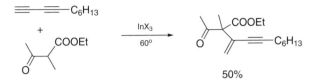

50%

Mixtures of epoxides and 3-substituted 2-methyl-4-penten-2-ols react to give homo-allylic alcohols, on treatment with In(OTf)$_3$ (or TfOH). Under the reaction conditions the tertiary alcohols are decomposed to generate allylating agents while the epoxides undergo rearrangement to aldehydes, the transformed species then combine to furnish the observed products.[6]

Retro-Claisen reaction.   Cleavage of 1,3-diketones by an alcohol (e.g., phenethyl alcohol) takes place when they are heated with In(OTf)$_3$.[7]

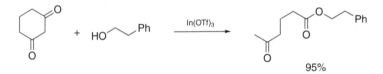

95%

Cycloisomerization.   Allenyl carbonyl compounds bearing a phenyl group at the γ-position are susceptible to cyclization to give products with a furan ring. Phenyl migration is involved.[8]

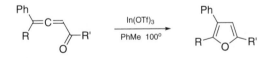

[1]Smith, B.M., Graham, A.E. *TL* **47**, 9317 (2006).
[2]Gregg, B.T., Golden, K.C., Quinn, J.F. *T* **64**, 3287 (2008).
[3]Mai, E., Schneider, C. *SL* 2136 (2007).
[4]Ramachandran, P.V., Pratihar, D. *OL* **9**, 2087 (2007).
[5]Endo, K., Hatakeyama, T., Nakamura, M., Nakamura, E. *JACS* **129**, 5264 (2007).
[6]Nokami, J., Maruoka, K., Soua, T., Tanaka, N. *T* **63**, 9016 (2007).
[7]Kawata, A., Takata, K., Kuninobu, Y., Takai, K. *ACIE* **46**, 7793 (2007).
[8]Dudnik, A.S., Gevorgyan, V. *ACIE* **46**, 5195 (2007).

Indium(III) triflimide.

Cyclization.[1]   Intramolecular alkenylation of β-keto esters is carried out with In(NTf₂)₃ in toluene at 100°, good yields of cyclic ketones (6- to 15-membered) are generally obtained.

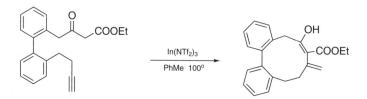

[1]Tsuji, H., Yamagata, K., Itoh, Y., Endo, K., Nakamura, M., Nakamura, E. *ACIE* **46**, 8060 (2007).

Iodine.

Iodination.   Arenes including some electron-poor members such as haloarenes, trifluoromethylbenzene, and benzoic acid, are iodinated by I_2 in an acidic medium (H_2SO_4-HOAc or CF_3COOH-CH_2Cl_2) with $K_2S_2O_8$ present.[1] It is also possible to activate iodine to iodinate electron-rich arenes by CAN,[2] and by $NaNO_2$ with air as oxidant to fully utilize I_2.[3] α-Iodination of alkyl aryl ketones is accomplished with I_2–CuO in refluxing MeOH.[4] N-Substituted 2,3-diiodoindoles are obtained when the 2-indolecarboxylic acids are treated with iodine and $NaHCO_3$.[5]

There is a review on electrophilic iodination with elemental iodine and other systems.[6]

Substitutions.   Iodine assists the S_N2 substitution of allylic alcohols to form sulfonamides and carbamates.[7] Allylic alcohols are sufficiently electrophilic toward 1,3-dicarbonyl compounds in the presence of iodine.[8]

Peroxidation.   Ketones are transformed into *gem*-bisperoxyalkanes using ROOH and iodine.[9] Bishydroperoxides are similarly prepared using H_2O_2.[10]

Friedel–Crafts reaction.   Iodine is a mild catalyst for the Pictet-Spengler reaction of tryptamine with ketones to generate tetrahydro-β-carbolines at room temperature.[11]

Cyclotrimerization of pyrrole-fused 1,3-dioxepanes occurs in the presence of iodine.[12]

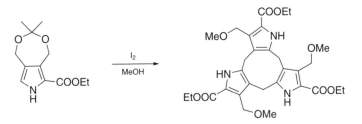

[1]Hossein, M.D., Oyamada, J., Kitamura, T. *S* 690 (2008).
[2]Das, B., Krishnaiah, M., Venkateswarlu, K., Reddy, V.S. *TL* **48**, 81 (2007).
[3]Iskra, J., Stavber, S., Zupan, M. *TL* **49**, 893 (2008).
[4]Yin, G., Gao, M., She, N., Hu, S., Wu, A., Pan, Y. *S* 3113 (2007).
[5]Putey, A., Popowycz, F., Joseph, B. *SL* 419 (2007).
[6]Stavber, S., Jereb, M., Zupan, M. *S* 1487 (2008).
[7]Wu, W., Rao, W., Er, Y.Q., Loh, J.K., Poh, C.Y., Chen, P.W.H. *TL* **49**, 2620 (2008).
[8]Rao, W., Tay, A.H.L., Goh, P.J., Choy, J.M.L., Ke, J.K., Chen, P.W.H. *TL* **49**, 122 (2008).
[9]Zmitek, K., Zupan, M., Stavber, S., Iskra, J. *JOC* **72**, 6534 (2007).
[10]Selvam, J.J.P., Suresh, V., Rajesh, K., Babu, D.C., Suryakiran, N., Venkateswarlu, Y. *TL* **49**, 3463 (2008).
[11]Lingam, Y., Rao, D.M., Bhowmik, D.R., Santu, P.S., Rao, K.R., Islam, A. *TL* **48**, 7243 (2007).
[12]Stepieri, M., Sessler, J.L. *OL* **9**, 4785 (2007).

Iodosuccinimide, NIS.

Iodination.   Practical and highly stereoselective iododesilylation of alkenylsilanes is performed with NIS in hexafluoroisopropanol.[1]

Treatment of 2-alkynyl-2-trimethylsiloxy carbonyl compounds with NIS leads to 4-iodo-3-furanones.[2] In some cases the presence of AuCl$_3$ (5 mol%) is beneficial.

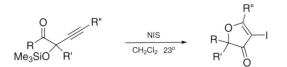

Interesting difference in iodocyclization that is participated by NIS and I$_2$ is noted, for which slight variation of the substrate structure is rather difficult to account.[3]

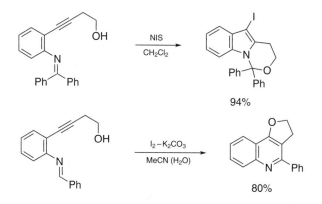

94%

80%

De-N-methylation. The *N*-methyl group of *N*-benzyl-*N*-methyl α-amino acid deriva-tives is selectively removed by treatment with NIS, then MeONH₃Cl in MeCN at room temperature.[4]

[1]Ilardi, E.A., Stivala, C.E., Zakarian, A. *OL* **10**, 1727 (2008).
[2]Crone, B., Kirsch, S.F. *JOC* **72**, 5435 (2007).
[3]Halim, R., Scammells, P.J., Flynn, B.L. *OL* **10**, 1967 (2008).
[4]Katoh, T., Watanabe, T., Nishitani, M., Ozeki, M., Kajimoto, T., Node, M. *TL* **49**, 598 (2008).

Iodosylbenzene.

Epoxidation. Epoxidation of alkenes by PhIO is co-catalyzed by a Mn-porphyrin and nanosized gold stabilized by RSH.[1]

Nitrenoids. In situ oxidation of $Cl_3CCH_2OSO_2NH_2$ by PhIO gives the nitrene that can be delivered to alkenes to form aziridines by azolecarbene-coordinated copper species.[2]

Cleavage of C=C bond. Adduct of PhI=O with tetrafluoroboric acid serves as ozone equivalent in its capacity of cleaving alkenes to dialdehyde in the presence of 18-crown-6.[3]

[1]Murakami, Y., Konishi, K. *JACS* **129**, 14401 (2007).
[2]Xu, Q., Appella, D.H. *OL* **10**, 1497 (2008).
[3]Miyamoto, K., Tada, N., Ochiai, M. *JACS* **129**, 2772 (2007).

o-Iodoxybenzoic acid, IBX.

Degradation. Carboxamides lose a one-carbon unit to give nitriles on heating with IBX and Et₄NBr.[1] The reaction involves generation of Br⁺ to induce a Hofmann rearrange-ment. Glycinamides give cyanamides.[2]

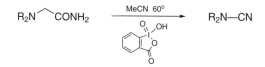

Oxidation.   Secondary α-amino nitriles afford α-imino nitriles upon reactin with IBX at room temperature.[3] Tetrahydroisoquinoline is carbamoylated at C-1 and *N*-acylated at the same time on exposure to IBX and treatment with RCOOH and R'NC.[4]

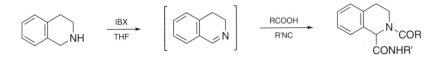

The method for conversion of benzylic bromide to aldehyde with IBX and DMSO has been applied to a molecule containing six such structural units, to deliver a precursor of the belt-shaped [6.8]₃cyclacene by a threefold intramolecular McMurry coupling.[5]

o-Quinones.   *o*-Oxygenation of a phenol to give a catechol intermediate constitutes a critical step in a synthesis of brazilin. Oxidation with IBX to give an *o*-quinone followed by reduction (with $Na_2S_2O_4$) accomplishes this transformation.[6]

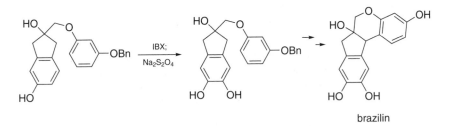

brazilin

Substituted IBX analogues.   Tetrafluoro-IBX (1) has been synthesized from 2,3,4,5-tetrafluorobenzoic acid. It is more soluble in organic solvents and shows higher reactivity comparing to the parent IBX.[7] Compound **2** is another analogue.[8]

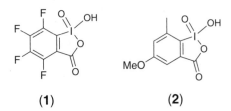

(1)   **(2)**

[1]Bhalerao, D.S., Mahajan, U.S., Chaudhari, K.H., Akamanchi, K.G. *JOC* **72**, 662 (2007).
[2]Chaudhari, K.H., Mahajan, U.S., Bhalerao, D.S., Akamanchi, K.G. *SL* 2815 (2007).
[3]Fontaine, P., Chiaroni, A., Masson, G., Zhu, J. *OL* **10**, 1509 (2008).
[4]Ngouansavanh, T., Zhu, J. *ACIE* **46**, 5775 (2007).
[5]Esser, B., Rominger, F., Gleiter, R. *JACS* **130**, 6716 (2008).
[6]Huang, Y., Zhang, J., Pettus, T.R.R. *OL* **7**, 5841 (2005).
[7]Richardson, R.D., Zayed, J.M., Altermann, S., Smith, D., Wirth, T. *ACIE* **46**, 6529 (2007).
[8]Moorthy, J.N., Singhal, N., Senapati, K. *TL* **49**, 80 (2008).

Ionic liquids.

Special ionic liquids.   A review of chiral ionic liquids is available.[1] A series of Lewis basic ionic liquids are prepared from DABCO by quaternization with RCl followed by anion exchange (to BF_4, PF_6).[2]

Some more significant applications.   Aldol reaction in ionic liquids is catalyzed by *O*-silylserines.[3] Michael reaction between malonitrile and chalcones proceeds without the usual catalysts using ionic liquids as reaction media, presumably the acidity of the carbon acid is enhanced.[4]

CAN oxidation in ionic liquids[5] is a useful development in view of the limitation in solvent systems for such a reagent. Depolymerization of nylon-6 to give caprolactam occurs when it is heated with DMAP in an ionic liquid at 300°.[6]

Phosphonium ionic liquids are good media for Pd-catalyzed carbonylation[7] and the Buchwald-Hartwig amination.[8] However, they must contain noncoordinating counteranions such as bistriflamide.

A synthesis of α-substituted acrylamides from 1-alkynes, amines and carbon monoxide based on catalysis by $Pd(OAc)_2$–DPPP is carried out in (bmim)NTf$_2$.[9]

While most synthetic applications have involved imidazolium ionic liquids, *N*-butylpyridinium salts are used in Sonogashira coupling.[10]

Ionic liquids are proposed as "designer solvents" for nucleophilic aromatic substitution.[11] Coating with a layer of ionic liquid onto silica-supported sulfonic acid improves its utility (such as acetalization) boasting selectivity in aqueous media.[12]

[1]Winkel, A., Reddy, P.V.G., Wilhelm, R. *S* 999 (2008).
[2]Wykes, A., MacNeil, S.L. *SL* 107 (2007).
[3]Teo, Y.-C., Chua, G.-L. *TL* **49**, 4235 (2008).
[4]Meciarova, M., Toma, S. *CEJ* **13**, 1268 (2007).
[5]Mehdi, H., Bodor, A., Lantos, D., Horvath, I.T., De Vos, D.E., Binnemans, K. *JOC* **72**, 517 (2007).
[6]Kamimura, A., Yamamoto, S. *OL* **9**, 2533 (2007).
[7]McNulty, J., Nair, J.J., Robertson, A. *OL* **9**, 4575 (2007).
[8]McNulty, J., Cheekoori, S., Bender, T.P., Coggan, J.A. *EJOC* 1423 (2007).
[9]Li, Y., Alper, H., Yu, Z. *OL* **8**, 5199 (2006).
[10]de Lima, P.G., Antunes, O.A.C. *TL* **49**, 2506 (2008).
[11]Newington, I., Perez-Arlandis, J.M., Welton, T. *OL* **9**, 5247 (2007).
[12]Gu, Y., Karam, A., Jerome, F., Barrault, J. *OL* **9**, 3145 (2007).

Iridium complexes.

Addition reactions.   Spiroacetal formation from dihydroxyalkynes[1] is the result of a triple bond activation by iridium complex (**1**). Prenylation of an aldehyde from its mixture with 1,1-dimethylallene under hydrogen is significant as the nucleophilic species is highly substituted.[2] Alcohols are dehydrogenated in the presence of proper iridium complexes (cf. **2**[3] and **3**[4]) therefore the same type of products are accessible while obviating hydrogen gas.[5]

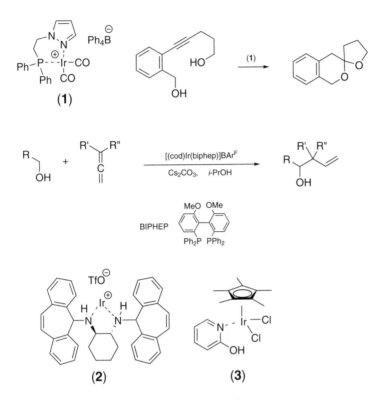

(1)

BIPHEP

(2)   **(3)**

An analogous complex of **1** (with a 3,5-diisopropylpyrazolylethyl group in the phosphine ligand and COD in place of two CO groups) is found to catalyze intramolecular hydroamination of properly constituted alkynes to afford cyclic imines.[6]

Alkenylation of activated ketones (e.g., α-keto esters[7]) and *N*-tosylimines[8] is similarly performed by the in situ reductive activation of alkynes (under H_2).

Hydrodehalogenation. Alkyl halides are reduced by Ir-catalyzed reaction with Et_3SiH.[9]

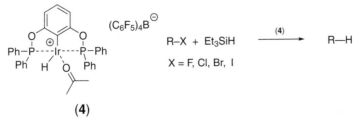

(4)

Nazarov cyclization.   2-Siloxy-4-alkenoylfurans fail to undergo Nazarov cyclization in the presence of conventional Lewis acids, but the reaction can be brought forth with addition of an iridium complex.[10] However, whether the true catalyst is a highly electrophilic silicon species cannot be excluded.

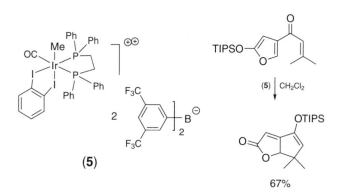

1Messerle, B.A., Vuong, K.Q. *OM* **26**, 3031 (2007).
2Skucas, E., Bower, J.F., Krische, M.J. *JACS* **129**, 12678 (2007).
3Königsmann, M., Donati, N., Stein, D., Schönberg, H, Harmer, J., Sreekanth, A., Grützmacher, H. *ACIE* **46**, 3567 (2007).
4Fujita, K., Tanino, N., Yamaguchi, R. *OL* **9**, 109 (2007).
5Bower, J.F., Skucas, E., Patman, R.L., Krische, M.J. *JACS* **129**, 15134 (2007).
6Field, L.D., Messerle, B.A., Vuong, K.Q., Turner, P., Failes, T. *OM* **26**, 2058 (2007).
7Ngai, M.-Y., Barchuk, A., Krische, M.J. *JACS* **129**, 280 (2007).
8Barchuk, A., Ngai, M.-Y., Krische, M.J. *JACS* **129**, 8432 (2007).
9Yang, J., Brookhart, M. *JACS* **129**, 12656 (2007).
10He, W., Huang, J., Sun, X., Frontier, A.J. *JACS* **130**, 300 (2008).

Iron(II) acetate.

Hydrosilylation.[1]   A protocol for reductive silylation of ketones by $(EtO)_2SiMeH$ includes addition of $Fe(OAc)_2$ and sodium 2-thienylcarboxylate.

1Furuta, A., Nishiyama, H. *TL* **49**, 110 (2008).

Iron(II) bromide.

2,6-Diacetylpyridine bis-N-(2,6-diisopropylphenyl)imine complex.   The readily synthesized air-stable complex is reduced in situ by $NaBEt_3H$ to catalyze intramolecular [2+2]cycloaddition.[1] Also the bromine atoms of the complex can be exchanged to dinitrogen so as to catalyze hydrogenation of aryl azides to give arylamines.[2]

1Bouwkamp, M.W., Bowman, A.C., Lobkovsky, E., Chirik, P.J. *JACS* **128**, 13340 (2006).
2Bart, S.C., Lobkovsky, E., Bill, E., Chirik, P.J. *JACS* **128**, 5302 (2006).

Iron(II) chloride.

Redox reactions. When complexed to a porphyrin ligand $FeCl_2$ mediates the reduction of α-alkoxy ketones with i-PrOH–NaOH.[1]

An activation system for the C—H bond constituted from $FeCl_2$ and $(t$-BuO$)_2$ enables the union of indan, tetralin and diphenylmethane with β-diketones.[2] Activation of cycloalkanes is more remarkable.[3]

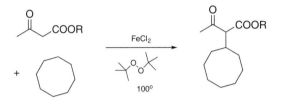

[1]Enthaler, S., Spilker, B., Erre, G., Junge, K., Tse, M.K., Beller, M. *T* **64**, 3867 (2008).
[2]Li, Z., Cao, L., Li, C.-J. *ACIE* **46**, 6505 (2007).
[3]Zhang, Y., Li, C.-J. *EJOC* 4654 (2007).

Iron(III) chloride.

Deacetalization. Diols protected as 1,2,-butanediacetals are released on treatment with $FeCl_3$ in HOAc at room temperature.[1]

Arylation. Traditionally, arylation of nucleophiles is carried out in the presence of copper catalysts, the use of $FeCl_3$ as an alternative, with its scope has now been delineated. In the synthesis of diaryl ethers, 1,3-di-t-butyl-1,3-propanedione serves an additive (ligand for the Fe^{3+} ion) and Cs_2CO_3 as base.[2] *N,N'*-Dimethylethylenediamine appears to be an excellent ligand in the reaction with *N*-nucleophiles (*N*-heterocycles,[3] amides[4]) in a nonpolar solvent (toluene) where a milder base (K_3PO_4) suffices, arylation of alkanethiols calls for t-BuONa.[5]

C-Arylation of 1-alkynes is similarly accomplished.[6]

$$R\!\!=\!\!= \; + \; ArI \quad \xrightarrow[\substack{Cs_2CO_3 \, / \, PhMe \\ 135^\circ}]{FeCl_3–DMEDA} \quad R\!\!=\!\!=\!\!-Ar$$

Friedel–Crafts reactions. Benzyl ethers are activated by $FeCl_3$ to react with arenes to provide diarylmethanes.[7] *N*-Tosylimines and aziridines also become electrophilic toward electron-rich arenes.[8]

Remarkably mild conditions are needed for the ring closure of *N*-aryl-*N*-hydroxypropargylamine derivatives to afford 4-allenylidene-1,2,3,4-tetrahydroisoquinolines.[9]

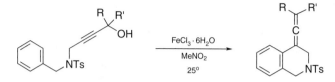

One of the difficult tasks in completing a synthesis of dimeric Vinca alkaloids such as vinblastine is the joining of two monomeric portions. Friedel–Crafts alkylation must be carried out in relatively mild conditions to avoid destruction of the many sensitive functional groups. An elegant solution to the synthetic problem involves oxidation of catharanthine and trapping the reactive species with vindoline. The intermolecular CC bond formation also implies fragmentation of the bridged ring system of catharanthine, most importantly in a stereoselective manner and having the emerging stereocenter in the natural configuration. Using FeCl$_3$ (in 0.1 N HCl) as oxidant and CF$_3$CH$_2$OH as cosolvent, such a task can be achieved at room temperature.[10]

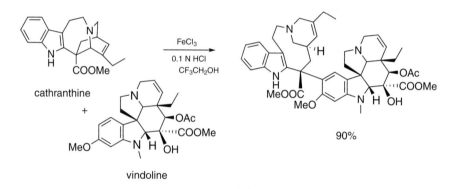

Alone or complexed to MeNO$_2$ or Ph$_2$CO as Friedel–Crafts alkylation catalyst to functionalize polystyrene resin with N-chloromethylphthalimide to produce aminomethylated polymer (for solid phase peptide synthesis), FeCl$_3$ performs well.[11]

Cyclization reactions. 3-Cyano-N-alkoxyindoles are formed when α-cyanobenzyl oxime ethers are oxidized with FeCl$_3$ (a one-electron oxidant). Cyclization follows the generation of the benzylic radicals.[12]

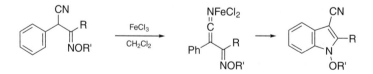

Addition of carboxylic acids to alkenes (e.g., norbornene) is promoted by FeCl$_3$–AgOTf in refluxing 1,2-dichloroethane. Unsaturated carboxylic acids give γ-lactones.[13]

Addition and cycloaddition. Styrenes are transformed into benzylamine derivatives by hydroamination with $TsNH_2$.[14] More unusual is the regioselective addition of arenes across the triple bond of an alkynylarene, as catalyzed by $FeCl_3$.[15]

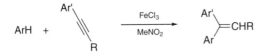

1-Oxallyl cations are readily generated from 2-alkenyl-1,3-dioxolanes. Bicyclo[n.2.0]cycloalkanes can be prepared by intramolecular trapping of such reactive intermediates.[16]

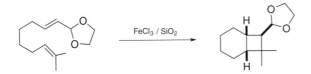

Substitution. Secondary benzylic and allylic alcohols are converted to carboxamido and sulfonamido derivatives by amides and sulfonamides, respectively, in the presence of $FeCl_3$.[17]

[1]Tzschucke, CC., Pradidphoe, N., Dieguez-Vazquez, A., Kongkathip, B., Kongkathip, N., Ley, S.V. *SL* 1293 (2008).

[2]Bistri, O., Correa, A., Bolm, C. *ACIE* **47**, 586 (2008).

[3]Correa, A., Bolm, C. *ACIE* **46**, 8862 (2007).

[4]Correa, A., Elmore, S., Bohm, C. *CEJ* **14**, 3527 (2008).

[5]Correa, A., Carril, M., Bolm, C. *ACIE* **47**, 2880 (2008).

[6]Carril, M., Correa, A., Bolm, C. *ACIE* **47**, 4862 (2008).

[7]Wang, B.-Q., Xiang, S.-K., Sun, Z.-P., Guan, B.-T., Hu, P., Zhao, K.-Q., Shi, Z.-J. *TL* **49**, 4310 (2008).

[8]Wang, Z., Sun, X., Wu, J. *T* **64**, 5013 (2008).

[9]Huang, W., Shen, Q., Wang, J., Zhou, X. *JOC* **73**, 1586 (2008).

[10]Ishikawa, H., Colby, D.A., Boger, D.L. *JACS* **130**, 420 (2008).

[11]Zikos, C., Alexiou, G., Ferderigos, N. *TL* **47**, 8711 (2006).

[12]Du, Y., Chang, J., Reiner, J., Zhao, K. *JOC* **73**, 2007 (2008).

[13]Komeyama, K., Mieno, Y., Yukawa, S., Morimoto, T., Takaki, K. *CL* **36**, 752 (2007).

[14]Michaux, J., Terrasson, V., Marque, S., Wehbe, J., Prim, D., Campagne, J.-M. *EJOC* 2601 (2007).

[15]Li, R., Wang, S.R., Lu, W. *OL* **9**, 2219 (2007).

[16]Ko, C., Feltenberger, J.B., Ghosh, S.K., Hsung, R.P. *OL* **10**, 1971 (2008).

[17]Jana, U., Maiti, S., Biswas, S. *TL* **49**, 858 (2008).

Iron(III) nitrate.

Hydroxymethylation. Formaldehyde is incorporated into β-dicarbonyl compounds by catalysis of $Fe(NO_3)_3 \cdot 9H_2O$. The reaction performed in water at room temperature is facilitated by sodium *p*-dodecylbenzenesulfate.[1]

Oxidation.   Heating an alcohol with $Fe(NO_3)_3 \cdot 9H_2O$ at $80°$ leads to its conversion into the carbonyl compound. However, the scope of the oxidation is not well studied, and its applicability (to give good yields) may be limited to benzylic alcohols.[2]

[1]Ogawa, C., Kobayashi, S. *CL* **36**, 56 (2007).
[2]Namboodiri, V.V., Polshettiwar, V., Varma, R.J. *TL* **48**, 8839 (2007).

Iron(III) perchlorate.

Transalkoxylation.   Facile exchange of the alkoxy group of an 2-alkoxytetrahydrofuran occurs on its treatment with ROH and $Fe(ClO_4)_3 \cdot 6H_2O$.[1]

[1]Yamanaka, D., Matsunaga, S., Kawamura, Y., Hosokawa, T. *TL* **49**, 53 (2008).

Iron(III) sulfate.

Modification of sugars.   Acid-sensitive sugars can be peracetylated using Ac_2O in the presence of $Fe_2(SO_4)_3 \cdot xH_2O$.[1] Ferrier rearrangement is accomplished by treatment with the Fe(III) salt.[2]

[1]Shi, L., Zhang, G., Pan, F. *T* **64**, 2572 (2008).
[2]Zhang, G., Liu, Q., Shi, L., Wang, J. *T* **64**, 339 (2008).

Iron(III) tosylate.

Allylation.   The title salt is another catalyst for allyl transfer from allylsilanes to aldehydes and acetals.[1]

[1]Spafford, M.J., Anderson, E.D., Lacey, J.R., Palma, A.C., Mohan, R.S. *TL* **48**, 8665 (2007).

Iron(II) triflate.

Oxygenation.   Hydrocarbons are oxygenated (e.g., *cis*-1,2-dimethylcyclohexane to *cis*-1,2-dimethylcyclohexan-1-ol) using a Fe(II) complex of amine **1** with catalytic amounts of H_2O_2. The oxygen atom introduced into the hydrocarbon molecules comes from water.[1]

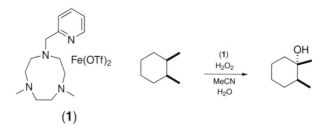

(1)

[1]Company, A., Gomez, L., Güell, M., Ribas, X., Luis, J.M., Que, Jr, L., Costas, M. *JACS* **129**, 15766 (2007).

L

Lanthanum.
Cyclopropanation.[1]   Styrenes are cyclopropanated with 1,1-dibromoalkanes in the presence of lanthanum powder and catalytic amounts of iodine in refluxing THF.

[1]Nishiyama, Y., Tanimizu, H., Tomita, T. *TL* **48**, 6405 (2007).

Lanthanum chloride.
Organolanthanum reagents.[1]   For nucleophilic addition to easily enolized ketones harder nucleophiles are preferred. Arylithiums, derived from *N*-Boc arylamines by *o,N*-dilithiation with *t*-BuLi, are treated with LaCl$_3$ · 2LiCl in THF before reaction with *N*-Boc 3-pyrrolidinone to give precursors of *N*$_b$-Boc tryptamines.[1]

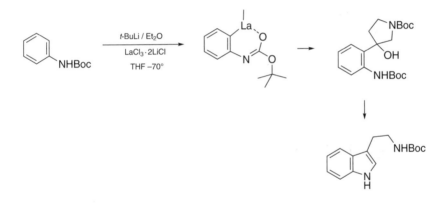

[1]Nicolaou, K.C., Krasovskiy, A., Trepanier, V.E., Chen, D.Y.-K. *ACIE* **47**, 4217 (2008).

Lanthanum tris(hexamethyldisilazide).
Carboxamides.[1]   A mixture of aldehydes and amines are converted into amides at room temperature by treatment with La[N(SiMe$_3$)$_2$]$_3$. The reaction proceeds without added oxidants, bases, and/or heat or light, while one portion of the aldehyde acts as oxidant.

Fiesers' Reagents for Organic Synthesis, Volume 25. By Tse-Lok Ho
Copyright © 2010 John Wiley & Sons, Inc.

Cyclization.   On treatment with La[N(SiMe$_3$)$_2$]$_3$, a 4-alkynol forms the alkoxide which cyclizes to give 2-methylenetetrahydrofuran; the isomer with endocyclic double bond is also formed from 3,4-pentadienol.[2] A modified tris(hexamethyldisilazide) of lanthanum (and of several other rare earth metals) (1) catalyzes intramolecular hydroamination of alkynylamines and the products also undergo hydrosilylation.[3]

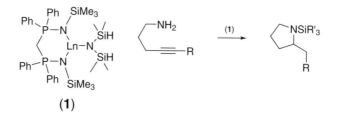

(1)

[1]Seo, S.Y., Marks, T.J. *OL* **10**, 317 (2008).
[2]Yu, X., Seo, S.Y., Marks, T.J. *JACS* **129**, 7244 (2007).
[3]Rastätter, M., Zulys, A., Roesky, P.W. *CEJ* **13**, 3606 (2007).

Lanthanum triflate.

Nucleophilic addition.[1]   Hydrated La(OTf)$_3$ is a useful Lewis acid catalyst for the addition of nucleophiles to *N*-phosphinylimines. In the presence of (CF$_3$CO)$_2$O the *N*-substitutent is also changed into a trifluoroacetyl group.

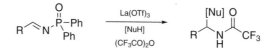

Imine formation.[2]   In the presence of La(OTf)$_3$ arylamines condense with aldehydes, further transformation such as electrocyclization may follow. Very poor results are obtained by using Bronsted acids.

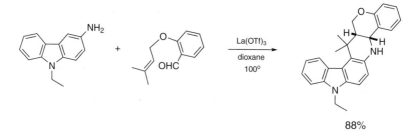

88%

[1]Ong, W.W., Beeler, A.B., Kesavan, S., Panek, J.S., Porco Jr, J.A. *ACIE* **46**, 7470 (2007).
[2]Gaddam, V., Nagarajan, R. *OL* **10**, 1975 (2008).

Lithium.

Lithium organoferrates.[1]   Ferrocene is reduced by lithium in DME in the presence of COD to form Li(dme)[CpFe(cod)]. The salt is useful for promoting intramolecular ene reaction and [5+2]cycloaddition.

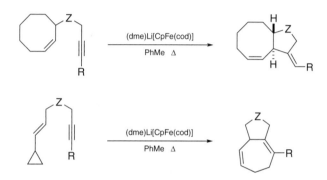

Reductive coupling.[2]   Cycloheptatriene gives a dimeric silylcycloheptadiene on treatment with Li and Me₃SiCl.

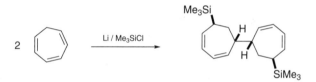

[1]Fürstner, A., Majima, K., Martin, R., Krause, H., Kattnig, E., Goddard, R., Lehmann, C.W. *JACS* **130**, 1992 (2008).
[2]Aouf, C., El Abed, D., Giorgi, M., Santelli, M. *TL* **48**, 4969 (2007).

Lithium – liquid ammonia.

Reductive cleavage.[1]   Certain cyclic sulfonamides suffer double cleavage of C—S and N—S bonds, but an *N*-benzyl group is more labile.

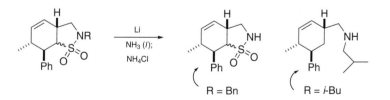

[1]Kelleher, S., Muldoon, J., Müller-Bunz, H., Evans, P. *TL* **48**, 4733 (2007).

Lithium aluminum hydride – niobium(IV) chloride.

Hydrodefluorination. A trifluoromethyl group of bis(trifluoromethyl)arenes is converted into the methyl group on heating with LAH (3 equiv.) and NbCl₅ (5 mol%) in DME. By increasing the amount of LAH to 10 equivalents both trifluoromethyl groups are reduced.[1]

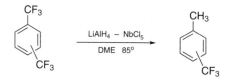

o-Trifluoromethylstyrenes cyclize to give indenes (yields around 60%) under similar conditions.[2]

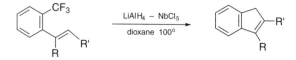

[1]Fuchibe, K., Ohshima, Y., Mitomi, K., Akiyama, T. *OL* **9**, 1497 (2007).
[2]Fuchibe, K., Mitomi, K., Akiyama, T. *CL* **36**, 24 (2007).

Lithium aluminum hydride – selenium.

Amide formation.[1] Carboxamides including peptides are synthesized from carboxylic acids and alkyl azides, after converting the acids into mixed anhydrides and then selenocarboxylates. Treatment of the mixed anhydrides with a suspension of freshly prepared from LAH and Se completes the first stage of the transformation.

[1]Wu, X., Hu, L. *JOC* **72**, 765 (2007).

Lithium borohydride.

Reductive amination.[1] LiBH₄ is said to be the reagent of choice for reductive amination of substituted cyclohexanones.

[1]Cabral, S., Hulin, B., Kawai, M. *TL* **48**, 7134 (2007).

Lithium chloride.

Mannich reaction. Mukaiyama-type Mannich reaction[1] can be effected by LiCl (0.2 equiv.) in DMF.

[1]Hagiwara, H., Iijima, D., Awen, B.Z.S., Hoshi, T., Suzuki, T. *SL* 1520 (2008).

Lithium di-*t*-butylbiphenylide.

Reductive lithiation. Phthalans are cleaved to give *O,C*-dilithio products which are readily quenched by electrophiles. The direction of cleavage is governed by substituents of the aromatic ring.[1]

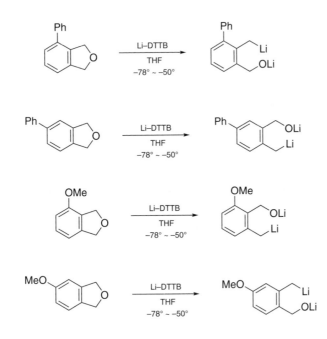

[1]Garcia, D., Foubelo, F., Yus, M. *T* **64**, 4275 (2008).

Lithium diisopropylamide, LDA.

Condensation. Condensation involving a carboxamide and an acylsilane (e.g., effected by LDA) proceeds via a Brook rearrangement to generate a β-siloxy homoenolate.[1] Trapping by a phosphinylimine leads to product that is convertible to γ-lactams.[2]

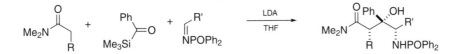

Quaterized α-alkylthiopropargylamines are dimerized to give enediynes that bear thio-substituents on the central double bond. Cyclic products are obtained from the intra-molecular version.[3]

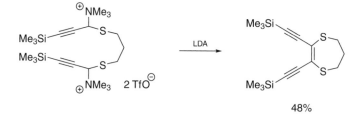

48%

Cleavage of allenylidenecyclopropanes.[4]   2,2-Diaryl-1-allenylidenecyclopropanes react in several different ways upon deprotonation at the cyclopropane. Hydroxyalkylation occurs at the central carbon of the original allene unit with aldehydes, but at a higher temp-erature the ring is ruptured accompanied by attack on ketones. Still a third type of products arises from reaction with conjugated carbonyl compounds. Electronic effects probably are responsible for the diverse results.

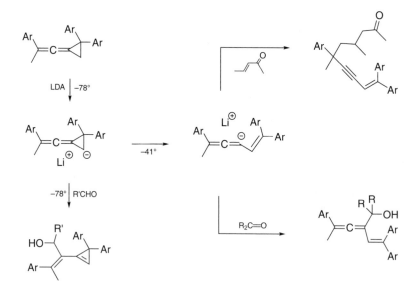

Cycloisomerization.[5]   2-Iodoethyl propargyl ethers give 3-iodomethylenetetra-hydrofurans after exposure to substoichiometric amounts of LDA at room temperature.

However, neither the all carbon analog nor 3-iodopropyl propargyl ethers show similar reactivity. Cycloisomerization of iodopropargyl ethers is induced by 1-hexynyllithium.

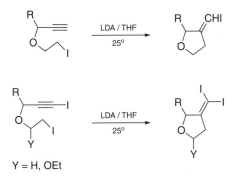

Annulated pyrroles.[6] 1-Dialkylamino-$(n+3,n+3)$-dichlorobicyclo[n.1.0]alkanes suffer dehydrochlorination, to generate, plausibly, allylic chlorocarbenes for subsequent H-abstraction, cyclization and aromatization.

Substitutions. Enolates generated from methyl ketones on treatment with LDA attack dimethyl 2-bromomethylfumarate in an S_N2' fashion, to give substitued itaconic esters.[7]

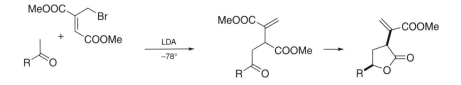

Deprotonation. Regioselective metallation of unsymmetrical 1,3-disubstituted allenes enables homologation. Thus lithiation followed by transmetallation with $ZnBr_2$ and Negishi coupling serves to introduce a new organic residue into the nonbenzylic position.[8]

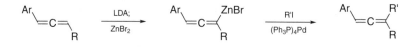

Twice deprotonated species from a 3-nitropropanoic ester is used in a double Michael reaction on 4,4-dimethoxy-2,5-cyclohexadienone to furnish bridged ring products (for a synthesis of gelsemine). The deprotonation is conveniently carried out with LDA in THF containing HMPA.[9]

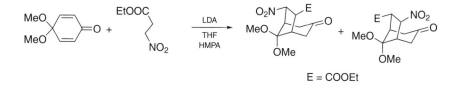

E = COOEt

The kinetic enolate of a conjugated ketone generated by LDA reacts with a conjugated acylsilane to form the monosilylated 1,3-cycloheptanedione. A Brook rearrangement following the initial C-acylation delivers an allyl anion that is poised to return an attack on the enone that re-emerges.[10]

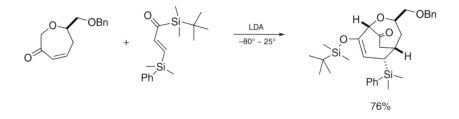

76%

[1]Lettan II, R.B., Reynolds, T.E., Galliford, C.V., Scheidt, K.A. JACS **128**, 15566 (2006).

[2]Lettan II, R.B., Woodward, C.C., Scheidt, K.A. ACIE **47**, 2294 (2008).

[3]Murai, T., Fukushima, K., Mutoh, Y. OL **9**, 5295 (2007).

[4]Lu, J.-M., Shi, M. OL **10**, 1943 (2008).

[5]Harada, T., Muramatsu, K., Mizunashi, K., Kitano, C., Imaoka, D., Fujiwara, T., Kataoka, H. JOC **73**, 249 (2008).

[6]Bissember, A.C., Phillis, A.T., Banwell, M.G., Willis, A.C. OL **9**, 5421 (2007).

[7]Baag, M.M., PurNIK, V.G., Argade, N.P. JOC **72**, 1009 (2007).

[8]Zhao, J., Liu, Y., Ma, S. OL **10**, 1521 (2008).

[9]Grecian, S., Aube, J. OL **9**, 3153 (2007).

[10]Sasaki, M., Hashimoto, A., Tanaka, K., Kawahata, M., Yamaguchi, K., Takeda, K. OL **10**, 1803 (2008).

Lithium hexamethyldisilazide, LHMDS.

Tandem cyclization. Using LHMDS as base the tetracyclic precursor of pareitropone is constructed from a biaryl. After addition of the tosylamide anion to an iodonioalkyne triple bond to generate an alkylidenecarbene, addition across an aromatic ring follows.[1]

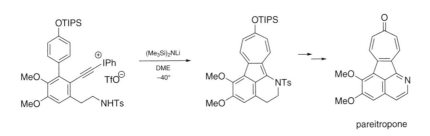

pareitropone

Wittig rearrangement. Functionalized 1,2-amino alcohols are acquired from 3-tosylamino-2-propen-1-yl ethers by Wittig rearrangement.[2] Interesting stereoselectivity of the reaction depending on the substituent of the alkyl component has been revealed.

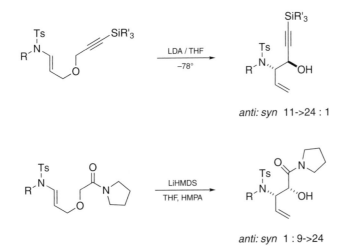

anti: syn 11->24 : 1

anti: syn 1 : 9->24

Dehydroalkylation. α-Keto amides are enolized and reaction of the resulting enediolate species with alkyl iodides furnishes 2-hydroxy-3-alkenamides.[3]

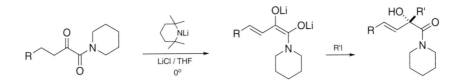

Cyclopropanation.   Ylides derived from 2-cyanoazetidinium triflates behave as stabilized carbenoids that cycloadd to enones.[4]

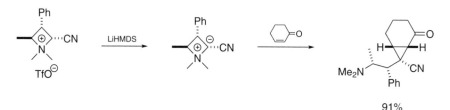

91%

[1]Feldman, K.S., Cutarelli, T.D. *JACS* **124**, 11600 (2002).
[2]Barbazanges, M., Meyer, C., Cossy, J. *OL* **9**, 3245 (2007).
[3]Marsden, S.P., Newton, R. *JACS* **129**, 12600 (2007).
[4]Couty, F., David, O., Larmanjat, B., Marrot, J. *JOC* **72**, 1058 (2007).

Lithium naphthalenide, LN.

Halogen/lithium exchange.   LN enables exclusive conversion of *sp*-hybridized C—Cl and C—Br bond in the presence of sp^3-hybridized C—Cl bond, but sp^3-hybridized C—I bond undergoes I/Li exchange preferentially to an *sp*-hybridized C-Cl bond.[1] It is also possible to exchange an sp^2-hybridized C-bonded Br in preference to an sp^3-hybridized C-bonded Cl.

[1]Abou, A., Foubelo, F., Yus, M. *T* **63**, 6625 (2007).

Lithium 2,2,6,6-tetramethylpiperidide, LiTMP.

Metallation.   5-Membered heterocycles are metallated by LiTMP, and transmetallated in the presence of $ZnCl_2$-TMEDA.[1] Remarkably regioselective lithiation of 3-methyl-thiophene at C-5 (79:1 over C-2) is observed with LiTMP in THF at −78°. (Cf. 3:1 and 5:1 in lithiation with MeLi and *t*-BuLi, respectively.)[2]

α-(*N*-Boc-*N*-bromoalkyl)amino esters cyclize on treatment with an alkali metal amide. Remarkably, enantiodivergence is observed on changing the base.[3]

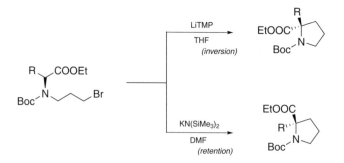

Alkene synthesis.   Exposure of epoxides to RM (M = Li, MgX) and LiTMP leads to alkenes.[4]

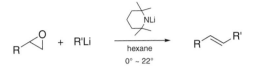

[1]L'Helgoual'ch, J.-M., Seggio, A., Chevallier, F., Yonehara, M., Jeanneau, E., Uchiyama, M., Mongin, F. *JOC* **73**, 177 (2008).
[2]Smith, K., Barratt, M.L. *JOC* **72**, 1031 (2007).
[3]Kawabata, T., Matsuda, S., Kawakami, S., Monguchi, D., Moriyama, K. *JACS* **128**, 15394 (2006).
[4]Hodgson, D.M., Fleming, M.J., Stanway, S.J. *JOC* **72**, 4763 (2007).

Lithium triethylborohydride.

Reduction.[1]   This borohydride reduces nitriles to afford stable aldimine-borane complexes, which can be used to prepare homoallylic amines on further reaction with $R_2BCH_2CH=CH_2$.

[1]Ramachandran, P.V., Biswas, D. *OL* **9**, 3025 (2007).

Lithium triflimide.

Aminolysis.   The conversion of lactones by reaction with amines to ω-hydroxy amides is catalyzed by $LiNTf_2$.[1]

[1]Lalli, C., Trabocchi, A., Menchi, G., Guarna, A. *SL* 189 (2008).

M

Magnesium.

Deoxygenation. Oxides of diorganochalcogenides (S, Se, Te) are deoxygenated by Mg in MeOH at room temperature. Both monoxides and dioxides are affected.[1]

Polysilanes.[2] Dichlorosilanes are reductively polymerized on treatment with Mg and a Lewis acid and LiCl.

Reformatsky reaction.[3] As an alternative promoter to zinc metal, the use of Mg-FeCl$_3$ or Mg-CuCl$_2 \cdot$ 2H$_2$O has been demonstrated.

Reductive acylation. Treatment of anthracene and a dicarboxylic acid chloride with magnesium in DMF leads to the dibenzo bridged diketone product.[4]

Indole synthesis.[5] Intramolecular sp^2-sp^3 coupling mediated by Mg is applied to synthesis of 2-fluoroalkylindoles.

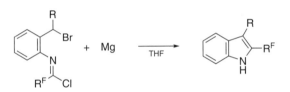

[1]Khurana, J.M., Sharma, V., Chacko, S.A. *T* **63**, 966 (2007).
[2]Kashimura, S., Tane, Y., Ishifune, M., Murai, Y., Hashimoto, S., Nakai, T., Hirose, R., Murase, H. *TL* **49**, 269 (2008).
[3]Chattopadhyay, A., Dubey, A.K. *JOC* **72**, 9357 (2007).
[4]Matsunami, M., Sakai, N., Morimoto, T., Maekawa, H., Nishiguchi, I. *SL* 769 (2007).
[5]Ge, F., Wang, Z., Wan, W., Hao, J. *SL* 447 (2007).

Magnesium bromide etherate.

C-Acylation. Condensation of ketones with acylating agents such as 1-acylbenzotriazoles is readily accomplished on treatment with MgBr$_2 \cdot$ OEt$_2$ and *i*-Pr$_2$NEt.

Additive aldol reaction.[2] An *anti*-selective synthesis of *S*-phenylthio esters of 3-hydroxy-2-phenylthiomethylalkanethiolates from a mixture of acryloyl chloride, PhSLi and aldehydes is mediated by MgBr$_2 \cdot$ OEt$_2$.

Fiesers' Reagents for Organic Synthesis, Volume 25. By Tse-Lok Ho
Copyright © 2010 John Wiley & Sons, Inc.

Radical addition.   A highly stereoselective addition of alkyl radicals to acrylic esters has been reported. The presence of MgBr$_2$ · OEt$_2$ seems critical.[3]

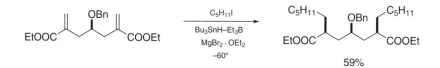

[1]Lim, D., Fang, F., Zhou, G., Coltart, D.M. *OL* **9**, 4139 (2007).
[2]Zhou, G., Yost, J.M., Sauer, S.J., Coltart, D.M. *OL* **9**, 4663 (2007).
[3]Nagano, H., Kuwahara, R., Yokoyama, F. *T* **63**, 8810 (2007).

Magnesium iodide.

Isomerization.   *N*-Aryl-2-alkylidenecyclopropanecarboxamides undergo isomerization to give β-alkylidene-γ-lactams on treatment with MgI$_2$ in THF.[1]

[1]Scott, M.E., Schwarz, C.A., Lautens, M. *OL* **8**, 5521 (2006).

Magnesium oxide.

Double Michael addition.   In the presence of MgO alkynyl ketones and 1,3-propane-dithiol react to afford 2-acylmethyl-1,3-dithianes.[1]

[1]Xu, C., Bartley, J.K., Enache, D.I., Knight, D.W., Lunn, M., Lok, M., Hutchings, G.J. *TL* **49**, 2454 (2008).

Magnesium perchlorate.

Alkylidenation.[1]   Condensation of β-diketones with aldehydes is catalyzed by Mg(ClO$_4$)$_2$.

Transfer hydrogenation.[2]   1-Acetyl-2,3-dimethylimidazolidine acts as H donor in MeCN-MeOH in the presence of Mg(ClO$_4$)$_2$ to reduce aromatic aldehydes, their imino derivatives (ArCH=NPh, ArCH=NTs) and cinnamaldehydes to give the corresponding alcohols and amines.

[1]Bartoli, G., Bosco, M., Carlone, A., Dalpozzo, R., Galzerano, P., Melchiorre, P., Sambri, L. *TL* **49**, 2555 (2008).
[2]Li, D., Zhang, Y., Zhou, G. *SL* 225 (2008).

Magnesium 2,2,6,6-tetramethylpiperidide.

Magnesiation.   The magnesium amide (as LiCl complex) is very useful for *o*-magnesiation of aroic esters[1] and aryl bis(dimethylamido)phosphates.[2] Reaction of the magnesiated species with electrophiles leads to products of substitution patterns otherwise difficult to achieve.

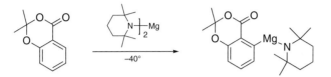

[1]Clososki, G.C., Rohbogner, C.J., Knochel, P. *ACIE* **46**, 7681 (2007).
[2]Rohbogner, C.J., Clososki, G.C., Knochel, P. *ACIE* **47**, 1503 (2008).

Mandelic acid.

Alcoholysis.   Styrene oxides open on treatment with ROH in the presence of mandelic acid and *N,N'*-di[3,5-bis(trifluoromethyl)phenyl]thiourea in a regioselective manner, due to cooperative interaction of the Bronsted acids.[1]

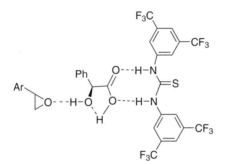

[1]Weil, T., Kleiner, C.M., Schreiner, P.R. *OL* **10**, 1513 (2008).

Manganese.

Conjugated esters.   Active Mn is prepared from $MnCl_2$ via Li_2MnCl_4 and treating the latter with Li and 2-phenylpyridine. The active metal promotes reaction of 2,2-dichloro-alkanoic esters with aldehydes to furnish conjugated esters.[1]

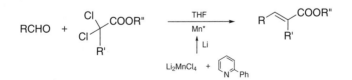

[1]Concellon, J.M., Rodriguez-Solla, H., Diaz, P., Llavona, R. *JOC* **72**, 4396 (2007).

Manganese(III) acetate.

Oxidation.   Aporphines undergo demethylative aromatization and further oxidation at ring C on exposure to Mn(OAc)₃.[1]

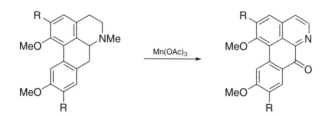

Oxidative cyclization.   4-Pentenylmalonic esters give bicyclic lactones in moderate to good yields when exposed to Mn(OAc)₃ and Cu(OTf)₂.[2]

A dramatic solvent effect has been observed in the following radical cyclization.[3] For reproducibly good yields 2,2,2-trifluoroethanol is used.

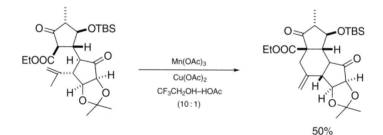

50%

No copper-based cooxidant is required in the oxidative cyclization of *N*-acylindoles such as shown below.[4]

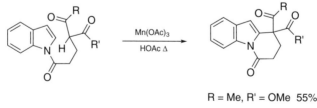

R = Me, R' = OMe 55%
R = R' = Me 60%

Phosphonation.[5] Heteroaryl compounds (furans, pyrroles, thiazoles, ...) undergo Mn(III)-mediated regioselective phosphonation with $HP(O)(OMe)_2$.

[1]Singh, O.V., Huang, W.-J., Chen, C.-H., Lee, S.-S. *TL* **48**, 8166 (2007).
[2]Powell, L.H., Docherty, P.H., Hulcoop, D.G., Kemmitt, P.D., Burton, J.W. *CC* 2559 (2008).
[3]Toueg, J., Prunet, J. *OL* **10**, 45 (2008).
[4]Magolan, J., Carson, C.A., Kerr, M.A. *OL* **10**, 1437 (2008).
[5]Mu, X.-J., Zou, J.-P., Qian, Q.-F., Zhang, W. *OL* **8**, 5291 (2006).

Manganese(II) triflate.

Epoxidation.[1] Various alkenes are epoxidized by peracetic acid with a complex of $Mn(OTf)_2$ to triazanonane **1** as catalyst.

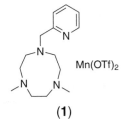

(1)

[1]Garcia-Bosch, I., Company, A., Fontrodona, X., Ribas, X., Costas, M. *OL* **10**, 2095 (2008).

Mercury(II) triflate.

Hydration-elimination. Addition of water (1 equiv.) to 3-acetoxy-1-ethoxyalkynes results in the formation of conjugated esters. A convenient catalyst is $Hg(OTf)_2$, which dictates the production of the (*E*)-isomers.[1]

Cyclodehydration. 6-Aryl-2-hexenols cyclize on brief heating with catalytic amounts of $Hg(OTf)_2$ in toluene.[2]

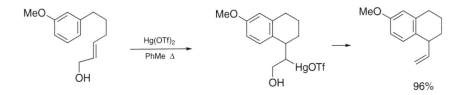

N-Tosylindolines.[3] Activation of the triple bond of *N*-tosyl-2-[2-(4-pentynoyloxy)-ethyl]anilines by $Hg(OTf)_2$ triggers cyclization. Benzologous pentynoic esters react similarly.

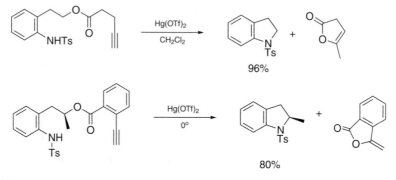

96%

80%

[1]Nishizawa, M., Hirakawa, H., Nakagawa, Y., Yamamoto, H., Namba, K., Imagawa, H. *OL* **9**, 5577 (2007).

[2]Namba, K., Yamamoto, H., Sasaki, I., Mori, K., Imagawa, H., Nishizawa, M. *OL* **10**, 1767 (2008).

[3]Yamamoto, H., Pandey, G., Asai, Y., Nakano, M., Kinoshita, A., Namba, K., Imagawa, H., Nishizawa, M. *OL* **9**, 4029 (2007).

Mesityltriphenylbismuthonium tetrafluoroborate.

Oxidation.[1]   The title compound is an oxidant for primary and secondary alcohols. Oxidation is carried out with the base tetramethylguanidine at room temperature, liberating mesitylene and Ph$_3$Bi as side products.

[1]Matano, Y., Suzuki, T., Shinokura, T., Imahori, H. *TL* **48**, 2885 (2007).

Methanesulfonic acid.

Isomerization.[1]   1-Alken-3-ols undergo isomerization to afford 2-alken-1-ols on treatment with MsOH in aq.THF.[1]

1-Amino-2-alkanols.[2]   Terminal epoxides are opened by NH$_3$ (in saturated EtOH) regioselectively. To render the procedure operationally simple and cost-effective, five equivalents of MsOH are added.

Cyclization.[3]   A critical step in a synthesis of (-)-cribrostatin-4 is the closure of the azabicyclo unit by treatment of an aldehyde precursor with MsOH. The reaction is very sensitive to the concentration of the acid catalyst and solvent.

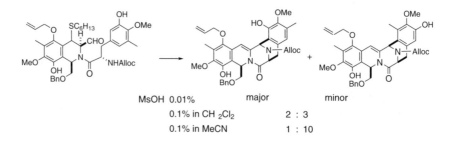

MsOH 0.01%   major   minor
0.1% in CH$_2$Cl$_2$ 2 : 3
0.1% in MeCN   1 : 10

[1]Leleti, R.R., Hu, B., Prashad, M., Repic, O. *TL* **48**, 8505 (2007).
[2]Kaburagi, Y., Kishi, Y. *TL* **48**, 8967 (2007).
[3]Chen, X., Zhu, J. *ACIE* **46**, 3962 (2007).

Methylaluminum bis(2,6-di-*t*-butyl-4-methylphenoxide), MAD.

Ring expansion.[1]   2-Methylenecyclopropanecarboxamides undergo expansion on reaction with *N*-tosylaldimines under the influence of a Lewis acid and with the assistance of iodide ion. Due to steric effects the reactions promoted by MAD–Bu$_4$NI and by MgI$_2$ lead to isomeric products.

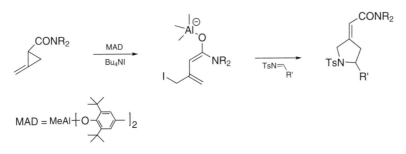

Rearrangement.[2]   (1*Z*,3*E*)-Alkadienyl acetals are convertd into 2-(α-alkoxy)alkyl-3-alkenals on treatment with MAD or ATPH. The rearrangement is α-regioselective.

[1]Taillier, C., Bethuel, Y., Lautens, M. *T* **63**, 8469 (2007).
[2]Tayama, E., Hashimoto, R. *TL* **48**, 7950 (2007).

N-Methyl-2-benzyloxypyridinium triflate.

Friedel–Crafts benzylation.[1]   The title salt is a stable precursor of benzylcarbenium ion. On thermolysis (at 80°) of its mixture with an electron-rich arene the cation is generated and trapped.

[1]Albiniak, P.A., Dudley, G.B. *TL* **48**, 8097 (2007).

Molybdenum(VI) dichloride dioxide.

β-Keto esters.[1]   For condensation of diazoacetic esters with aldehydes to furnish β-keto esters MoO$_2$Cl$_2$ is an effective catalyst.

[1]Jeyakumar, K., Chand, D.K. *S* 1685 (2008).

Molybdenum hexacarbonyl.

Acylation. Formation of bicyclic lactams related to the Pauson–Khand reaction is realized from 5,6-alkadienal hydrazones by heating with $Mo(CO)_6$ and DMSO in CH_2Cl_2 at 100°.[1] With thiocarbonyl compounds the reaction leads to thiolactones.[2]

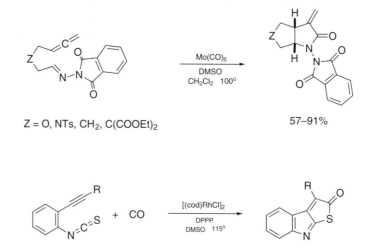

Z = O, NTs, CH_2, $C(COOEt)_2$

57–91%

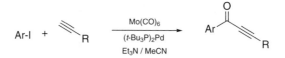

Coupling of electron-deficient ArI and 1-alkynes while incorporating CO to provide alkynyl aryl ketones is accomplished at room temperature by a Pd-catalyzed reaction with $Mo(CO)_6$.[3]

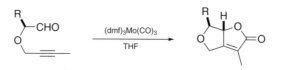

Ketone formation from aryl and alkenyl halides via carbonylation and addition to alkenes is accomplished by heating the substrates with $Mo(CO)_6$ in DMF at 160°.[4]

The $(dmf)_3Mo(CO)_3$ modification furnishes CO to combine with an aldehyde and a triple bond to form a butenolide unit.[5]

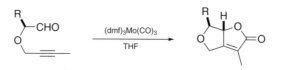

Chain elongation of ArCHO. Transformation of ArCHO to 2-arylidene derivatives of Meldrum's acid followed by reduction with $PhSiH_3$ accomplishes homologation of the aldehydes by two carbon units. The second step is catalyzed by $Mo(CO)_6$.[6]

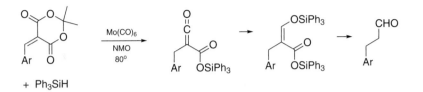

+ Ph₃SiH

Reductive cleavage.   Pentafluorophenyl isoxazolidine-4-sulfonates are converted into α-hydroxymethyl-β-sultams on treatment with $Mo(CO)_6$ in aqueous MeCN at 90°. Cyclization occurs after the N-O bond is cleaved.[7]

[1]Kim, S.-H., Kang, E.S., Yu, C.-M. *SL* 2439 (2007).
[2]Saito, T., Nihei, H., Otani, T., Suyama, T., Furukawa, N., Saito, M. *CC* 172 (2008).
[3]Iizuka, M., Kondo, Y. *EJOC* 5180 (2007).
[4]Sangu, K., Watanabe, T., Takaya, J., Iwasawa, N. *SL* 929 (2007).
[5]Adrio, J., Carretero, J.C. *JACS* **129**, 778 (2007).
[6]Frost, C.G., Hartley, B.C. *OL* **9**, 4259 (2007).
[7]Lewis, A.K.deK., Mok, B.J., Tocher, D.A., Wilden, J.D., Caddick, S. *OL* **8**, 5513 (2006).

N

Nafion resin.

Photochemical reactions. The Na^+-form resin is found to be an excellent reaction medium for photo-induced cyclization of α-pyridone derivatives to provide bicyclic β-lactams.[1]

Deprotection. Terminal acetonides and trityl ethers are selectively cleaved on exposure to Nafion-H in MeOH at room temperature.[2]

[1]Arumugam, S. *TL* **48**, 2461 (2008).
[2]Rawal, G.K., Rani, S., Kumar, A., Vankar, Y.D. *TL* **46**, 9117 (2006).

Nickel.

Reductions. In refluxing isopropanol carbonyl compounds are reduced in the presence of nickel nanoparticles.[1] Reductive amination is also performed under the same conditions.[2]

Catalytic hydrogenation of alkenes and alkynes is achieved with nickel nanoparticles prepared from $NiCl_2$ by reduction with Li and catalytic amounts of DTBB and an alcohol.[3]

[1]Alonso, F., Riente, P., Yus, M. *T* **64**, 1847 (2008).
[2]Alonso, F., Riente, P., Yus, M. *SL* 1289 (2008).
[3]Alonso, F., Osante, I., Yus, M. *T* **63**, 93 (2007).

Nickel, Raney.

Pyrazoles. Isoxazoles are readily converted into pyrazoles on treatment with hydrazine at room temperature in the presence of Raney nickel.[1]

[1]Sviridov, S.I., Vasil'ev, A.A., Shorshnev, S.V. *T* **63**, 12195 (2007).

Nickel(II) acetate.

Aminocarbonylation.[1] A method for preparation of $ArCONMe_2$ from ArX involves heating with DMF and NaOMe in dioxane. The catalyst system contains $Ni(OAc)_2 \cdot 4H_2O$ and $[2,4-(t-Bu)_2C_6H_3O]_3P$.

[1]Ju, J., Jeong, M., Moon, J., Jung, H.M., Lee, S. *OL* **9**, 4615 (2007).

Fiesers' Reagents for Organic Synthesis, Volume 25. By Tse-Lok Ho
Copyright © 2010 John Wiley & Sons, Inc.

Nickel(II) acetylacetonate.

Coupling. A route to biaryls entails the use of Ni(acac)$_2$ to couple ArX with Ar'Ti(OEt)$_3$.[1] A hindered phosphine or carbene ligand is also added.

N-Arylation. Secondary amines are arylated by ArCl under the following conditions: heating with Ni(acac)$_2$, 3,5,6,8-tetrabromo-1,10-phenanthroline, sodium *t*-butoxide, and PMHS in toluene at 130°.[2]

Addition. Diorgano dichalcogenides are split and add to 1-alkynes to afford (Z)-1,2-bis(organochalcogeno)-1-alkenes. Better results are obtained on using Ni(acac)$_2$ – PhPMe$_2$ than (dba)$_3$Pd$_2$ – PhPCy$_2$ although excess alkynes are required.[3]

[1]Manolikakes, G., Dastbaravardeh, N., Knochel, P. *SL* 2077 (2007).
[2]Manolikakes, G., Gavryushin, A., Knochel, P. *JOC* **73**, 1429 (2008).
[3]Ananikov, V.P., Gayduk, K.A., Beletskaya, I.P., Krustalev, V.N., Antipin, M.Yu. *CEJ* **14**, 2420 (2008).

Nickel(II) acetylacetonate–diorganozinc.

Reductive aldol reaction.[1] Conjugated amides with *N*-acylalkyl substituents are liable to cyclize on treatment with Ni(acac)$_2$ – Et$_2$Zn, with the organozinc reagent serving as a reducing agent to generate the amide enolates. Often inferior results are obtained when Co(acac)$_2$ – Et$_2$Zn is employed.

Cyclization.[2] 1,6-Enynes afford alkylidenecyclopentanes as exemplified by the following equation.

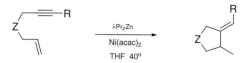

[1]Joensuu, P.M., Murray, G.J., Fordyce, E.A.F., Luebbers, T., Lam, H.W. *JACS* **130**, 7328 (2008).
[2]Chen, M., Weng, Y., Guo, M., Zhang, H., Lei, A. *ACIE* **47**, 2279 (2008).

Nickel bromide.

Propargylarenes.[1] Secondary benzylic bromides couple with trialkynylindium reagents in the presence of (diglyme)NiBr$_2$. An enantioselective version employing a Pybox ligand leads to chiral products in about 80% ee.

[1]Caeiro, J., Sestelo, J.P., Sarandeses, L.A. *CEJ* **14**, 741 (2008).

Nickel bromide–zinc.

   Cycloaddition.   Formation of [*a,e*]-fused cyclooctatetraenes by cyclodimerization of terminal diynes in a formal [2+2+2+2]cycloaddition is effected by (dme)NiBr$_2$–Zn.[1]

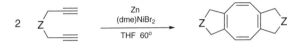

   Reductive coupling.   Certain β-amino esters are synthesized from imines and acrylic esters, by mediation of (phenanthroline)NiBr$_2$–Zn.[2] Conjugated sulfones, nitriles also couple with the imines.

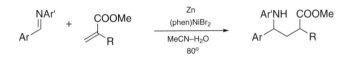

[1]Wender, P.A., Christy, J.P. *JACS* **129**, 13402 (2007).
[2]Yeh, C.-H., Korivi, R.P., Cheng, C.-H. *ACIE* **47**, 4892 (2008).

Nickel chloride.

   Coupling reactions.   Modified procedures for the Stille coupling,[1] Hiyama coupling,[2] and Negishi coupling[3] involving aliphatic substrates are based on (dme)NiCl$_2$ promotion. C_2-symmetric *vic*-diamines ligands are employed in the first two reaction types and chiral α-branched carboxylic esters are accessible by the method. Allylic chlorides are transformed into chiral products.

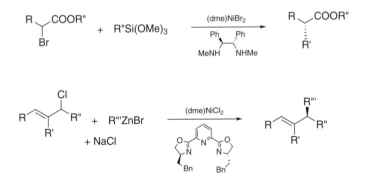

   From alkenyl triflates (or iodides), diorganozincs and CO, conjugated ketones are assembled with the aid of NiCl$_2$.[4]

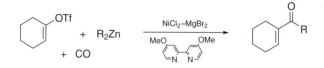

(Py)$_4$NiCl$_2$ is a precatalyst for Negishi coupling that is preceded by a cyclization process.[5]

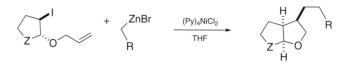

The complex of NiCl$_2$ to both Ph$_3$P and 1,3-bis(2,6-diisopropylphenyl)imidazol-2-ylidene promotes α-arylation of alkyl aryl ketones and N-arylation of ArNH$_2$.[6]

[1]Saito, B., Fu, G.C. JACS **129**, 9602 (2007).
[2]Dai, X., Strotman, N.A., Fu, G.C. JACS **130**, 3302 (2008).
[3]Son, S., Fu, G.C. JACS **130**, 2756 (2008).
[4]Wang, Q., Chen, C. TL **49**, 2916 (2008).
[5]Phapale, V.B., Bunuel, E., Garcia-Iglesias, M., Cardenas, D.J. ACIE **46**, 8790 (2007).
[6]Matsubara, K., Ueno, K., Koga, Y., Hara, K. JOC **72**, 5069 (2007).

Nickel iodide.

Coupling reactions.[1]   The Ph$_3$P-complex of NiI$_2$ is found to succor the reaction of o-bromobenzylzinc bromides with alkenes and alkynes, delivering indanes and indenes, respectively.

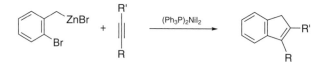

[1]Deng, R., Sun, L., Li, Z. OL **9**, 5207 (2007).

Nickel perchlorate.

Cycloaddition.[1]   Azomethine imines constituting a segment of heteroaromatic systems are activated toward cycloaddition with cyclopropanes.

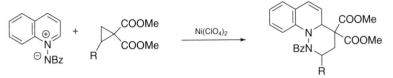

[1]Perreault, C., Goudreau, S.R., Zimmer, L.E., Charette, A.B. *OL* **10**, 689 (2008).

2-Nitrophenyl isocyanide.

Ugi reaction.[1]   As a participant of the Ugi reaction, the title compound provides an active acyl unit for further synthetic transformations such as for the elaboration of fused γ-lactam/β-lactones. By reduction and diazotization, the *N*-(2-nitrophenyl)carboxamide products are converted into *N*-acylbenzotriazoles.

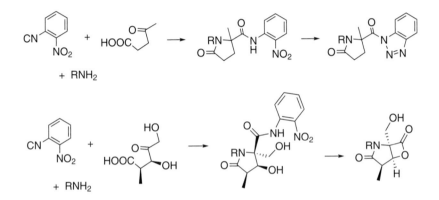

[1]Gilley, C.B., Kobayashi, Y. *JOC* **73**, 4198 (2008).

2-Nitro-5-piperidinylbenzyl alcohol.

Carboxyl protection.[1]   Esters of the title reagent **1** are cleaved photochemically under specific conditions. The nitro group, while existing mainly in the nitronate form due to resonance interaction of the *p*-amino group, is therefore photoinactive. By addition of TfOH or Cu(OTf)$_2$ to remove the resonance the susceptibility to decomposition by UV light revives.

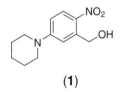

(1)

[1]Riguet, E., Bochet, C.G. *OL* **9**, 5453 (2007).

O

Organoaluminum reagents.

Addition reactions. The 1,2- and 1,4-additions of organoalanes to conjugated carbonyl compounds have been reviewed.[1] There is a report on reaction of R_3Al with *N*-diphenylphosphonylketimines.[2] Alkynyldimethylalanes are superior reagents for transferring an alkynyl group to *N*-toluenesulfinylimines in comparison with the corresponding lithium and magnesium reagents.[3]

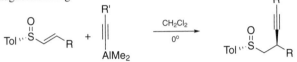

Conjugated sulfoxides.[4] A method for the preparation of conjugated sulfoxides involves reaction of the pyridine-coordinated alkenyldiisobutylalanes with $ArSO_2Cl$ in the presence of Ph_3P.

Aluminum iminoxides.[5] Nucleophilic amides are formed upon condensation of esters with amines, which are liable to undergo conjugate addition to enones when catalyzed by $Sn(OTf)_2$. The reaction has found an application in the synthesis of (-)-cephalotaxine.

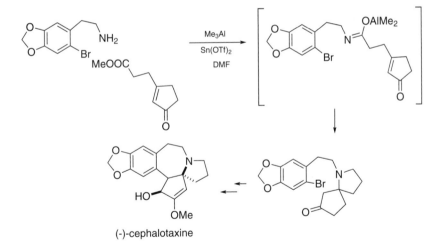

(-)-cephalotaxine

Fiesers' Reagents for Organic Synthesis, Volume 25. By Tse-Lok Ho
Copyright © 2010 John Wiley & Sons, Inc.

Reduction.[6] Meerwein–Ponndorf–Verley reduction of 4-pyranones with a reagent derived from Me_3Al and benzhydrol favors hydride delivery from tris(diphenylmethoxy)-aluminum on the equatorial side to give axial alcohols.

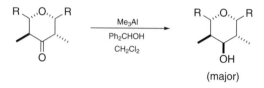

(major)

Cycloadditions. With Et_2AlOEt present, α-chloroalkenyl acetates react with carbonyl compounds and imines to furnish β-lactones and β-lactams, respectively. The ketene equivalent also combines with oximes to give 5-isoxazolidones.[7]

A synthesis of 4-substituted 1,2,3-triazoles involves reaction of 1-alkynes with diorgano-aluminum azides. The inexpensive and nontoxic azide reagents are available from mixing R_2AlCl with NaN_3 in toluene.[8]

The carbon acid Tf_2CH_2 reacts with Me_3Al to form $Me_2AlCHTf_2$, which is an excellent catalyst for the highly *endo*-selective Diels–Alder reaction between cyclopentadiene and conjugated lactones.[9]

Modified Claisen condensation.[10] Silyl ketene acetals and methyl esters (important!) react in the presence of Me_3Al in toluene to afford mixed methyl/silyl acetals of β-ketoesters.

1,4-Diynes.[11] A synthesis of the skipped diynes from 1-alkynes and propargylic mesylates involves prior conversion of the former compounds into aluminum derivatives.

[1]Von Zezschwitz, P. *S* 1809 (2008).
[2]Reingruber, R., Bräse, S. *CC* 105 (2008).
[3]Turcaud, S., Berhal, F., Royer, J. *JOC* **72**, 7893 (2007).
[4]Signore, G., Calderisi, M., Malanga, C., Menicagli, R. *T* **63**, 177 (2007).
[5]Tietze, L.F., Braun, H., Steck, P.L., El Bialy, S.A.A., Tölle, N., Düfert, A. *T* **63**, 6437 (2007).
[6]Dilger, A.K., Gopalsamuthiram, V., Burke, S.D. *JACS* **129**, 16273 (2007).
[7]Bejot, R., Anjaiah, S., Falck, J.R., Mioskowski, C. *EJOC* 101 (2007).
[8]Aureggi, V., Sedelmeier, G. *ACIE* **46**, 8440 (2007).
[9]Yanai, H., Takahashi, A., Taguchi, T. *TL* **48**, 2993 (2007).
[10]Iwata, S., Hamura, T., Matsumoto, T., Suzuki, K. *CL* **36**, 538 (2007).
[11]Kessabi, J., Beaudegnies, R., Jung, P.M.J., Martin, B., Montel, F., Wendeborn, S. *OL* **8**, 5629 (2006).

Organocerium reagents.

Tertiary amines from thioamides.[1] Alkyldichlorocerium reagents (alkyl group being primary) react with imino thioethers, which are formed by treatment of thioamides with MeOTf.

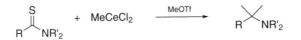

[1]Agosti, A., Britto, S., Renaud, P. *OL* **10**, 1417 (2008).

Organocopper reagents.

Arylcuprates.[1]   Mixed aryl(2-thienyl)cuprate reagents can be prepared from ArTeBu and Me(2-Th)Cu(CN)Li$_2$.

Additions.   Lithium diorganocuprates R$_2$CuLi (e.g., R = *n*-Bu) add to the double bond of cyclopropenyl carbinols.[2] The stereochemical course of the reaction is different from that employing RMgBr-CuI.

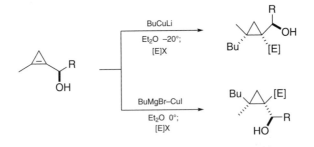

Benzaldimines are readily cuprated at the α-position of the *N*-alkyl substituent via lithiation and treatment with PhSCu. When an *o*-position of the benzene ring carries an alkynyl group, cyclization to give benzazepines occurs.[3]

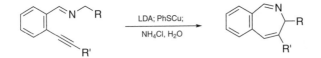

A new application of the conjugate addition of lithium diorganocuprates to enals is found in a synthesis of (-)-grandisol.[4]

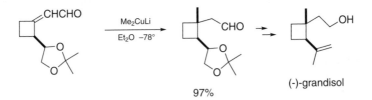

97%

(-)-grandisol

A synthesis of 3,4-disubstituted β-prolines from α-(allylaminomethyl)acrylic esters is initiated by conjugate addition.[5] Participation of the unactivated double bond at the second stage of the transformation is remarkable.

Cupration. Arylcopper reagents are formed with the Cu atom attached to a functional group (e.g., alkoxy, amide, cyano, . . .) of an arene, on treatment with $R(TMP)Cu(CN)Li_2$ derived from (**1**). Such copper reagents can engage in hydroxylation, phenylation, alkylation, acylation, silylation, and dimerization.[6]

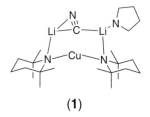

(1)

Substitution. Reaction of the diepoxides derived from allenes with organocuprates to afford α-hydroxy ketones is stereoselective.[7]

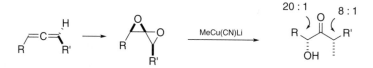

[1] Toledo, F., Cunha, R.L.O.R., Raminelli, C., Comasseto, J.V. *TL* **49**, 873 (2008).
[2] Simaan, S., Marek, I. *OL* **9**, 2569 (2007).
[3] Lyaskovskyy, V., Bergander, K., Fröhlich, R., Würthwein, E.-U. *OL* **9**, 1049 (2007).
[4] Bernard, A.M., Frongia, A., Ollivier, J., Piras, P.P., Secci, F., Spiga, M. *T* **63**, 4968 (2007).
[5] Denes, F., Perez-Luna, A., Chemla, F. *JOC* **72**, 398 (2007).
[6] Usui, S., Hashimoto, Y., Morey, J.V., Wheatley, A.E.H., Uchiyama, M. *JACS* **129**, 15102 (2007).
[7] Ghosh, P., Lotesta, S.D., Williams, L.J. *JACS* **129**, 2438 (2007).

Organogallium reagents.

Condensation reactions. 1-Alkynes combine with aldehydes to afford propargylic alcohols. The condensation can be mediated by Me_3Ga in CH_2Cl_2 at room temperature.[1]

Wittig reaction of propargylphosphonium salts with aldehydes is (Z)-selective using Me₃Ga as base. From the analogous sulfonium salts, alkynyl epoxides are obtained (base: t-Bu₃Ga).[2]

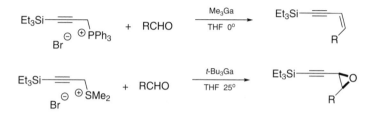

[1]Jia, X., Yang, H., Fang, L., Zhu, C. *TL* **49**, 1370 (2008).
[2]Nishimura, Y., Shiraishi, T., Yamaguchi, M. *TL* **49**, 3492 (2008).

Organoindium reagents.

Cyclization. Organoindium reagents participate in a Pd-catalyzed reaction of an allylic ester that contains a triple bond at some distance by addition, leading to a cyclic compound (with a 5- or 6-membered ring).[1]

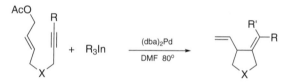

[1]Metza J.T., Jr, Terzian, R.A., Minehan, T. *TL* **47**, 8905 (2006).

Organolithium reagents.

2-Arylethanols. A straightforward preparation of ArCH₂CH₂OH consists of mixing ArLi with ethylene sulfate.[1]

Alkenylsilanes. Bis(trimethylsilyl)methyllithium is formed by Cl/Li exchange from reaction of (Me₃Si)₂CHCl with s-BuLi. A Peterson reaction leads to (E)-alkenylsilanes.[2]

Nucleophilic reactions. 2-Lithio-1,3-dithianes are found to add to styrenes and stilbenes.[3] Access to mononuclear Fischer carbene complexes is realized by the reaction of organolithiums to the CO ligand followed by O-alkylation.[4]

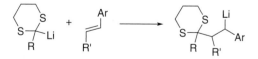

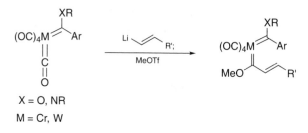

X = O, NR

M = Cr, W

Based on the reaction of chiral α-lithiated carbamates it is convenient to prepare alcohols with two contiguous stereocenters from secondary boronic esters.[5]

Difluoromethyl phenyl sulfone is lithiated by LiHMDS at −78°, and the lithio species attacks cyclic sulfates of 1,2-diols and 1,2-amino alcohols to afford α,α-difluoroalkyl sulfones.[6] The benzenesulfonyl group can be reductively removed (Mg, HOAc, NaOAc) or eliminated to provide 1,1-difluoroalkenes with an allylic OH or NH_2 group.

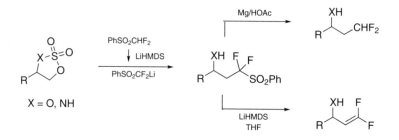

X = O, NH

Ketone synthesis.   The reaction of 1-chloroalkenyl *p*-tolyl sulfoxides with lithio-acetonitrile leads to 2-amino-1-cyanocyclopentadienes, which on acid hydrolysis give 2-cyclopentenones. The method has been employed in a formal synthesis of acorone.[7]

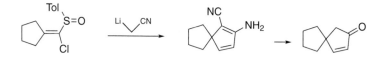

Alkynones are prepared from esters by treatment with lithioalkynes and lithium morpho-linide in the presence of $BF_3 \cdot OEt_2$ at −78°. More rapid aminolysis of the esters precedes an attack of the alkynides.[8]

Pentamethylcyclopentadienyllithium attacks ArCOCl to give Cp*COAr. On further reaction with allyldimethylaluminum, tertiary alcohols are formed. Thermolysis in toluene generates allyl aryl ketones.[9]

The elegant application of the reaction between an α-sulfonyllithium reagent and acyl-silane to afford an enol silyl ether is described in a synthetic route to limonoids.[10] Brook rearrangement following the nucleophilic addition of the organolithium protects the oxy functionality.

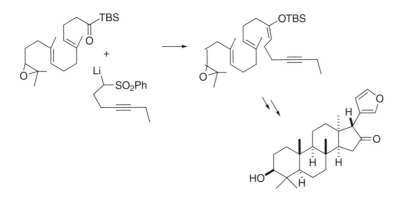

Reaction with 1,3-dithianes.   The special reactivity of 2-thiomethyl-1,3-dithianes has been exploited. The thio group is eliminated to provide 2-methylene-1,3-dithiane which is susceptible to attack by organolithium reagents. Trapping of the dimeric dithianyllithium species leads to precursors of various 1,3-diketones.[11]

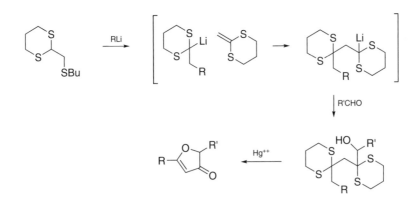

Addition reactions.   Organolithium reagents add to α,β-unsaturated amides. If the reagents are mixed with *t*-BuOK the regiochemical sense is reversed (to the α-carbon).[12]

Organolithium RLi and Grignard reagent R′MgX mixtures add to thioformamides to give amines RR′CHNR″.[2] The metal sulfide LiSMgX created from such a mixture is a thiolating agent, for example, for converting RCOCl to RCOSH.[13]

A controlled opening of α,β-epoxy ketones involves reaction with Ph(Me)$_2$SiLi and mild hydrolysis. Attack of the silyllithium reagent on the ketone groups is followed by a Brook rearrangement and β-elimination.[14]

[1]Schläger, T., Oberdorf, C., Tewes, B., Wünsch, B. *S* 1793 (2008).
[2]McNulty, J., Das, P. *CC* 1244 (2008).
[3]Tang, S., Han, J., He, J., Zheng, J., He, Y., Pan, X., She, X. *TL* **49**, 1348 (2008).

[4]Barluenga, J., Trabano, A.A., Perez-Sanchez, I., De la Campa, R., Florez, J., Garcia-Granda, S., Aguirre, A. *CEJ* **14**, 5401 (2008).

[5]Stymiest, J.L., Dutheuil, G., Mahmood, A., Aggarwal, V.K. *ACIE* **46**, 7491 (2007).

[6]Ni, C., Liu, J., Zhang, L., Hu, J. *ACIE* **46**, 786 (2007).

[7]Satoh, T., Kawashima, T., Takahashi, S., Sakai, K. *T* **59**, 9599 (2003).

[8]Yim, S.J., Kwon, C.H., An, D.K. *TL* **48**, 5393 (2007).

[9]Iwasaki, M., Morita, E., Uemura, M., Yorimitsu, H., Oshima, K. *SL* 167 (2007).

[10]Behenna, D.C., Corey, E.J. *JACS* **130**, 6720 (2008).

[11]Valiulin, R.A., Halliburton, L.M., Kutateladze, A.G. *OL* **9**, 4061 (2007).

[12]Hinago, T., Teshima, N., Kenmoku, S., Kamata, T., Terauchi, N., Chiba, N., Satoh, C., Nakamura, A., Asaoka, M. *CL* **36**, 54 (2007).

[13]Murai, T., Asai, F. *JACS* **129**, 780 (2007).

[14]Reynolds, S.C., Wengryniuk, S.E., Hartel, A.M. *TL* **48**, 6751 (2007).

Organomagnesium reagents.

Deprotonation. Enolization of ketones with t-Bu$_2$Mg and treatment of the enolates with Me$_3$SiCl–LiCl in THF at $0°$ afford silyl enol ethers.[1]

[1]Kerr, W.J., Watson, A.J.B., Hayes, D. *SL* 1386 (2008).

Organozinc reagents.

Zincation. Sensitive arenes and heteroarenes undergo zincation using as base zinc bis(2,2,6,6-tetramethylpiperidide), [complexed with MgCl$_2$], which is prepared from the corresponding chloromagnesium amide and ZnCl$_2$ in THF.[1] Zincation at C-2 of benzofurans, benzothiophenes, N-Boc indoles and the like (to generate the corresponding diarylzincs) is also achieved via lithiation with LiTMP and treatment with (tmeda)ZnCl$_2$.[2]

Lithium dialkyl(2,2,6,6-tetramethylpiperidino)zincates are readily formed from R$_2$Zn and lithium 2,2,6,6-tetramethylpiperidide. The alkyl group in such reagents is critical to their utility in zincation of haloarenes. Rapid elimination occurs after reaction with the dimethylzincate, but o-bromaryldi-t-butylzincates persist and they can used to react with electrophiles.[3]

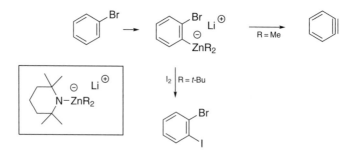

2,3-Disubstituted succinic esters.[4] Substituted succinic esters differentiated at the two termini and alkyl substituents (one being an allyl group) are readily prepared from diallyl fumarate. Copper-catalyzed conjugate addition of R_2Zn in the presence of Me_3SiCl leads to silyl ketene acetals in which the other O-substituent is allyl. On warming, an Ireland–Claisen rearrangement occurs, furnishing succinic esters.

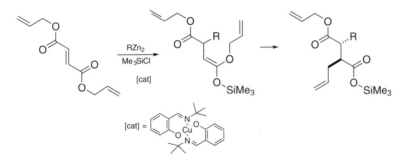

Alkenylzinc reagents. Hydroboration of pinacolatoborylalkynes with dicyclohexylborane affords 1,1-diboryl-1-alkenes. The dicyclohexylboryl group is selectively exchanged on treatment with Me_2Zn, and the resulting species show differentiated chemoselectivity such that homologation/functionalization proceed in a desired manner.[5]

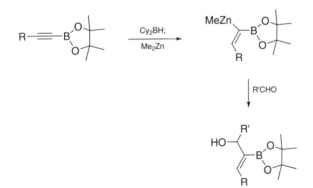

1-Bromo-1-dibromoborylalkenes are useful for the synthesis of trisubstituted alkenes. Reaction with R_2Zn (as demonstrated by Me_2Zn) affords alkenylzinc species which can be converted to the corresponding alkenyl iodides for further coupling.[6]

Allenylarenes undergo lithiation at a far sp^2-terminus of the diene system on exposure to LDA. Negishi coupling enables transforming the parent compounds into 1,3-diarylallenes upon conversion of the lithio derivatives to zincio species.[7]

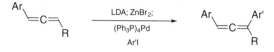

Addition reactions.   Allylzinc reagents add to *t*-butanesulfinylimines in a straight-forward fashion.[8] Interestingly, the products can be different from those using the corresponding Grignard reagents.

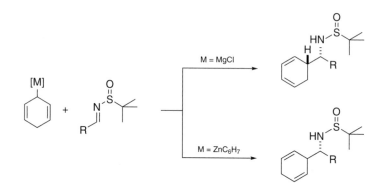

For hydroamination a co-catalyst system contains {PhNMe$_2$H[B(C$_6$F$_5$)$_4$]} and **1**, which is made from Me$_2$Zn.[9] Hydrosilylation of terminal alkenes employing *t*-butylzinciosilanes and lithium biphenyl-2,2'-dioxide in the presence of CuCN proceeds nicely.[10]

(1)

Reformatsky reagents are found to perform conjugate addition to alkylidenemalonic esters.[11]

Allylic substitution.[12]   Catalyzed by CuCl, the reaction of RZnBr (R = aryl, alkenyl) with *cis*-4-cyclopentene-1,3-diol monoesters proceeds regioselectively along an S$_N$2' pathway to provide *trans*-2-substituted 3-cyclopentenols.

Homologation.[13]   Ethyl(iodomethyl)zinc contributes a methylene group while effectuating a debrominative dimerization of bromomethyl ketones. Formation of the reaction products involves a rearrangement step.

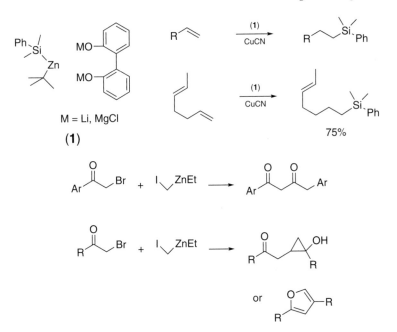

(1)

75%

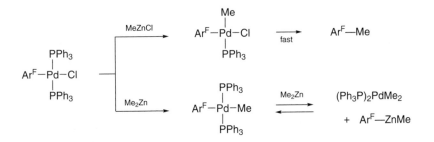

or

Negishi coupling.[14]   Differences in reactivity on using R_2Zn and $RZnX$ are due to formation of isomeric four-coordinate Pd complexes.

[1]Wunderlich, S.H., Knochel, P. *ACIE* **46**, 7685 (2007).
[2]L'Helgonal'ch, J.-M., Seggio, A., Chevallier, F., Yonehara, M., Jeanneau, E., Uchiyama, M., Mongin, F. *JOC* **73**, 177 (2008).
[3]Uchiyama, M., Kobayashi, Y., Furuyama, T., Nakamura, S., Kajihara, Y., Miyoshi, T., Sakamoto, T., Kondo, Y., Morokuma, K. *JACS* **130**, 472 (2008).
[4]Bausch, C.C., Johnson, J.S. *JOC* **73**, 1575 (2008).
[5]Li, H., Carroll, P.J., Walsh, P.J. *JACS* **130**, 3521 (2008).
[6]Huang, Z., Negishi, E. *JACS* **129**, 14788 (2007).

[7]Zhao, J., Liu, Y., Ma, S. *OL* **10**, 1521 (2008).
[8]Maji, M.S., Fröhlich, R., Studer, A. *OL* **10**, 1847 (2008).
[9]Dohlnahl, M., Löhnwitz, K., Pissarek, J.-W., Biyikal, M., Schulz, S.R., Schön, S., Meyer, N., Roesky, P.W., Blechert, S. *CEJ* **13**, 6654 (2007).
[10]Nakamura, S., Uchiyama, M. *JACS* **129**, 28 (2007).
[11]Benz, E., Moloney, M.G., Westaway, S.M. *SL* 733 (2007).
[12]Nakata, K., Kiyotsuka, Y., Kitazume, T., Kobayashi, Y. *OL* **10**, 1345 (2008).
[13]Li, L., Cai, P., Xu, D., Guo, Q., Xue, S. *JOC* **72**, 8131 (2007).
[14]Casares, J.A., Espinet, P., Fuentes, B., Salas, G. *JACS* **129**, 3508 (2007).

Osmium tetroxide.

Osmylation.   A review of recent works on osmylation of alkenes has been published.[1]

[1]Francais, A., Bedel, O., Haudrechy, A. *T* **64**, 2495 (2008).

Oxalic acid.

Cleavage of dithioacetals.   Use of oxalic acid in $MeNO_2$ to regenerate carbonyl compounds represents a mild method.[1]

[1]Miyake, H., Nakao, Y., Sasaki, M. *CL* **36**, 104 (2007).

1-Oxo-4-acetamino-2,2,6,6-tetramethylpiperidinium tetrafluoroborate.

Allylic oxidation.   An ene-type reaction occurs when alkenes are treated with the title reagent in MeCN at room temperature.[1]

[1]Pradhan, P.P., Bobbitt, J.M., Bailey, W.F. *OL* **8**, 5485 (2006).

Oxygen.

Dehydrogenation.   2-Arylimidazoles are formed on heating the 4,5-dihydro derivatives with activated carbon in xylene under oxygen.[1]

Oxidations   Metal-catalyzed aerobic oxidation of organic compounds has been reviewed.[2] Aerial oxidation of primary and secondary alcohols is mediated by TEMPO in the presence of HCl and $NaNO_2$.[3] Secondary benzylic alcohols undergo aerial oxidation (or with *t*-BuOOH) based on catalysis by AuCl – neocuproine,[4] but another report describes the oxidation of both primary and secondary alcohols (to acids and ketones, respectively) using nanoclusters of gold that are stabilized by poly(*N*-vinyl-2-pyrrolidone).[5]

Allylic oxidation, for example, of cyclohexene to 2-cyclohexenone, and oxidative cleavage of styrene to benzaldehyde are readily accomplished with oxygen; such reaction systems contain *N*-hydroxyphthalimide and 1,4-diamino-2,3-dichloro-9,10-anthraquinone.[6]

Aldehydes are converted into carboxylic acids with Pd/C, KOH and catalytic amounts of $NaBH_4$ in the air.[7] Very similar conditions (K_2CO_3 instead of KOH) are described for oxidation of benzylic and allylic alcohols.[8]

Arylacetamides undergo aerial oxidation to yield the corresponding α-keto amides without the need of a transition metal salt. The transformation is carried out in the presence of a base (Cs$_2$CO$_3$) and Bu$_4$NBr.[9] Primary amines are converted into oximes in an aerobic oxidation employing 1,1-diphenyl-2-picrylhydrazyl (1) and WO$_3$/Al$_2$O$_3$ as catalyst.[10]

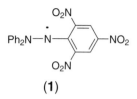

(1)

Perhaps of more practical value is the oxygenation of alkanes, and a reaction catalyzed by VO(acac)$_2$ is notable.[11]

Oxidative coupling.   *N*-Alkenylation of carboxamides to acquire enamides is accomplished by potassium alkenyltrifluoroborates. It involves treatment of the reaction components with Cu(OAc)$_2$ under oxygen in the presence of 4A-molecular sieves.[12]

Certain ynamides are obtained from 1-alkynes and amine derivatives (e.g., 3-acylindoles, imidazolidinones, oxazolidinones, and sulfonamides) by coupling under oxygen, using the CuCl$_2$-pyridine catalyst.[13]

Aldehydes and isonitriles undergo hydrative condensation. Depending on the hydrophilicity of the isonitriles, either α-hydroxycarboxamides or α-acyloxycarboxamides are formed.[14]

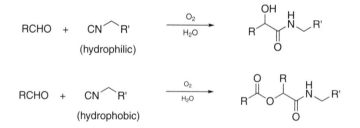

o-Arylation of anilides[15] proceeds on heating with arenes and Pd(OAc)$_2$, DMSO, and CF$_3$COOH under oxygen (1 atm.) at 100°. Twofold C—H activation is involved in the coupling.

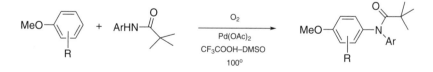

Dimerization of 1-alkynes is accomplished with 1% Pd/C, CuI in DMSO under oxygen at room temperature.[16] This protocol is base-free and ligand-free. Homocoupling of Grignard reagents also takes place in dry air, with some $MnCl_2$ or $FeCl_3$ as catalyst.[17,18]

Oxidative cross-coupling of 2-naphthols is catalyzed by Cu(OH)Cl-TMEDA and assisted by Yb(OTf)$_3$. Under such conditions BINOL containing a single methyl ester at C-3 is formed with >99% selectivity.[19]

Short chain ω-alkynamides undergo Wacker oxidation and coupling with 1-alkenes in tandem, on treatment with Na_2PdCl_4, $CuCl_2$ in $MeCN-H_2O$ (5 : 1) under O_2.[20]

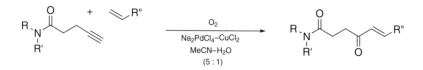

Addition reactions. Under oxygen *B*-propylcatecholborane catalyzes anti-Markovnikov hydrophosphorylation of 1-alkenes.[21]

[1]Haneda, S., Okui, A., Ueba, C., Hayashi, M. *T* **63**, 2414 (2007).
[2]Piera, J., Bäckvall, J.-E. *ACIE* **47**, 3506 (2008).
[3]Wang, X., Liu, R., Jin, Y., Liang, X. *CEJ* **14**, 2679 (2008).
[4]Li, H., Guan, B., Wang, W., Xing, D., Fang, Z., Wan, X., Yang, L., Shi, Z. *T* **63**, 8430 (2007).
[5]Tsunoyama, H., Tsukuda, T., Sakurai, H. *CL* **36**, 212 (2007).
[6]Tong, X., Xu, J., Miao, H., Yang, G., Ma, H., Zhang, Q. *T* **63**, 7634 (2007).
[7]Lim, M., Yoon, C.M., An, G., Rhee, H. *TL* **48**, 3835 (2007).
[8]An, G., Lim, M., Chun, K.-S., Rhee, H. *SL* 95 (2007).
[9]Song, B., Wang, S., Sun, C., Deng, H., Xu, B. *TL* **48**, 8982 (2007).
[10]Suzuki, K., Watanabe, T., Murahashi, S.-I. *ACIE* **47**, 2079 (2008).
[11]Kobayashi, H., Yamanaka, I. *CL* **36**, 114 (2007).
[12]Bolshan, Y., Batey, R.A. *ACIE* **47**, 2109 (2008).
[13]Hamada, T., Ye, X., Stahl, S.S. *JACS* **130**, 833 (2008).
[14]Shapiro, N., Vigalok, A. *ACIE* **47**, 2849 (2008).
[15]Brasche, G., Garcia-Fortanet, J., Buchwald, S.L. *OL* **10**, 2207 (2008).
[16]Kurita, T., Abe, M., Maegawa, T., Monguchi, Y., Sajiki, H. *SL* 2521 (2007).
[17]Cahiez, G., Moyeux, A., Buendia, J., Duplais, C. *JACS* **129**, 13788 (2007).
[18]Liu, W., Lei, A. *TL* **49**, 610 (2008).
[19]Yan, P., Sugiyama, Y., Takahashi, Y., Kinemuchi, H., Temma, T., Habaue, S. *T* **64**, 4325 (2008).
[20]Momiyama, N., Kanan, M.W., Liu, D.R. *JACS* **129**, 2230 (2007).
[21]Montgomery, I., Parsons, A.F., Ghelfi, F., Roncaglia, F. *TL* **49**, 628 (2008).

Oxygen, singlet.

Oxygenation. Photooxidation of primary alcohols to RCOOH and ArMe to ArCOOH is accomplished in EtOAc, in the presence of Ph_3P and CBr_4.[1] The light source of the reaction is a fluorescent lamp.

A new electron-transfer mediator for photooxygenation of alkenes (to give allylic hydro-peroxides) and methylarenes (to give aromatic aldehydes) is 9-mesityl-10-methylacridinium perchlorate (**1**).[2]

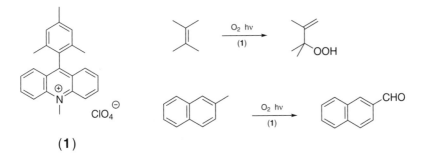

(**1**)

N-Benzyl carboxamides undergo photooxidation to deliver aroylimides, which is cata-lyzed by iodine.[3] 2-Allyl-5-silylfurans are converted into spiroperoxy lactones on treatment with singlet oxygen and then an acid.[4]

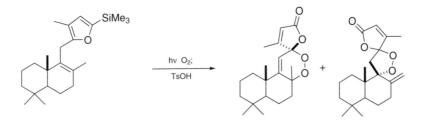

Degradation. α-Substituted mandelic acids are oxidatively degraded to aryl ketones by singlet oxygen in the presence of iodine. Unsubstituted analogs give aroic acids.[5] Actually, many alkylarenes are converted into aroic acids by singlet oxygen using allyl bromide as catalyst, except those carrying nitro group(s) in the aromatic nucleus.[6]

[1]Sugai, T., Itoh, A. *TL* **48**, 9096 (2007).
[2]Griesbeck, A.G., Cho, M. *OL* **9**, 611 (2007).
[3]Nakayama, H., Itoh, A. *SL* 675 (2008).
[4]Margaros, I., Montagnon, T., Vassilikogiannakis, G. *OL* **9**, 5585 (2007).
[5]Nakayama, H., Itoh, A. *TL* **49**, 2792 (2008).
[6]Sugai, T., Itoh, A. *TL* **48**, 2931 (2007).

Ozone.

Cleavage of alkenes.[1]   Ozonization of alkenes in aqueous acetone (H_2O : Me_2CO = 5:95) at $0°$ gives carbonyl products directly, additional reagents for decomposition of ozonides are not needed.

Unsaturated organotrifluoroborate salts are cleaved at the CC multiple bond to yield carbonyl compounds. The trifluoroborate group is resistant to attack by ozone.[2]

Ozonolysis of 1-triorganosilyl-1-alken-3-ols afford α-formyl-β-hydroxy silyl peroxides.[3]

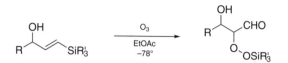

[1]Schiaffo, C.E., Dussault, P.H. *JOC* **73**, 4688 (2008).
[2]Molander, G.A., Cooper, D.J. *JOC* **72**, 3558 (2007).
[3]Igawa, K., Sakita, K., Murakami, M., Tomooka, K. *S* 1641 (2008).

P

Palladacycles.

Coupling.   In the synthesis of ArCH=CH$_2$ and ArCH=CHAr' from ethylene by the Heck-reaction both steps can be catalyzed by palladacycle **1**.[1]

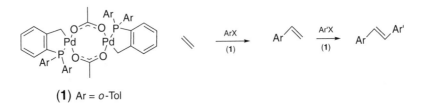

(1) Ar = o-Tol

Pincer complexes **2** are active catalysts for the Heck reaction. Interestingly, their catalytic activities are controlled by the central metal atom of the porphyrin nucleus.[2]

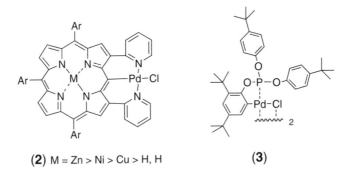

(2) M = Zn > Ni > Cu > H, H   **(3)**

In a preparation of ArCN from ArX and K$_4$Fe(CN)$_6$, the employment of a palladacycle[3] as a catalyst under rather drastic conditions (NMP, 140°) is perhaps of questionable value.

Coupling reactions catalyzed by Pd complexes of the Se-C-Se pincer type are accelerated when the *p*-position to the metal atom is substituted by an electron donor (e.g., MeO group).[4]

Allylation.   In the addition of allylstannanes to aldehydes and *N*-tosylimines, **3** and **4** act in the promoter role, respectively.[5,6] Noteworthy is that the pincer complex **4** features two different but synergistic donor groups.

Fiesers' Reagents for Organic Synthesis, Volume 25. By Tse-Lok Ho
Copyright © 2010 John Wiley & Sons, Inc.

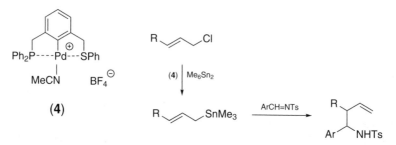

(4)

The S,S-pincer **5**, cocatalyzed by TsOH, converts allylic alcohols to allylboronic acids for direct allylation of aldehydes.[7]

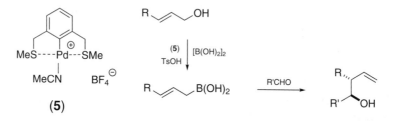

(5)

[1] Kormos, C.M., Leadbeater, N.E. *JOC* **73**, 3854 (2008).
[2] Katoh, T., Shinokubo, H., Osuka, A. *JACS* **129**, 6392 (2007).
[3] Cheng, Y., Duan, Z., Li, T., Wu, Y. *SL* 543 (2007).
[4] Aydin, J., Selander, N., Szabo, K.J. *TL* **47**, 8999 (2006).
[5] Bedford, R.B., Pilarski, L.T. *TL* **49**, 4216 (2008).
[6] Gagliardo, M., Selander, N., Mehendale, N.C., van Koten, G., Gebbink, R.J.M.K., Szabo, K.J. *CEJ* **14**, 4800 (2008).
[7] Selander, N., Kipke, A., Sebelius, S., Szabo, K.J. *JACS* **129**, 13723 (2007).

Palladium.

Hydrogenation.   The polyethyleneimine complex of Pd is useful for partial hydrogenation of alkynes.[1] Selective reduction of alkenes under transfer hydrogenation conditions (hydrogen source: HCOOH) is accomplished with Pd and t-Bu$_3$P.[2] Numerous substrates including styrene, stilbene, allylic alcohols, enals, enones, enoic acids, conjugated nitriles are susceptible to reduction, although the pinenes do not react.

A recyclable system of Pd nanoparticulates in water for hydrogenation of alkenes such as conjugated carbonyl compounds, esters, nitriles, allylic alcohols and ethers, styrenes, has been developed.[3]

Coupling reactions.   Pd nanoparticles supported on polyaniline fibers are catalytically active for Suzuki coupling. 2-Chlorobiaryls are obtainable from a reaction of ArB(OH)$_2$ with 1,2-dichlorobenzene.[4] The products can be used to prepare 2-hydroxybiaryls. There are many other types of Pd nanoparticles on solid supports, the one version (from reduction of Na$_2$PdCl$_4$ with hydrazine) deposited on NiFe$_2$O$_4$, proved active in Heck and Suzuki coupling, has the advantage of magnetic recovery.[5]

It is possible to synthesize diarylamines or triarylamines from ArBr by reaction with BnONH$_2$. The catalyst system contains Pd, t-Bu$_3$P, and t-BuONa.[6]

[1]Sajiki, H., Mori, S., Ohkubo, T., Ikawa, T., Kume, A., Maegawa, T., Monguchi, Y. *CEJ* **14**, 5109 (2008).
[2]Brunel, J.M. *T* **63**, 3899 (2007).
[3]Callis, N.M., Thiery, E., Le Bras, J., Muzart, J. *TL* **48**, 8128 (2007).
[4]Gallon, B.J., Kojima, R.W., Kaner, R.B., Diaconescu, P.L. *ACIE* **46**, 7251 (2007).
[5]Baruwati, B., Guin, D., Manorama, S.V. *OL* **9**, 5377 (2007).
[6]Bedford, R.B., Betham, M. *TL* **48**, 8947 (2007).

Palladium/alumina.

Suzuki coupling.[1]   The catalyst is prepared from impregnating Pd(OAc)$_2$ in alumina and calcined. With KF in EtOH, Suzuki coupling is conducted.

[1]Kudo, D., Masui, Y., Onaka, M. *CL* **36**, 918 (2007).

Palladium/calcium carbonate.

Stille coupling.[1]   With this catalyst reservoir the Stille coupling is performed with ligand-free Pd in aq. EtOH.

[1]Coelho, A.V., de Souza, A.L.F., de Lima, P.G., Wardell, J.L., Antunes, O.A.C. *TL* **48**, 7671 (2007).

Palladium/carbon.

Deprotection.   Aryl ketones are recovered from 2-aryl-1,3-dithianes on heating with Pd/C and the resin Amberlite IR-120 in MeOH.[1] The presence of the aryl group is critical.

Hydrogenation.   Transfer hydrogenation (Pd/C, HCOONH$_4$) affects alkenylfurans and benzyloxy substituents. Enones are reduced to saturated alcohols.[2]

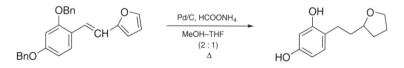

The formyl group of activated aromatic aldehydes such as p-anisaldehyde is deoxygenated on hydrogenation. The hydrogenolysis is promoted by HCl.[3]

From nitroarenes hydrogenation in the presence of aldehydes leads to N-monoalkyl arylamines.[4] Hydrogenation of 2-(o-nitroaryl)-2-cycloalkenones gives annulated indoles (e.g., tetrahydrocarbazoles).[5]

Reduction with nascent hydrogen generated from Et$_3$SiH has been reported.[6]

Coupling reactions. Studies of coupling reactions that are catalyzed by Pd/C continue. Ligand-free Pd-catalyzed Suzuki coupling is an obvious advantage, and Pd/C is a reusable catalyst showing no leaching of the metal into solution (i.e., <1 ppm).[7,8] Another report regarding Suzuki coupling with ArCl with assistance of a X-Phos ligand.[9] In the aerogel form, Pd/C is also capable of catalyzing the Sonogashira coupling.[10]

N,N-Dimethylbenzamides are obtained from ArI, DMF on heating with Pd/C and POCl$_3$ at 140°.[11] For transformation of ArBr into ArCN in NMP, the cyanide source is K$_4$Fe(CN)$_6$ and a catalyst-additive system is made from Pd/C (1 mol%), Na$_2$CO$_3$, and Bu$_3$N.[12]

[1] Wang, E.-C., Wu, C.-H., Chien, S.-C., Chiang, W.-C., Kuo, Y.-H. *TL* **48**, 7706 (2007).
[2] Nandy, S.K., Liu, J., Padmapriya, A.A. *TL* **48**, 2469 (2007).
[3] Xing, L., Wang, X., Cheng, C., Zhu, R., Liu, B., Hu, Y. *T* **63**, 9382 (2007).
[4] Sydnes, M.O., Isobe, M. *TL* **49**, 1199 (2008).
[5] Scott, T.L., Burke, N., Carrero-Martínez, G., Söderberg, B.C.G. *T* **63**, 1183 (2007).
[6] Mandal, P.K., McMurray, J.S. *JOC* **72**, 6599 (2007).
[7] Maegawa, T., Kitamura, Y., Sako, S., Udzu, T., Sakurai, A., Tanaka, A., Kobayashi, Y., Endo, K., Bora, U., Kurita, T., Kozaki, A., Monguchi, Y., Sajiki, H. *CEJ* **13**, 5937 (2007).
[8] Kitamura, Y., Sakurai, A., Udzu, T., Maegawa, T., Monguchi, Y., Sajiki, H. *T* **63**, 10596 (2007).
[9] Simeone, J.P., Sowa Jr, J.R. *T* **63**, 12646 (2007).
[10] Soler, R., Cacchi, S., Fabrizi, G., Forte, G., Martin, L., Martinez, S., Molins, E., Moreno-Manas, M., Petrucci, F., Roig, A., Sebastian, R.M., Vallribera, A. *S* 3068 (2007).
[11] Tambade, P.J., Patil, Y.P., Bhanushali, M.J., Bhanage, B.M. *TL* **49**, 2221 (2008).
[12] Zhu, Y.-Z., Cai, C. *EJOC* 2401 (2007).

Palladium(II) acetate.

Coupling reactions. Various coupling reactions involving nonpolar substrates may be carried out with hydrophobicized silica sol-gel matrix in which Pd(OAc)$_2$ is trapped. The substrates form microemulsions in water with SDS and ROH. The catalyst is leach-proof and reusable.[1]

Heck reaction with ArB(OH)$_2$ is achieved in the air under mild conditions when 2,9-dimethyl-1,10-phenanthroline is present.[2] A carboxylate ion directs palladation of a C–H bond for coupling with organoboronates, and the reaction is useful for attaching a carbon fragment to an o-position of aroic acids and homologation at the methyl group of t-alkanoic acids.[3]

Suzuki coupling catalyzed by Pd(OAc)$_2$ and assisted by (Me$_2$N)$_2$C=NBu is very efficient. In aqueous solvent at room temperature a typical reaction of PhB(OH)$_2$ and 4-O$_2$NC$_6$H$_4$I shows TON of up to 850,000.[4] Also reported are methods employing cryptand-22,[5] N-phenylurea,[6] and a polymer N-linked to DABCO.[7]

A ligand-free Suzuki coupling protocol indicates employment of Pd(OAc)$_2$ in PEG-400, in which nanoparticles of Pd are generated in situ.[8] More conventionally, NaOMe is used as a base for coupling at room temperature.[9] Under certain coupling reaction conditions reduction of nitro group(s) also occurs.[10]

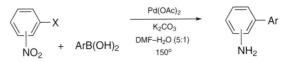

Under oxygen and in the presence of Pd(OAc)$_2$ organoboronic acids/esters couple with electron-deficient alkenes, no base is needed.[11]

A conjugated polyene chain can be assembled stereoselectively by iterative Suzuki coupling. The strategy is based on using a chelated boronato group to moderate reactivity.[12]

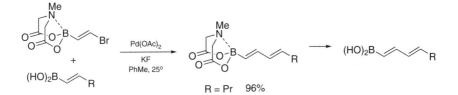

Indoles are synthesized from *o*-iodoaroic acids, via Curtius rearrangement and subsequent coupling reactions of the aryl isocyanates.[13]

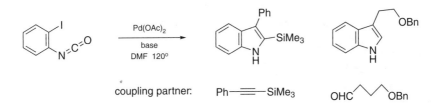

There is no need to protect the N—H of indoles and pyrroles during coupling (at C-2) with ArI, while employing Pd(OAc)$_2$ as catalyst and a mild base of CsOAc in DMA at 125°.[14] A rather unusual catalyst system for Heck and Suzuki couplings constitutes a salen complex that is formed by adding Pd(OAc)$_2$ to the ligand,[15] and for Heck reaction in water, sodium 2-[(pyrid-3-yl)ethyl]aminoethanesulfonate serves as a base and ligand.[16]

Intramolecular coupling involving a Fischer carbene unit is particularly interesting. While both quinoline and indole derivatives are produced in the reaction, it is arrested upon change of the catalyst to Pd(MeCN)$_2$Cl$_2$.[17]

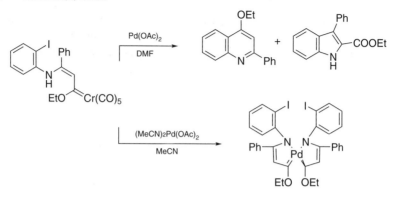

In the formation of indoles from *o*-haloarylamines and alkynes, regioselectivity is attained to some degrees. The larger substituent of the alkyne appears at C-2 of the product.[18]

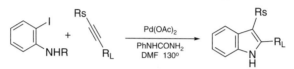

When ArBr, an alkyne, and $K_4Fe(CN)_6$ are heated with Pd(OAc)$_2$ in DMA, *cis*-arylcyanation of the alkyne occurs. Substituted (*Z*)-cinnamonitriles are obtained in moderate to good yields.[19]

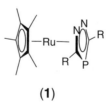

Air-stable hexaarylcyclotrisiloxanes, which are generated from Ar$_2$Si(OH)$_2$ on contact with TsOH, are activated by KOH to undergo cross-coupling with ArX.[20] Hydrosilanes can be used to couple with electron-deficient ArI at room temperature. Additives to complement Pd(OAc)$_2$ are LiCl and pyridine.[21] *cis*-Hydroarylation by ArN$_2$BF$_4$ and Ph$_3$SiH proceeds at room temperature.[22]

A rather unusual ligand is the ruthenocene **1** that is used in Heck reaction catalyzed by Pd(OAc)$_2$.[23]

(1)

In the Pd-catalyzed arylation of unactivated arenes such as benzene with ArBr, pivalic acid is a key cocatalyst.[24]

Cyclization. Two different modes of cyclization are available to 2,2-difluoro-3-alkynamides. The 4-exo-dig mode is favored by the Pd-catalysis, while 5-endo-dig cyclization occurs in the presence of TBAF.[25]

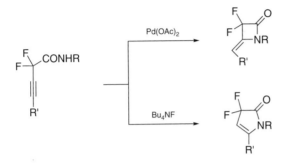

Halogenation.[26] Aroic acids undergo *o*-iodination by IOAc in the presence of Pd(OAc)$_2$. Thus 2,6-diiodobenzoic acid is produced. However, if R$_4$NX (X = Br, I) is added the anion is rapidly oxidized and it becomes the electrophile.

[1]Tsvelikhovsky, D., Blum, J. *EJOC* 2417 (2008).

[2]Lindh, J., Enquist, P.-A., Pilotti, A., Nilsson, P., Larhed, M. *JOC* **72**, 7957 (2007).

[3]Giri, R., Maugel, N., Li, J.-J., Wang, D.-H., Breazzano, S.P., Saunders, L.B., Yu, J.-Q. *JACS* **129**, 3510 (2007).

[4]Li, S., Lin, Y., Cao, J., Zhang, S. *JOC* **72**, 4067 (2007).

[5]Hsu, M.-H., Hsu, C.-M., Wang, J.-C., Sun, C.-H. *T* **64**, 4268 (2008).

[6]Cui, X., Zhou, Y., Wang, N., Liu, L., Guo, Q.-X. *TL* **48**, 163 (2007).

[7]Li, J.-H., Hu, X.-C., Xie, Y.-X. *TL* **47**, 9239 (2006).

[8]Han, W., Liu, C., Jin, Z.-L. *OL* **9**, 4005 (2007).

[9]Deng, C.-L., Guo, S.-M., Xie, Y.-X., Li, J.-H. *EJOC* 1457 (2007).

[10]Wang, H.-S., Wang, Y.-C., Pan, Y.-M., Zhao, S.-L., Chen, Z.-F. *TL* **49**, 2634 (2008).

[11]Yoo, K.S., Yoon, C.H., Jung, K.W. *JACS* **128**, 16384 (2006).

[12]Lee, S.J., Gray, K.C., Paek, J.S., Burks, M.D. *JACS* **130**, 466 (2008).

[13]Leogane, O., Lebel, H. *ACIE* **47**, 350 (2008).

[14]Wang, X., Gribkov, D.V., Sames, D. *JOC* **72**, 1476 (2007).

[15]Borhade, S.R., Waghmode, S.B. *TL* **49**, 3423 (2008).

[16]Pawar, S.S., Dekhane, D.V., Shingare, M.S., Thore, S.N. *TL* **49**, 4252 (2008).

[17]Lopez-Alberca, M.P., Mancheno, M.J., Fernandez, I., Gomez-Gallego, M., Sierra, M.A., Torrs, R. *OL* **9**, 1757 (2007).

[18]Cui, X., Li, J., Fu, Y., Liu, L., Guo, Q.-X. *TL* **49**, 3458 (2008).

[19]Cheng, Y., Duan, Z., Yu, L., Li, Z., Zhu, Y., Wu, Y. *OL* **10**, 901 (2008).

[20]Endo, M., Sakurai, T., Ojima, S., Katayama, T., Unno, M., Matsumoto, H., Kowase, S., Sano, H., Kosugi, M., Fugami, K. *SL* 749 (2007).

[21]Iizuka, M., Kondo, Y. *EJOC* 1161 (2008).

[22]Cacchi, S., Fabrizi, G., Goggiamani, A., Persiani, D. *OL* **10**, 1597 (2008).

[23]Yorke, J., Wan, L., Xia, A., Zhang, W. *TL* **48**, 8843 (2007).

[24]Lafrance, M., Fagnou, K. *JACS* **128**, 16496 (2006).

[25]Fustero, S., Fernandez, B., Bello, P., del Pozo, C., Arimitsu, S., Hammond, G.B. *OL* **9**, 4251 (2007).

[26]Mei, T.-S., Giri, R., Maugel, N., Yu, J.-Q. *ACIE*, **47**, 5215 (2008).

Palladium(II) acetate – imidazol-2-ylidene.

Coupling reactions. Using complexes in which Pd is coordinated with polymer-supported azolecarbene ligands Suzuki coupling between ArB(OH)$_2$ and Ar'N$_2$BF$_4$ or haloarenes has been studied.[1,2]

A ligand (1) containing three carbene units has been prepared and used in conjunction with Pd(OAc)$_2$ in the Heck reaction.[3]

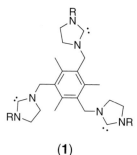

(1)

Aryl ketones and esters. A Pd-carbene complex is useful for synthesis of heteroaryl ketones under CO.[4] The coupling involving ArCHO in the air leads to aryl aroates.[5]

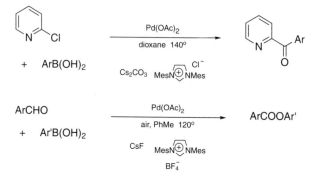

[1]Qin, Y., Wei, W., Luo, M. *SL* 2410 (2007).
[2]Lee, D.-H., Kim, J.-H., Jun, B.-H., Kang, H., Park, J., Lee, Y.-S. *OL* **10**, 1609 (2008).
[3]özdemir, I., Demir, S., Centinkaya, B. *SL* 889 (2007).
[4]Maerten, E., Sauthier, M., Mortreux, A., Castanet, Y. *T* **63**, 682 (2007).
[5]Qin, C., Wu, H., Chen, J., Liu, M., Cheng, J., Su, W., Ding, J. *OL* **10**, 1537 (2008).

Palladium(II) acetate – copper salts.

Cyclization. Intramolecular addition of a guanidino group to a double bond is promoted by Pd(OAc)$_2$ (10 mol%) and Cu(OAc)$_2$ (2.1 equiv.). Bicyclic products containing a bridgehead nitrogen atom are usually obtained in good yields.[1]

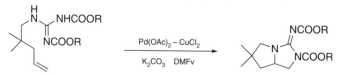

Coupling reactions. α-Arylation of cyclic enaminones (vinylogous lactams) is accomplished by coupling with ArBF$_3$K, but *N*-Boc derivatives fail to follow suit.[2]

A carboxyl group in the five-membered ring of benzannulated heteroles plays an *o*-directing role and its is detached at the end of the coupling reaction.[3]

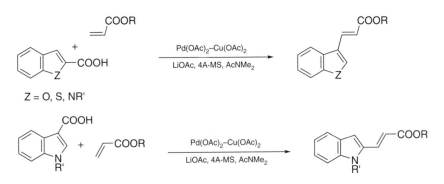

N-Acylindoles are arylated at either C-3 or C-2, by changing the auxiliary metal salt, i.e., from Cu(OAc)$_2$ to AgOAc.[4] Such phenomenon had been reported in the previous year using catalyst systems based on Pd(OCOCF$_3$)$_2$. [*loc. cit.*]

N-Acylated indolines and 1,2,3,4-tetrahydroquinolines free of *o*-substituent to the heteroatom are arylated by reaction with ArB(OH)$_2$. Besides Pd(OAc)$_2$, the catalyst system also contains equivalents of Cu(OTf)$_2$ and Ag$_2$O.[5] Similarly, a protocol for the Pd-catalyzed *o*-arylation of acetanilides employs Cu(OTf)$_2$ and AgF, with ArSi(OR)$_3$ as aryl group donors.[6]

Further variants of coupling conditions involving oxygen in a carboxylic acid solvent enable the use of ArH as reaction partners.[7,8] Cyclization of *N*-aroylindoles proceeds via double C—H activation.[9]

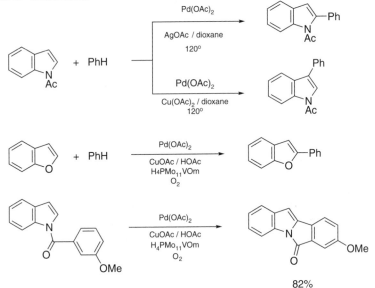

Oxidative amination carried out under improved catalyst reoxidation conditions permits the use of alkenes as limiting reagents.[10]

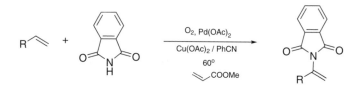

[1]Hövelmann, C.H., Streuff, J., Brelot, L., Muniz, K. *CC* 2334 (2008).
[2]Ge, H., Niphakis, M.J., Georg, G.I. *JACS* **130**, 3708 (2008).
[3]Maehara, A., Tsurugi, H., Satoh, T., Miura, M. *OL* **10**, 1159 (2008).
[4]Potavathri, S., Dumas, A.S., Dwight, T.A., Naumiec, G.R., Hammann, J.M., DeBoef, B. *TL* **49**, 4050 (2008).
[5]Shi, Z., Li, B., Wan, X., Cheng, J., Fang, Z., Cao, B., Qin, C., Wang, Y. *ACIE* **46**, 5554 (2007).
[6]Yang, S., Li, B., Wan, X., Shi, Z. *JACS* **129**, 6066 (2007).
[7]Li, B.-J., Tian, S.-L., Fang, Z., Shi, Z.-J. *ACIE* **47**, 1115 (2008).
[8]Yang, S.-D., Sun, C.-L., Fang, Z., Li, B.-J., Li, Y.-Z., Shi, Z.-J. *ACIE* **47**, 1473 (2008).
[9]Dwight, T.A., Rue, N.R., Charyk, D., Josselyn, R., DeBoef, B. *OL* **9**, 3137 (2007).
[10]Rogers, M.M., Kotov, V., Chatwichien, J., Stahl, S.S. *OL* **9**, 4331 (2007).

Palladium(II) acetate – oxidants.

Oxidative cleavage. Alkynes are cleaved to provide two ester fragments on reaction with Pd(OAc)$_2$, ZnCl$_2 \cdot$ 2H$_2$O under O$_2$ in an alcohol at 100°.[1] Aerobic oxidation of α-aminoarylacetamides leads to aroylformamides.[2]

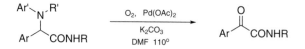

Allylic amination. 1-Alkenes can be functionalized with double bond migration that results in derivatives of 1-amino-2-alkenes, when they are treated with TsNHCOOR and catalytic amounts of Pd(OAc)$_2$ under O$_2$. Maleic anhydride, 4A-molecular sieves and NaOAc are the proper additives for this reaction.[3] Alternatively, the same transformation is accomplished with benzoquinone as the oxidant, together with 1,2-bis(benzenesulfinyl)-ethane and (salen)Cr^{III}Cl.[4] A heterobimetallic catalytic system is involved. However, only the Pd(II) catalyst is needed for oxidative cyclization of homoallylic *N*-tosylcarbamates.[5]

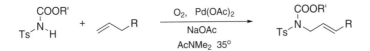

Cyclization of alkenylamines usually proceeds via *cis*-aminopalladation.[6]

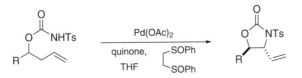

Functionalization at sp-carbons.   Several different oxidants have been employed in oxidative functionalization. For *o*-acetoxylation of acylaminoarenes $K_2S_2O_8$ is the oxidant,[7] and oxone is present to facilitate the transfer of an ethoxycarbonyl group from DEAD to organopalladium intermediates.[8]

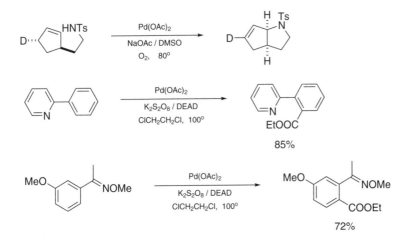

One of the silyl groups in *N*-bis(trimethylsilyl)methylcarboxamides is activated by coordination to the carbonyl and under the Heck reaction conditions it can submit a methyl group.[9] Among *N*-directed reactions methyl group transfer from dicumyl peroxide, involving Pd insertion into the O—O bond and elimination of acetophenone to afford the transfer reagent, is also a relatively new discovery.[10]

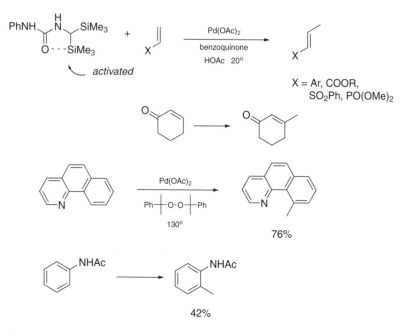

42%

Intramolecular oxidative addition to the double bond of *o,o'*-bifunctional stilbenes is accomplished using PhI(OAc)$_2$ as the oxidant. This method is applicable to forming head-to-tail fused biindolines and furoindolines.[11]

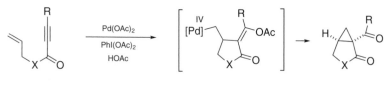

Allyl 2-alkynoates and amides undergo formal hydration and cycloaddition under similar conditions.[12,13] The net result is equivalent to transforming the triple bond into an acylcarbenoid.

X = O, NTs

[1]Wang, A., Jiang, H. *JACS* **130**, 5030 (2008).
[2]El Kaim, L., Gamez-Montano, R., Grimaud, L., Ibarra-Rivera, T. *CC* 1350 (2008).

[3]Liu, G., Yin, G., Wu, L. *ACIE* **47**, 4733 (2008).
[4]Reed, S.A., White, M.C. *JACS* **130**, 3316 (2008).
[5]Fraunhoffer, K.J., White, M.C. *JACS* **129**, 7274 (2007).
[6]Liu, G., Stahl, S.S. *JACS* **129**, 63294 (2007).
[7]Wang, G.-W., Yuan, T.-T., Wu, X.-L. *JOC* **73**, 4717 (2008).
[8]Yu, W.-Y., Sit, W.N., Lai, K.-M., Zhou, Z., Chan, A.S.C. *JACS* **130**, 3304 (2008).
[9]Rauf, W., Brown, J.M. *ACIE* **47**, 4228 (2008).
[10]Zhang, Y., Feng, J., Li, C.-J. *JACS* **130**, 2900 (2008).
[11]Muniz, K. *JACS* **129**, 14542 (2007).
[12]Tong, X., Beller, M., Tse, M.K. *JACS* **129**, 4906 (2007).
[13]Welbes, L.L., Lyons, T.W., Cychosz, K.A., Sanford, M.S. *JACS* **129**, 5836 (2007).

Palladium(II) acetate – phase-transfer catalyst.

Coupling reactions. Structural limitation of the Baylis–Hillman reaction is amended by the Heck reaction.[1]

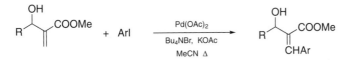

Elaboration of a Baylis–Hillman adduct to give a benzoazepino[2,1-*a*]isoindole[2] serves to demonstrate the power of the coupling method.

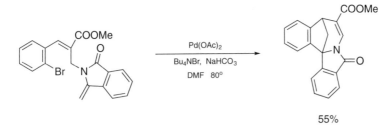

55%

A convenient preparation of indene-1,1-dicarboxylic esters involves coupling of *o*-iodoarylmalonic esters with alkynes. The *sp*-carbon bearing the larger group becomes C-2, and that bearing the smaller group, C-3.[3]

R^L = Ph, *t*-Bu, SiMe₃

Electrooxidation. A mixture of Pd(OAc)₂ and Bu₄NX is electrooxidized in MeCN to provide (MeCN)₄PdX₂ for use in Wacker oxidation.[4]

[1]Kim, J.M., Kim, K.H., Kim, T.H., Kim, J.N. *TL* **49**, 3248 (2008).
[2]Gowrisankar, S., Lee, H.S., Lee, K.Y., Lee, J.-E., Kim, J.N. *TL* **48**, 8619 (2007).
[3]Zhang, D., Liu, Z., Yum, E.K., Larock, R.C. *JOC* **72**, 251 (2007).
[4]Mitsudo, K., Kaide, T., Nakamoto, E., Yoshida, K., Tanaka, H. *JACS* **129**, 2246 (2007).

Palladium(II) acetate – silver salts.

Coupling reactions.   A catalyst system for coupling of indoles (at C-2) with ArI is made up of Pd(OAc)$_2$, Ag$_2$O and the ArCOOH additive, no phosphine ligand is needed.[1] Cinnamyl esters are synthesized from ArI and allylic esters by the Heck reaction, the traditional leaving group (acyloxy) is retained.[2]

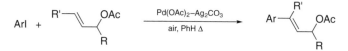

N-Methoxy-2,2-dimethylalkanamides are homologated at one of the methyl groups on reaction with organoboronic acids. The solvent of choice for arylation is *t*-BuOH, and for alkylation, 2,2,5,5-tetramethyltetrahydrofuran.[3]

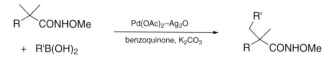

When neopentylpalladium intermediates are generated under non-nucleophilic conditions C—H activation is the course they pursue.[4]

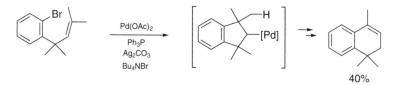

Benzylamines are *o*-arylated in CF$_3$COOH when catalyzed by Pd(OAc)$_2$ – AgOAc.[5]

[1]Lebrasseur, N., Larrosa, I. *JACS* **130**, 2926 (2008).
[2]Pan, D., Chen, A., Su, Y., Zhou, W., Li, S., Jia, W., Xiao, J., Liu, Q., Zhang, L., Jiao, N. *ACIE* **47**, 4729 (2008).
[3]Wang, D.-H., Wasa, M., Giri, R., Yu, J.-Q. *JACS* **130**, 7190 (2008).
[4]Liron, F., Knochel, P. *TL* **48**, 4943 (2007).
[5]Lazareva, A., Daugulis, O. *OL* **8**, 5211 (2006).

Palladium(II) acetate – tertiary phosphine.

Coupling reactions.   Using ligand **1** in Pd(OAc)$_2$-catalyzed C—N bond coupling at room temperature the scope encompasses RX (R = aryl, alkenyl) containing base-sensitive groups.[1] The catalyst system is actually an extension from a success in Suzuki coupling.[2] In preparing pinacolatoborylarenes from ArCl and bis(pinacolato)diboron the biarylphosphine ligand **2** in the Pd-catalyzed reaction plays a special role; with it PdL species is favored over PdL$_2$ and the oxidative addition to ArCl is facilitated.[3]

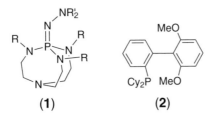

(1)   **(2)**

Ligand **2** has also been employed to advantage in the coupling reaction of β,β-dihalo-*o*-aminostyrenes leading to 2-substituted indoles.[4] (With addition of Cu(OAc)$_2$ to promote *N*-arylation the process runs better.)

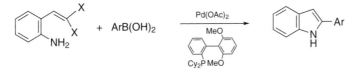

A report describes an application of ferrocenylphosphine ligand **3** to Suzuki coupling involving ArCl.[5] A great number of α-acetaminostyrenes are readily prepared by Heck reaction in which high regioselectivity is observed with Pd(OAc)$_2$, DPPP, Et$_3$N, and *i*-Pr$_2$NH$_2$BF$_4$ in isopropanol.[6] Other styrenes bearing electron-rich substituents at the α-position are similarly accessed (the superiority of alcohol solvents for the reaction is noted).[7]

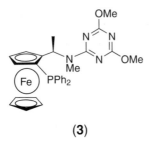

(3)

Sonogashira coupling is promoted by Pd(OAc)$_2$, Ph$_3$P in refluxing THF, where Cp*Li adequately serves as a base.[8]

Tertiary homoallylic alcohols undergo CC bond scission during Heck reaction.[9,10] The emerging carbonyl fragment is lost from the alicyclic substrates except in the case of a 1-substituted 3-cycloalkenol.

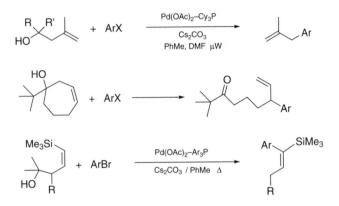

Cleavage of a cyclobutanol subunit also creates a site for CC coupling,[11] ring strain is the cause for such reactivity.

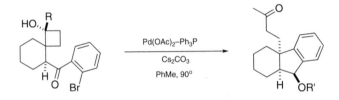

Hydronaphthacenes[12] and benzofluorenes[13] can be elaborated in one step, based on sequential allylic substitution and Heck reaction.

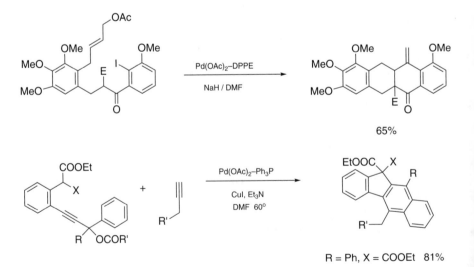

Diarylamines and *o*-dihaloarenes condense to give *N*-arylcarbazoles.[14] Exposure of *o*-bromobenzamides to Pd(OAc)$_2$, Cs$_2$CO$_3$ and a phosphine leads to debrominative coupling and cyclization with elimination of one amide unit, to give phenanthridinones.[15]

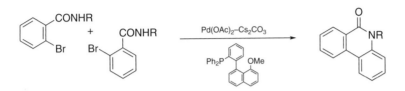

A remarkable change of the second-stage coupling described in the following equations is apparently the effect of the base employed.[16]

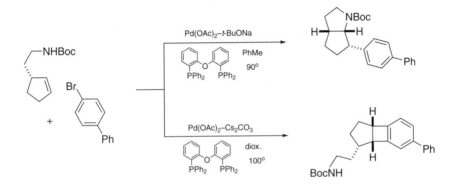

The possibility of Pd shift, as exemplified by cases of *o*-iodobiaryls,[17] must be heeded. On the other hand, the phenomenon can be used to synthetic advantage, as shown by an unusual route to fluorenones, xanthones, and acridones.[18]

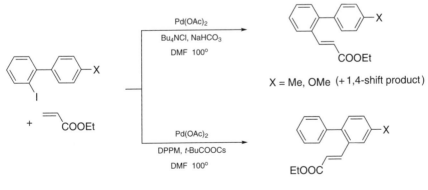

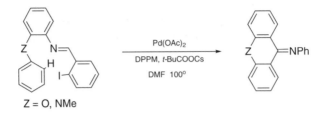

Z = O, NMe

The transient involvement of norbornene in Heck reaction via its carbopalladation and help to functionalization of the ortho position(s) is highly profitable. Its further exploitation includes o-alkylation at the site of Heck reaction[19] and amination.[20]

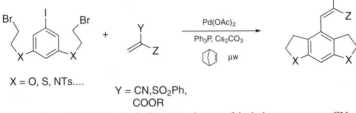

X = O, S, NTs....

Y = CN, SO$_2$Ph, COOR

The process is applicable to completing an exchange of the halogen atom to a CN group by reaction with K$_4$Fe(CN)$_6$, after performing o-functionalization.[21] In another report on the preparation of ArCN, the bulky butylbis(1-adamantyl)phosphine is the ligand employed.[22]

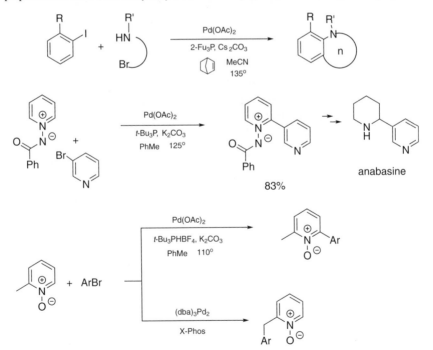

anabasine

83%

Directed activation of the C—H bond adjacent to the nuclear nitrogen atom of *N*-iminopyridinium ylides enables a rapid elaboration of a synthetic intermediate of anabasine.[23] 2-Picoline-*N*-oxide undergoes coupling with ArBr at C-6. Interestingly, the reaction is shifted to the methyl group by changing the catalyst system to (dba)$_3$Pd$_2$/X-Phos,[24] or Pd(OAc)$_2$/DavePhos.[25]

Cross-coupling of two different types of alkynes in a controlled and atom-economical fashion is of great synthetic value. By this method, 2-alken-4-ynoic esters are readily elaborated.[26]

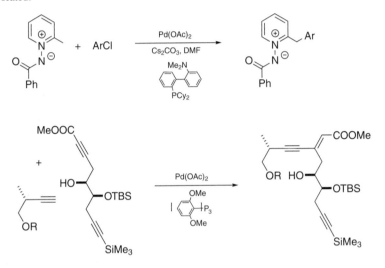

Alkenylation and arylation. Alkenyl triflates are highly active in intramolecular alkenylation of both electron-rich and electron-poor arenes.[27] 2-Alkynylbiaryls give 9-alkylidenefluorenes via palladation of the *o'*-position [the benzene ring bearing electron-withdrawing group(s)].[28]

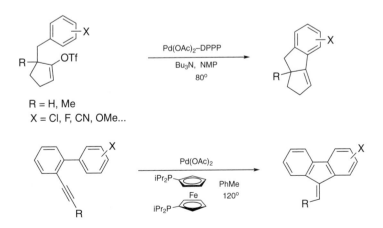

α-Methallyl-γ-bromocinnamyl alcohols undergo cyclization to provide 3-aryl-4,4-dimethyl-2-cyclopentenones.[29] Some members of this series of compounds are useful precursors of cuparan sesquiterpenes.

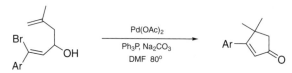

Phenanthrenes spanned at C-4 and C-10 by an aminomethyl bridge emerge from cyclization of N-(m-bromoaryl)-β-bromocinnamylamines.[30] The reaction involves double C—H activation.

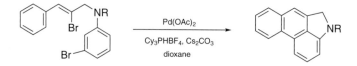

Many metal-catalyzed reactions of propargylic esters proceed as if alkylidene-metal carbenoids are involved. In the Pd-catalyzed reaction such species add to norbornenes/norbornadienes to give ring expansion products.[31]

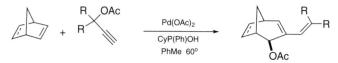

For α-arylation of aldehydes catalytic systems with wide-scoped applicability have been determined.[32] Excellent phosphine ligand for assisting Pd(OAc)$_2$ in the reaction with ArCl is 1, and with ArBr, rac. BINAP or 2.

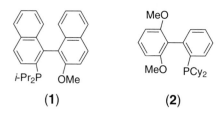

i-Pr$_2$P   OMe

MeO   MeO   PCy$_2$

(1)   (2)

Coupling via C(sp³)-H activation.   In addition to some examples mentioned in the previous section (e.g., 2-picoline derivatives), sp^3-hybridized C—H can be activated with Pd catalysts under proper conditions. Such a process enables formation of indolines[33] and dihydrobenzofurans.[34]

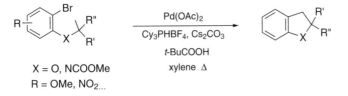

X = O, NCOOMe
R = OMe, NO₂...

A benzylic alkyl group of *o*-bromoarylacetonitriles or the corresponding arylacetic esters is either dehydrogenated with one hydrogen atom to replace the bromine atom, or engaged in ring formation. The catalyst system to effect the transformation(s) consists of Pd(OAc)₂ and tris(*m*-fluorophenyl)phosphine.[35]

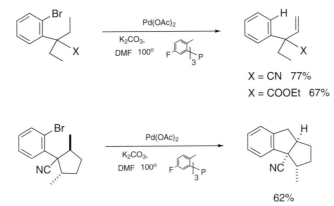

X = CN 77%
X = COOEt 67%

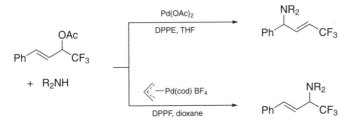

62%

Allylic substitution.   α-Trifluoromethylcinnamyl acetate undergoes substitution with R₂NH. Regiochemical contrasts are observed in reactions with different catalyst systems.[36]

In the presence of (*i*-PrO)₄Ti, an allylic alcohol unit serves as a leaving group and a secondary hydroxyl group as a nucleophile in Pd-catalyzed cyclization.[37]

Catalysts for diallylation of active methylene compounds (e.g., malonate esters) and RCM reaction are compatible, therefore 3-cyclopentene-1,1-dicarboxylic esters can be prepared in one-pot.[38]

Two stereogenic centers are established on allylation of α-substituted β-keto esters and cyanoacetic esters with secondary cinnamyl acetates. The reaction is directed toward a regioselective and diastereoselective pathway when o-diphenylphosphinobenzoic acid is used as a ligand.[39]

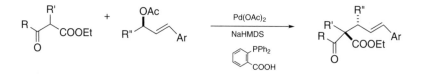

Epimerization. Treatment of (Z)-γ-(2-bromovinyl)-γ-butyrolactones with Pd(OAc)₂ and Ph₃P simultaneously inverts the less stable stereocenter and change the double bond configuration via formation of π-allylpalladium species from which reclosure of the lactone ring follows CC bond rotation. This stereochemical readjustment is an important operation in an approach to 4-epiethisolide.[40]

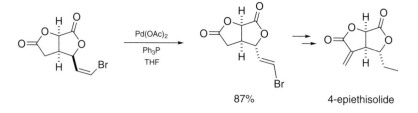

87%   4-epiethisolide

Reduction. Hydrodehalogenation of ArX and α-bromo ketones is accomplished by heating the halides with Pd(OAc)₂, Ph₃P and K₂CO₃ in an alcohol at 100°.[41] Enol triflates are defunctionalized by the Pd-catalyzed reduction with HCOOH – Bu₃N in DMF.[42]

Selective hydrogenation of alkenes such as the strained double bond of dicyclopenta-diene is achieved by heating with Pd(OAc)$_2$, t-Bu$_3$P and HCOOH in THF.[43]

Addition reactions.   Acrylic esters and analogues (amides, nitriles) are receptive to aryl group transfer to their β-position from ArB(OH)$_2$, under the influence of Pd(OAc)$_2$.[44]

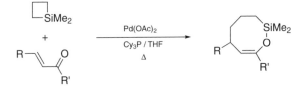

Complementary to copper catalyst, a Pd(OAc)$_2$ – Ph$_3$P system also promotes the conjugate addition of R$_2$Zn to enals.[45]

With Pd insertion into a C—Si bond of a siletane elements of the small ring add to conjugated carbonyl compounds, eight-membered unsaturated O,Si-heterocycles are produced.[46]

[1]Reddy, C.V., Kingston, J.V., Verkade, J.G. *JOC* **73**, 3047 (2008).

[2]Kingston, J.V., Verkade, J.G. *JOC* **72**, 2816 (2007).

[3]Billingsley, K.L., Barder, T.E., Buchwald, S.L. *ACIE* **46**, 5359 (2007).

[4]Fang, Y.-Q., Lautens, M. *JOC* **73**, 538 (2008).

[5]Yu, S.-B., Hu, X.-P., Deng, J., Huang, J.-D., Wang, D.-Y., Duan, Z.-C., Zheng, Z. *TL* **49**, 1253 (2008).

[6]Liu, Z., Xu, D., Tang, W., Xu, L., Mo, J., Xiao, J. *TL* **49**, 2756 (2008).

[7]Hyder, Z., Ruan, J., Xiao, J. *CEJ* **14**, 5555 (2008).

[8]Uemura, M., Yorimitsu, H., Oshima, K. *T* **64**, 1829 (2008).

[9]Iwasaki, M., Hayashi, S., Hirano, K., Yorimitsu, H., Oshima, K. *T* **63**, 5200 (2007).

[10]Hayashi, S., Hirano, K., Yorimitsu, H., Oshima, K. *JACS* **129**, 12650 (2007).

[11]Ethirajan, M., Oh, H.-S., Cha, J.K. *OL* **9**, 2693 (2007).

[12]Tietze, L.F., Redert, T., Bell, H.P., Hellkamp, S., Levy, L.M. *CEJ* **14**, 2527 (2008).

[13]Guo, L.N., Duan, X.-H., Liu, X.-Y., Hu, J., Bi, H.-P., Liang, Y.-M. *OL* **9**, 5425 (2007).

[14]Ackermann, L., Althammer, A. *ACIE* **46**, 1627 (2007).

[15]Furuta, T., Kitamura, Y., Hashimoto, A., Fujii, S., Tanaka, K., Kan, T. *OL* **9**, 183 (2007).

[16]Bertrand, M.B., Wolfe, J.P. *OL* **9**, 3073 (2007).

[17]Campo, M.A., Zhang, H., Yao, T., Ibdah, A., McCulla, R.D., Huang, Q., Zhao, J., Jenks, W.J., Larock, R.C. *JACS* **129**, 6298 (2007).

[18]Zhao, J., Yue, D., Campo, M.A., Larock, R.C. *JACS* **129**, 5288 (2007).

[19]Alberico, D., Rudolph, A., Lautens, M. *JOC* **72**, 775 (2007).

[20]Thansandote, P., Raemy, M., Rudolph, A., Lautens, M. *OL* **9**, 5255 (2007).

[21]Mariampillai, B., Alliot, J., Li, M., Lautens, M. *JACS* **129**, 15372 (2007).

[22]Schareina, T., Zapf, A., Mägerlein, W., Müller, N., Beller, M. *TL* **48**, 1087 (2007).

[23]Larivee, A., Mousseau, J.L., Charette, A.B. *JACS* **130**, 52 (2008).

[24]Campeau, L.-C., Schipper, D.J., Fagnou, K. *JACS* **130**, 3266 (2008).

[25]Mousseau, J.J., Larivee, A., Charette, A.B. *OL* **10**, 1641 (2008).

[26]Trost, B.M., Ashfeld, B.L. *OL* **10**, 1893 (2008).

[27]Cruz, A.C.F., Miller, N.D., Willis, M.C. *OL* **9**, 4391 (2007).

[28]Chernyak, N., Gevorgyan, V. *JACS* **130**, 5636 (2008).

[29]Ray, D., Ray, J.K. *OL* **9**, 191 (2007).

[30]Ohno, H., Iuchi, M., Fujii, N., Tanaka, T. *OL* **9**, 4813 (2007).

[31]Bigeault, J., de Riggi, I., Gimbert, Y., Giordano, L., Buono, G. *SL* 1071 (2008).

[32]Martin, R., Buchwald, S.L. *ACIE* **46**, 7236 (2007).

[33]Watanabe, T., Oishi, S., Fujii, N., Ohno, H. *OL* **10**, 1759 (2008).

[34]Lafrance, M., Gorelsky, S.I., Fagnou, K. *JACS* **129**, 14570 (2007).

[35]Hitce, J., Retailleau, P., Baudoin, O. *CEJ* **13**, 792 (2007).

[36]Kawatsura, M., Hirakawa, T., Tanaka, T., Ikeda, D., Hayase, S. Itoh, T. *TL* **49**, 2450 (2008).

[37]Brioche, J.C.R., Goodenough, K.M., Whatrup, D.J., Harrity, J.P.A. *JOC* **73**, 1946 (2008).

[38]Kammerer, C., Prestat, G., Gaillard, T., Madec, D., Poli, G. *OL* **10**, 405 (2008).

[39]Kawatsura, M., Ikeda, D., Komatsu, Y., Mitani, K., Tanaka, T., Uenishi, J. *T* **63**, 8815 (2007).

[40]Hon, Y.-S., Chen, H.-F. *TL* **48**, 8611 (2007).

[41]Chen, J., Zhang, Y., Yang, L., Zhang, X., Liu, J., Li, L., Zhang, H. *T* **63**, 4266 (2007).

[42]Pandey, S.K., Greene, A.E., Poisson, J.-F. *JOC* **72**, 7769 (2007).

[43]Brunel, J.M. *SL* 330 (2007).

[44]Horiguchi, H., Tsurugi, H., Satoh, T., Miura, M. *JOC* **73**, 1590 (2008).

[45]Marshall, J.A., Herold, M., Eidam, H.S., Eidam, P. *OL* **8**, 5505 (2006).

[46]Hirano, K., Yorimitsu, H., Oshima, K. *OL* **10**, 2199 (2008).

Palladium(II) acetate – tertiary phosphine – carbon monoxide.

Carbonylation. The conversion of ArOTf to ArCHO with syngas (reductive carbonylation) simply uses the Pd(OAc)$_2$ – DPPP system as catalyst.[1] Homologation of cycloalkenyl triflates to the corresponding enals is perhaps favored by the hindered bis-phosphine **1**.[2]

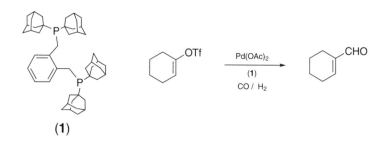

(1)

Without the reducing agent (e.g., H_2) aryl sulfonates are converted into esters in ROH under basic conditions (a protocol indicates addition of the hydrofluoroborate salt of 1,3-biscyclohexylphosphinopropane).[3] In an analogous synthesis of amides in DMSO, in which amines replace ROH, the PhONa base appears to play a special and critical role.[4]

With $Al(OTf)_3$ cocatalyst 1-alkenes are transformed into saturated esters favoring the linear isomer (against the branched isomer in a ratio of $>2:1$).[5]

A general method for preparation of ArCOAr' from $ArB(OH)_2$, Ar'Br and CO employs $Pd(OAc)_2$, butylbis(1-adamantyl)phosphine and TMEDA.[6]

Carboxylation.   Arylzinc bromides undergo carboxylation with CO_2 readily.[7] The Pd catalyst (with Cy_3P as ligand) performs better than a similar Ni complex.

[1]Brennführer, A., Neumann, H., Beller, M. *SL* 2537 (2007).
[2]Neumann, H., Sergeev, A., Beller, M. *ACIE* **47**, 4887 (2008).
[3]Munday, R.H., Martinelli, J.R., Buchwald, S.L. *JACS* **130**, 2754 (2008).
[4]Martinelli, J.R., Clark, T.P., Watson, D.A., Munday, R.H., Buchwald, S.L. *ACIE* **46**, 8460 (2007).
[5]Williams, D.B.G., Shaw, M.L., Green, M.J., Holzapfel, C.W. *ACIE* **47**, 560 (2008).
[6]Neumann, H., Brennführer, A., Beller, M. *CEJ* **14**, 3645 (2008).
[7]Yeung, C.S., Dong, V.M. *JACS* **130**, 7826 (2008).

Palladium(II) acetylacetonate.

Coupling reactions.   Electron-deficient aroic acids undergo decarboxylative coupling with ArBr on heating with $Pd(acac)_2$, $CuCO_3$.[1]

Allylsilanes are prepared from allyl ethers (silyl ethers or phenyl ethers) and R_3SiCl; coupling takes place in the presence of $Pd(acac)_2$ and PhMgBr (240 mol%).[2]

Stereoselective cyclization of 6-hydroxy-4-hexenals is observed. 2-Vinylcyclobutanols are obtained.[3]

[1]Goossen, L.J., Rodriguez, N., Melzer, B., Linder, C., Deng, G., Levy, L.M. *JACS* **129**, 4824 (2007).
[2]Naitoh, Y., Bando, F., Terao, J., Otsuki, K., Kuniyasu, H. *CL* **36**, 236 (2007).
[3]Kimura, M., Mukai, R., Tamaki, T., Horino, Y., Tamaru, Y. *JACS* **129**, 4122 (2007).

Palladium(II) bis(trifluoroacetate).

Decarboxylation.   Aroic acids bearing electronic-rich substituents are found to undergo decarboxylation on warming with $Pd(OCOCF_3)_2$, CF_3COOH in DMF (containing 5% DMSO) at 70°.[1]

Dehydrogenation. Cycloalkanones are converted into the corresponding enones by the Pd-catalyzed aerial oxidation in the presence of a bipyridyl ligand.[2]

Coupling reactions. Aryl cyanides are formed in a reaction with Zn and Zn(CN)$_2$, using t-Bu$_3$P to ligate with Pd(OCOCF$_3$)$_2$ [or (t-Bu$_3$P)$_2$Pd].[3]

The coupling of o-borylaroic esters with alkynes is immediately followed by acylation, which results in the formation of indenones. The homologous arylacetic esters give 2-naphthols.[4]

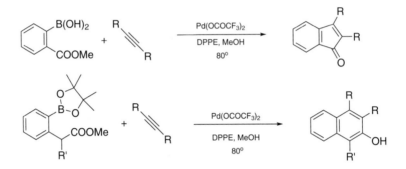

Cyclization of o-allyl-N-tosylarylamines to afford 2-vinylindolines is accomplished by Pd(OCOCF$_3$)$_2$ and a carbene ligand (**1**) in oxygen (or air).[5]

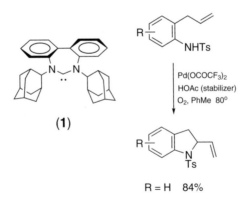

Regiochemical dependency in the oxidative coupling of N-acetylindole with arenes on the added oxidant [AgOAc vs. Cu(OAc)$_2$] is a remarkable phenomenon.[6]

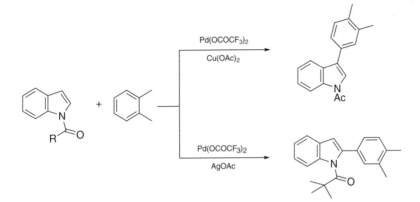

Using benzoquinone as oxidant to regenerate Pd(II) species in situ in the reaction catalyzed by Pd(OCOCF₃)₂, a 3,5-bridged tetrahydropyran is formed from 2,2-diprenyl-1,3-propanediol monobenzoate. Asymmetric induction by a spiro(isoxazole-isoxazoline) ligand has also been scrutinized.[7]

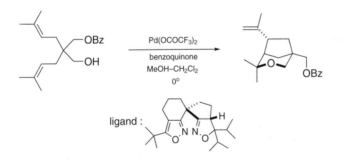

Aryl(2-pyridyl)methanes are obtained from coupling of picolyl diisopropyl carbinol with ArCl. The reaction involves fragmentation to form aryl(2-methylene-1,2-dihydropyridyl)-palladium intermediates.[8]

[1]Dickstein, J.S., Mulrooney, C.A., O'Brien, E.M., Morgan, B.J., Kozlowski, M.C. *OL* **9**, 2441 (2007).
[2]Tokunaga, M., Harada, S., Iwasawa, T., Obora, Y., Tsuji, Y. *TL* **48**, 6860 (2007).
[3]Littke, A., Soumeillant, M., Kaltenbach III, R.F., Cherney, R.J., Tarby, C.M., Kiau, S. *OL* **9**, 1711 (2007).
[4]Tsukamoto, H., Kondo, Y. *OL* **9**, 4227 (2007).
[5]Rogers, M.M., Wendlandt, J.E., Guzei, I.A., Stahl, S.S. *OL* **8**, 2257 (2006).
[6]Stuart, D.R., Villemure, E., Fagnou, K. *JACS* **129**, 12072 (2007).
[7]Koranne, P.S., Tsujihara, T., Arai, M.A., Bajracharya, G.B., Suzuki, T., Onitsuka, K., Sasai, H. *TA* **18**, 919 (2007).
[8]Niwa, T., Yorimitsu, H., Oshima, K. *ACIE* **46**, 2643 (2007).

Palladium carbene complexes.

Coupling reactions.   A PdCl$_2$ complex with 1,3-bis(2,6-diisopropylphenyl)imidazol-2-ylidene and 3-chloropyridine ligands is active for catalyzing the Suzuki coupling.[1]

Hydrogenation.   In a polar solvent, the Pd species ligated by both 1,3-dimesitylimidazol-2-ylidene and maleic anhydride promotes hydrogenation of alkynes to give (Z)-alkenes by HCOOH−NEt$_3$ without over-reduction.[2]

On the other hand, complexes **1** with a very basic and electron-rich Pd center serves as a hydrogenation catalyst for alkenes.[3] In comparison, the less basic **1a** is a poor catalyst.

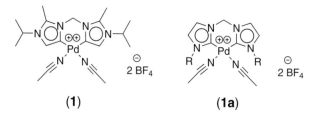

(1)   **(1a)**

[1]Valente, C., Baglione, S., Candito, D., O'Brien, C.J., Organ, M.G. *CC* 735 (2008).
[2]Hauwert, P., Maestri, G., Sprengers, J.W., Catellani, M., Elsevier, C.J. *ACIE* **47**, 3223 (2008).
[3]Heckenroth, M., Kluser, E., Neels, A., Albrecht, M. *ACIE* **46**, 6293 (2007).

Palladium(II) chloride.

Benzhydryl ethers.   Alcohols and benzhydrol condense to form ROCHPh$_2$ on warming with PdCl$_2$ at 80°.[1]

Alkylation.   Viologen-supported nanoparticles of palladium prepared from PdCl$_2$ show an activity of promoting alkylation of methyl ketones with alcohols in the air at 100°. A 1,3-cycloalkanedione undergoes alkylation but the ring is opened. The reaction also requires a base such as Ba(OH)$_2$ · H$_2$O or Sr(OH)$_2$ · 8H$_2$O.[2]

Allylation of 1,3-dicarbonyl compounds catalyzed by PdCl$_2$−Bu$_4$NBr is subject to solvent effects. Diallylation occurs in THF, but in water, monoallylation.[3]

Coupling reactions.   The coupling of 1,2-diiodoalkenes with 1-alkenes affords conjugated enynes, as dehydroiodination also occurs. The (E)-isomer is produced from acrylonitrile, (Z)-isomers from other alkenes.[4]

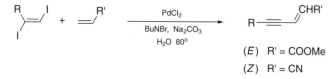

(E) R' = COOMe
(Z) R' = CN

For Suzuki coupling, amine ligands tested to support PdCl$_2$ include polyaniline,[5] 2-(2-pyridyl)-6-isopropylpiperidine,[6] and piperazine **1**.[7] The naphthidine **2** actually reduces PdCl$_2$ to Pd(0) nanoparticles and stabilizes such to perform catalysis.[8]

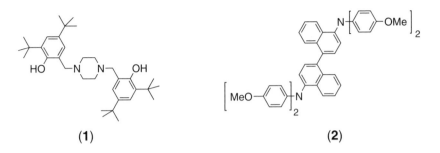

(1)   **(2)**

With a cationic 2,2′-bipyridyl ligand the Suzuki coupling in water proceeds very efficiently (TOF up to 81,000 per hr, TON up to 395,000).[9]

Exploration of metal-carbene complexes in catalysis for various reactions is in vogue. Performing Heck reaction in the ionic liquid *N*-methyl-*N*′-(2-diisopropylaminoethyl)imidazolium triflimide must involve a carbene-complexed Pd species.[10] An *N*-arylation method to derivatize amines has been developed in that spirit.[11]

Cyclopentenones.   3-Acetoxy-4-alken-1-ynes are converted into cyclopentenones on treatment with PdCl₂ in MeCN at 60° (or with in CH₂Cl₂ at room temperature).[12]

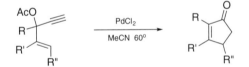

[1]Bikard, Y., Weibel, J.-M., Sirlin, C., Dupuis, L., Loeffler, J.-P., Pale, P. *TL* **48**, 8895 (2007).

[2]Yamada, Y.M.A., Uozumi, Y. *T* **63**, 8492 (2007).

[3]Ranu, B.C., Chattopadhyay, K., Adak, L. *OL* **9**, 4595 (2007).

[4]Ranu, B.C., Chattopadhyay, K. *OL* **9**, 2409 (2007).

[5]Kantam, M.L., Roy, M., Roy, S., Sreedhar, B., Madhavendra, S.S., Choudary, B.M., De, R.L. *T* **63**, 8002 (2007).

[6]Puget, B., Roblin, J.-P., Prim, D., Troin, Y. *TL* **49**, 1706 (2008).

[7]Mohanty, S., Suresh, D., Balakrishna, M.S., Mague, J.T. *T* **64**, 240 (2008).

[8]Desmarets, C., Omar-Amrani, R., Walcarius, A., Lambert, J., Champagne, B., Fort, Y., Schneider, R. *T* **64**, 372 (2008).

[9]Wu, W.-Y., Chen, S.-N., Tsai, F.-Y. *TL* **47**, 9267 (2006).

[10]Ye, C., Xiao, J.-C., Twamley, B., LaLonde, A.D., Norton, M.G., Shreeve, J.M. *EJOC* 5095 (2007).

[11]Organ, M.G., Abdel-Hadi, M., Avola, S., Dubovyk, I., Hadei, N., Assen, E., Kantchev, B., O'Brien, C.J., Sayah, M., Valente, C. *CEJ* **14**, 2443 (2008).

[12]Caruana, P.A., Frontier, A.J. *T* **63**, 10646 (2007).

Palladium(II) chloride – di-*t*-butylphosphinous acid.

Hydroarylation. The transfer of the aryl group of $ArSi(OMe)_3$ to the β-position of conjugated ketones, esters, nitriles, and nitroalkenes can be carried out in water with a Pd–Cu salt combinant such as $(t\text{-BuPOH})PdCl_2$ and $CuBF_4$.[1]

[1] Lerebours, R., Wolf, C. *OL* **9**, 2737 (2007).

Palladium(II) chloride – metal salts.

Coupling reactions. To form biaryls through cross-coupling of ArCOOH with Ar′I, one of the aryl groups in the products comes from decarboxylation of the acids. The catalyst system used to effect this reaction contains $PdCl_2$, Ph_3As and Ag_2CO_3.[1]

Arenes undergo coupling with $ArSnCl_3$ in the presence of a bimetallic catalyst of $PdCl_2$ and $CuCl_2$.[2] An *o*-position of *N,N*-dimethylbenzylamines is also activated toward coupling with acrylic acid derivatives.[3]

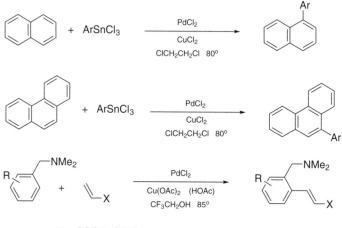

X = COOR, CONH₂…

A method for preparing unsymmetrical 1,2-diarylethyne based on Sonogashira coupling, 1-ethynylcyclohexanol is employed. The two-stage process is intervened by the addition of KOH (after completion of the first coupling) to release arylethynes.[4]

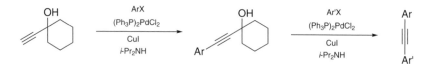

3-Aminoindolizines are readily obtained from the coupling of 2-bromopyridines and pro-pargylamines,[5] whereas precursors of indolines can be assembled from an *N*-protected pro-pargylamine by extending the carbon chain to an enyne and completing *N*-alkynylation. Thermolysis of such products gives substituted indolines.[6]

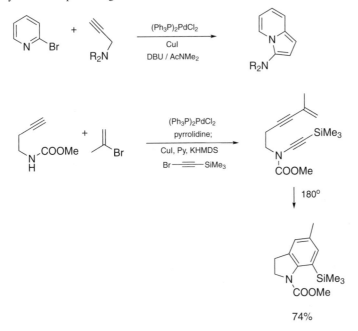

74%

2-Arylethylamines form *N*-(2-arylethyl)-3,4-diarylpyrroles in a trimerizative conden-sation on treatment with PdCl$_2$–Cu(OAc)$_2$.[7]

Ring expansion.[8]   2-Arylidenecyclopropane-1-methanols possessing electron-rich aryl groups undergo ring expansion to afford cyclobutenemethanols under argon, when they are exposed to PdCl$_2$–CuBr$_2$. On the other hand, electron-poor congeners give hydrofuran derivatives under air.

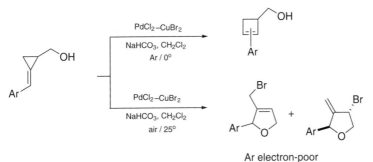

Ar electron-poor

Wacker oxidation.   Directed by an allylic trifluoroacetamido group a terminal double bond is converted into an aldehyde as the major product. The transformation has been exploited in a synthesis of daunosamine.[9]

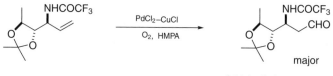

(aldehyde:ketoene 82:18)

Under reaction conditions essentially those for the Wacker oxidation, electron-deficient 1-alkenes (e.g., acrylic esters) are trimerized to provides 1,3,5-trisubstituted benzenes.[10]

Pauson–Khand reaction.   In the Pd-catalyzed reaction tetramethylthiourea and LiCl are important additives.[11]

[1]Becht, J.-M., Catala, C., Le Drian, C., Wagner, A. *OL* **9**, 1781 (2007).
[2]Kawai, H., Kobayashi, Y., Oi, S., Inoue, Y. *CC* 1464 (2008).
[3]Cai, G., Fu, Y., Li, Y., Wan, X., Shi, Z. *JACS* **129**, 7666 (2007).
[4]Csekei, M., Novak, Z., Kotschy, A. *T* **64**, 975 (2008).
[5]Liu, Y., Song, Z., Yan, B. *OL* **9**, 409 (2007).
[6]Dunetz, J.R., Danheiser, R.L. *JACS* **127**, 5776 (2005).
[7]Wan, X., Xing, D., Fang, Z., Li, B., Zhao, F., Zhang, K., Yang, L., Shi, Z. *JACS* **128**, 12046 (2006).
[8]Tian, G.-Q., Yuan, Z.-L., Zhu, Z.-B., Shi, M. *CC* 2668 (2008).
[9]Friestad, G.K., Jiang, T., Mathies, A.K. *OL* **9**, 777 (2007).
[10]Jiang, H.-F., Shen, Y.-X., Wang, Z.-Y. *TL* **48**, 7542 (2007).
[11]Deng, L.-J., Liu, J., Huang, J.-Q., Hu, Y., Chen, M., Lan, Y., Chen, J.-H., Lei, A., Yang, Z. *S* 2565 (2007).

Palladium(II) chloride – tertiary phosphine.

Hydrodehalogenation.   Removal of halogen atom(s) from haloarenes is accomplished by heating with PdCl$_2$, DPPF, NaHCO$_3$ in DMF, the solvent is a convenient hydride source.[1]

Arylation.   With the highly active and air-stable ligand, 1,1′-bis(di-*t*-butylphosphino)-ferrocene, to assist PdCl$_2$, α-arylation of ketones can be accomplished in good yields even with the hindered 2,6-dimethylchlorobenzene.[2] More unusual is the arylation that converts enol acetates into aryl ketones as shown below.[3] The reaction employs Bu$_3$SnOMe as base.

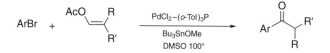

2-Arylthiazoles are found to undergo arylation at C-5. Cleaner and faster reactions are performed in water than in other common solvents.[4] Arylation of 4-aryl-1,2,3-triazoles has also been reported.[5]

An alkynyl group is introduced into C-3 of the indolizine nucleus by the Pd-catalyzed reaction with a 1-alkyne.[6]

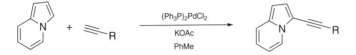

Coupling reactions. The PdCl$_2$–DPPF system placed in a microemulsion environment (SDS, NaHCO$_3$) can be used to catalyze Suzuki coupling of long-chain alkyl and oxyalkyl substrates.[7]

Heck reaction in the presence of the aminoethylimidazolium salt **1** requires no additional base.[8] Using the 1,1'-bis(di-*t*-butylphosphino)ferrocene ligand Heck reaction is accomplished at room temperature in water to prepare styrenes and cinnamate esters.[9] The same conditions are conducive to performing Suzuki coupling.[10]

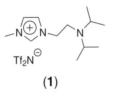

(1)

Azolecarbene ligands are shown to be favorable to Suzuki coupling involving polyfluorinated aryltrifluoroborate salts (e.g., C$_6$F$_5$BF$_3$K) catalyzed by (Ph$_3$P)$_2$PdCl$_2$ under anaerobic conditions.[11]

Formation of 3-spiroannulated oxindoles involving a Heck reaction is the result of subsequent activation of a proximal aromatic C—H bond by the organopalladium complex.[12]

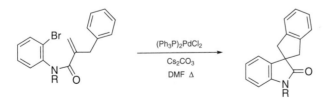

1,2-Migration of alkenyl-Pd(II) intermediates may occur during Heck reaction of enol phosphates, but it may be suppressed by using the X-Phos ligand.[13]

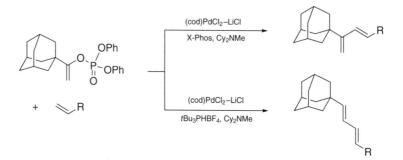

α-Substituted styrenes are synthesized from 1-alkynes by hydroarylation, using NaBAr$_4$ (e.g., Ar=Ph) as reagents.[14] Hydroalkenylation of alkynes, via hydropalladation, proceeds well with acrylic esters and amides.[15]

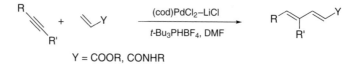

$$Y = COOR, CONHR$$

Another copper-free protocol of Sonogashira coupling instructs heating ArBr, 1-alkynes with PdCl$_2$, Ph$_3$P in pyrrolidine containing water at 120°.[16]

Silylalkynes and propargylic chlorides couple with loss of both functional groups, affording alka-1,2-dien-4-ynes.[17]

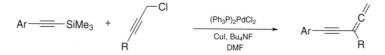

Ring formation.   Imines derived from *o*-alkynylaraldehydes react with chloroform to give 1-trichloromethyl-1,2-dihydroisoquinolines.[18]

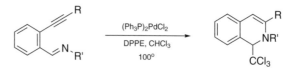

Carbonylation follows the formal alkenylation of 1,3-dicarbonyl compounds, therefore 5-acyl-3,4-dihydro-2-pyrones can be assembled from the Pd-catalyzed reaction involving alkynes and CO.[19]

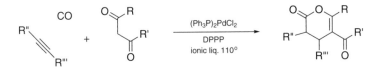

[1]Zawisza, A.M., Muzart, J. *TL* **48**, 6738 (2007).
[2]Grasa, G.A., Colacot, T.J. *OL* **9**, 5489 (2007).
[3]Jean, M., Renault, J., Uriac, P., Capet, M., van de Weghe, P. *OL* **9**, 3623 (2007).
[4]Turner, G.L., Morris, J.A., Greaney, M.F. *ACIE* **46**, 7996 (2007).
[5]Chuprakov, S., Chernyak, N., Dudnik, A.S., Gevorgyan, V. *OL* **9**, 2333 (2007).
[6]Seregin, I.V., Ryabova, V., Gevorgyan, V. *JACS* **129**, 7742 (2007).
[7]Vashehenko, V., Krivoshey, A., Knyazeva, I., Petrenko, A., Goodby, J.W. *TL* **49**, 1445 (2008).
[8]Ye, C., Xiao, J.-C., Twamley, B., LaLonde, A.D., Norton, M.G., Shreeve, J.M. *EJOC* 5095 (2007).
[9]Lipshutz, B.H., Taft, B.R. *OL* **10**, 1329 (2008).
[10]Lipshutz, B.H., Petersen, T.B., Abela, A.R. *OL* **10**, 1333 (2008).
[11]Adonin, N.Yu., Babushkin, D.E., Parmon, V.N., Bardin, V.V., Kostin, G.A., Mashukov, V.I., Frohn, H.-J. *T* **64**, 5920 (2008).
[12]Ruck, R.T., Huffman, M.A., Kim, M.M., Shevlin, M., Kandur, W.V., Davies, I.W. *ACIE* **47**, 4711 (2008).
[13]Ebran, J.-P., Hansen, A.L., Gogsig, T.M., Skrydstrup, T. *JACS* **129**, 6931 (2007).
[14]Zeng, H., Hua, R. *JOC* **73**, 558 (2008).
[15]Lindhardt, A.T., Mantel, M.L.H., Skrydstrup, T. *ACIE* **47**, 2668 (2008).
[16]Guan, J.T., Weng, T.Q., Yu, G.-A., Liu, S.H. *TL* **48**, 7129 (2007).
[17]Girard, D., Broussons, S., Provot, O., Brion, J.-D., Alami, M. *TL* **48**, 6022 (2007).
[18]Nakamura, H., Saito, H., Nanjo, M. *TL* **49**, 2697 (2008).
[19]Li, Y., Yu, Z., Alper, H. *OL* **9**, 1647 (2007).

Palladium(II) hydroxide/carbon.

Deprotection.[1]   Cleavage of benzyl ethers and cyclic acetals by heating with large amounts of this catalyst [20% $Pd(OH)_2/C$] in MeOH is totally impractical. It does not seem to offer any advantage over the less expensive Pd/C.

Hydrogenation.   An interesting transfer hydrogenation (with double bond migration) that converts 1,3-dihydroindoles to 4,5,6,7-tetrahydroisoindoles is catalyzed by $Pd(OH)_2/C$.[2]

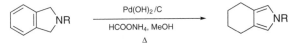

A benzylic nitro group is hydrogenolyzed in the presence of $Pd(OH)_2$, whereas homogeneous catalysts are ineffective and only low conversion is observed with Pt/C (and with Rh/C hydrogenation to amines occurs).[3] In conjunction with an enantioselective Henry reaction, chiral homobenzylic alcohols are accessible.

[1]Murali, C., Shashidhar, M.S., Gopinath, C.S. *T* **63**, 4149 (2007).
[2]Hou, D.-R., Wang, M.-S., Chung, M.-W., Hsieh, Y.-D., Tsai, H.-H.G. *JOC* **72**, 9231 (2007).
[3]Fesard, T.C., Motoyoshi, H., Carreira, E.M. *ACIE* **46**, 2078 (2007).

Palladium(II) iodide.

Oxidation. Benzils are formed on heating diarylethynes with PdI_2 and DMSO at $140°$.[1]

Coupling.[2] Palladation of allenyl carbinols leads to 2,5-dihydrofuran intermediates that contain a C—Pd bond at C-3. Such intermediates can undergo coupling with unreacted allenyl carbinols. Because substitution at the β-carbon affects rates of palladation two different carbinols can be joined.

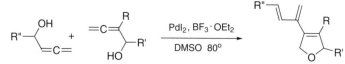

Coumarin synthesis.[3] Double carbonylation is involved when α-(o-hydroxyaryl)parpargyl alcohols are placed in an autoclave with PdI_2, KI, and CO in MeOH.

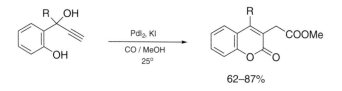

62–87%

[1]Mousset, C., Provot, O., Hamze, A., Bignon, J., Brion, J.-D., Alami, M. *T* **64**, 4287 (2008).
[2]Deng, Y., Li, J., Ma, S. *CEJ* **14**, 4263 (2008).
[3]Gabriele, B., Mancuso, R., Salerno, G., Plastina, P. *JOC* **73**, 756 (2008).

Palladium(II) triflate.

1-Indenols. Coupling between o-formylarylboronic acids with alkynes leads to indenols. The catalyst system includes $Pd(OTf)_2 \cdot 2H_2O$ and Me_4-SEGPHOS ligand.[1]

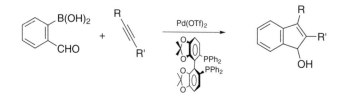

[1]Yang, M., Zhang, X., Liu, X. *OL* **9**, 5131 (2007).

Pentafluoroanilinium triflate.

Acylation.[1]   Silyl enol ethers and ketene silyl acetals are acylated to give β-keto carbonyl compounds with $C_6F_5NH_3OTf$ as catalyst.

[1]Iida, A., Osada, J., Nagase, R., Misaki, T., Tanabe, Y. *OL* **9**, 1859 (2007).

Pentamethylcyclopentadienylbis(vinyltrimethylsilane)cobalt.

Hydrogen transfer.[1]   Cyclic amines are dehydrogenated via hydrogen transfer to the *N*-vinyldimethylsilyl group, attached by heating with the title catalyst.

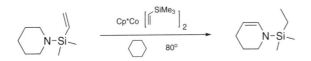

[1]Bolig, A.D., Brookhart, M. *JACS* **129**, 14544 (2007).

Perfluorooctanesulfonic acid.

Pictet–Spengler reaction.[1]   Serving as both a Bronsted acid and a surfactant the sulfonic acid enables Pictet–Spengler reaction in water or aqueous hexafluoroisopropanol.

[1]Saito, A., Numaguchi, J., Hanzawa, Y. *TL* **48**, 835 (2007).

Perrhenic acid.

Phosphorylation.[1]   Alcohols are phosphorylated by phosphoric acid on heating with $ReO_3(OH)$ and Bu_2NH in NMP.

[1]Sakakura, A., Katsukawa, M., Ishihara, K. *ACIE* **46**, 1423 (2007).

Phase-transfer catalysts.

Robust catalyst.   Didecyldimethylammonium bromide is proposed as a universal potent PTC because it is stable to heat and alkali.[1]

Dehydrochlorination.   The ready availability of β-chloro enamides makes them useful precursors of ynamides. Thus phase-transfer conditions (e.g., with Bu_4NHSO_4, 50% NaOH in PhMe) are generally applied to complete the preparation.[2]

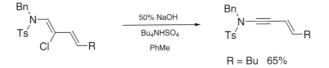

Isomerization.[3] Alkenylidenecyclopropanes can be converted into the conjugated alkenylcyclopropenes by heating with Bu_4NHSO_4 and NaOH in toluene at 60°.

Alkylation.[4] Fischer carbene complexes can be allylated at room temperature under phase transfer conditions.

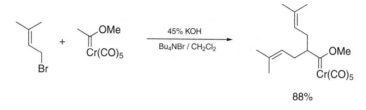

88%

[1]Chidambaram, M., Sonavane, S.U., de la Zerda, J., Sasson, Y. *T* **63**, 7696 (2007).
[2]Couty, S., Barbazanges, M., Meyer, C., Cossy, J. *SL* 905 (2005).
[3]Shao, L.-X., Zhang, Y.-P., Qi, M.-H., Shi, M. *OL* **9**, 117 (2007).
[4]Menon, S., Sinha-Mahapatra, D., Herndon, J.W. *T* **63**, 8788 (2007).

Phenyliodine(III) bis(trifluoroacetate).

Resin oxidant. By mixing $PhI(OCOCF_3)_2$ with a resin containing trimethylammonium iodide subunits, iodolysis occurs and the resin becomes attached to trifluoroacetoxyiodate anions. The new resin is capable of mediating the oxidative hydrolysis of dithioacetals.[1]

Addition reactions. Electron-rich styrenes undergo twofold addition to give arylacetaldehyde 1,1-bis(trifluoroacetates).[2] The phenyliodonio intermediates are subject to displacement, for example by *N*-acylhydrazines.[3]

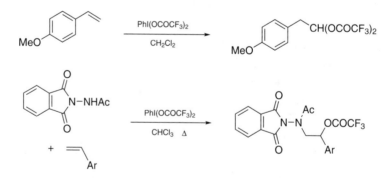

4-Alkynoic acids and amides cyclize to provide γ-acyl-γ-butyrolactones and lactams, respectively.[4] But $PhI(OCOCF_3)_2$ plays a different role when the substrates are changed to 3-alkenoic acids (with the presence of PhSeSePh in 5 mol%).[5]

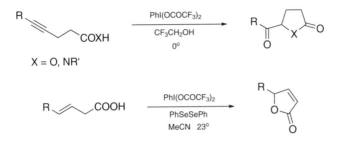

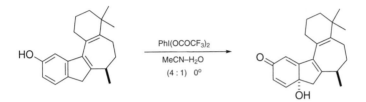

Oxidation of aromatic compounds. The completion of a synthesis of frondosin-C relied on the conversion of the phenolic moiety to a 4-hydroxy-2,5-cyclohexadienone by PhI(OCOCF$_3$)$_2$.[6] The transformation cannot be done with PhI(OAc)$_2$

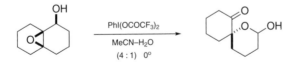

Naphthalene is oxidized by PhI(OCOCF$_3$)$_2$, becoming an arylating agent for mesitylene and other polymethylbenzenes in the presence of BF$_3$ · OEt$_2$.[7]

The highly reactive π-bond of an indole nucleus can be protected by the addition of an ethylenedioxy unit. The adducts are formed in a reaction with ethanediol in the presence of PhI(OCOCF$_3$)$_2$. Reversion of the process is accomplished by NaBH$_3$CN.[8]

Oxidative cleavage and rearrangement. Epoxy carbinols of the oxapropellane-type are driven by ring strain to undergo hydrative devolution. Formation of phenyliodoniodi-oxolanes provokes an intramolecular redox decomposition and further transformations.[9]

[1]Luiken, S., Kirschning, A. *JOC* **73**, 2018 (2008).
[2]Tellitu, I., Dominguez, E. *T* **64**, 2465 (2008).
[3]Murata, K., Tsukamoto, M., Sakamoto, T., Saito, S., Kikugawa, Y. *S* 32 (2008).
[4]Tellitu, I., Serna, S., Herrero, M.T., Moreno, L., Dominguez, E., SanMartin, R. *JOC* **72**, 1526 (2007).
[5]Browne, D.M., Niyomura, O., Wirth, T. *OL* **9**, 3169 (2007).
[6]Li, X., Kyne, R.E., Ovaska, T.V. *T* **63**, 1899 (2007).
[7]Dohi, T., Ito, M., Morimoto, K., Iwata, M., Kita, Y. *ACIE* **47**, 1301 (2008).

[8]Takayama, H., Misawa, K., Okada, N., Ishikawa, H., Kitajima, M., Hatori, Y., Murayama, T., Wongseripipatana, S., Tashima, K., Matsumoto, K., Horie, S. *OL* **8**, 5705 (2006).
[9]Fujioka, H., Matsuda, S., Horai, M., Fujii, E., Morishita, M., Nishiguchi, N., Hata, K., Kita, Y. *CEJ* **13**, 5238 (2007).

Phenyliodine(III) diacetate.

Generation of electrophilic halogen.   The title reagent has found use in oxidation of KBr to initiate cyclization of homoallylic sulfonamides.[1] The use of the *p*-anisyl congener to oxidize powdered KBr for benzylic bromination is also realized. 4-Arylbutanoic acids are converted into γ-aryl-γ-butyrolactones by this method.[2]

Oxidative functionalization.   *p*-Substituted phenols are oxidized and trapped by various nucleophiles. For example, reaction carried out in the presence of MeCN furnishes 4-acetamino-2,5-cyclohexadienones.[3] *N*-Tosylimines of 4-fluoro-2,5-cyclohexadienones are obtained when HF-pyridine is added to the corresponding reaction of *N*-tosylanilines.[4] Hydroxylation of 4-arylphenols (in aqueous MeCN) proceeds much better from the corresponding trimethylsilyl ethers.[5]

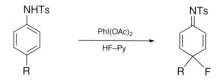

Using an allylsilane as trapping agent for the reactive species derived from phenols, formation of 2-silylmethyldihydrobenzofurans is observed.[6]

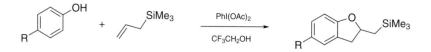

Thioureas suffer desulfurization and *N*-acetylation on treatment with PhI(OAc)$_2$. Acetylation occurs at the nitrogen atom with a lower pKa.[7]

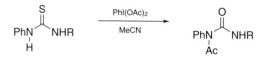

Oxidation of guanidines can lead to cyclization, i.e., functionalization at an unactivated carbon atom.[8] Such a reaction is of obvious synthetic interest.

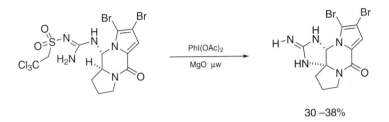

30 –38%

In the last step of a dibromophakellstatin synthesis,[9] PhI(OAc)₂ serves well to effect closing of the central piperazinone moiety in what is actually a vicinal diamination reaction. [Pd(OAc)₂ – DMSO can also be employed.]

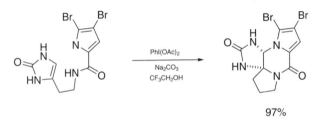

97%

vic-Dioxygenation of double bonds has been carried out at room temperature with PhI(OAc)₂ and catalytic amounts of the (dppp)Pd(2H₂O)(OTf)₂ complex.[10]

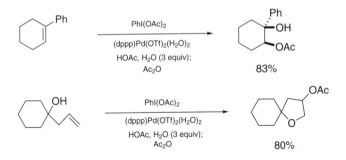

83%

80%

3-Butenols are converted into 3-phthalimidotetrahydrofurans in an oxidative cyclization process in the presence of phthalimide.[11] The oxidation is mediated by PhI(OAc)₂ while a combination of Pd(OAc)₂ and AgBF₄ provides the necessary catalyst.

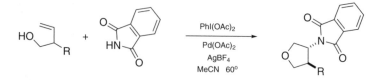

α-Substituted benzylidenecyclopropanes and cyclobutanes undergo ring-enlarging rearrangement on reaction with the nitrenoid generated from N-aminophthalimide [with PhI(OAc)$_2$] to provide homologous cycloalkanone hydrazones.[12]

Trapping of oxidized phenols by reaction with furan features a formal [3 + 2]cyclo-addition.[13]

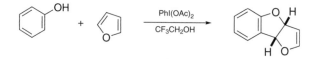

For cleavage of 1,2-diols, a polymer-linked aryliodine(III) diacetate has been developed.[14]

[1]Fan, R., Wen, F., Qin, L., Pu, D., Wang, B. *TL* **48**, 7444 (2007).

[2]Dohi, T., Takenaga, N., Goto, A., Maruyama, A., Kita, Y. *OL* **9**, 3129 (2007).

[3]Liang, H., Ciufolini, M.A. *JOC* **73**, 4299 (2008).

[4]Basset, L., Martin-Mingot, A., Jouannetaud, M.-P., Jacquesy, J.-C. *TL* **49**, 1551 (2008).

[5]Felpin, F.-X. *TL* **48**, 409 (2007).

[6]Berard, D., Racicot, L., Sabot, C., Canesi, S. *SL* 1076 (2008).

[7]Singh, C.B., Ghosh, H., Murru, S., Patel, B.K. *JOC* **73**, 2924 (2008).

[8]Wang, S., Romo, D. *ACIE* **47**, 1284 (2008).

[9]Lu, J., Tan, X., Chen, C. *JACS* **129**, 7768 (2007).

[10]Li, Y., Song, D., Dong, V.M. *JACS* **130**, 2962 (2008).

[11]Desai, L.V., Sanford, M.S. *ACIE* **46**, 5737 (2007).

[12]Liang, Y., Jiao, L., Wang, Y., Chen, Y., Ma, L., Xu, J., Zhang, S., Yu, Z.-X. *OL* **8**, 5877 (2006).

[13]Berard, D., Jean, A., Canesi, S. *TL* **48**, 8238 (2007).

[14]Chen, F.-E., Xie, B., Zhang, P., Zhao, J.-F., Wang, H., Zhao, L. *SL* 619 (2007).

Phenyliodine(III) diacetate – copper salts.

Iodoalkynes.   Iodination agent for 1-alkynes is generated from a mixture of PhI(OAc)$_2$ and KI, and the iodination is catalyzed by CuI while the liberated proton is removed by Et$_3$N.[1]

Homocoupling of 1-alkynes is achieved. Dialkynylpalladium intermediates are formed from alkynes in the presence of PdCl$_2$, CuI, Ph$_3$P, and Et$_3$N, and the Pd(0) species liberated on reductive elimination is reoxidized with PhI(OAc)$_2$ in situ.[2]

C−H Functionalization.   Allylic and benzylic positions are oxygenated by reaction with N-hydroxyphthalimide, and the reaction is realized with the assistance of PhI(OAc)$_2$−CuCl.[3]

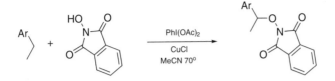

Cyclic ethers are also activated at an α-position on treatment with PhI(OAc)$_2$–Cu(OTf)$_2$ and tosylamination can be accomplished by introducing TsNH$_2$ into the reaction media.[4]

Three-membered rings. The combination of PhI(OAc)$_2$ and Cu(II) trifluoro-acetylacetonate is particularly useful for generating nitrenoid from 6-methylpyridine-2-sulfonamide.[5]

A cyclopropanation agent is created from MeNO$_2$ by PhI(OAc)$_2$ and Rh$_2$(esp)$_4$.[6]

[1]Yan, J., Li, J., Cheng, D. *SL* 2442 (2007).
[2]Yan, J., Lin, F., Yang, Z. *S* 1301 (2007).
[3]Lee, J.M., Park, E.J., Cho, S.H., Chang, S. *JACS* **130**, 7824 (2008).
[4]He, L., Yu, J., Zhang, J., Yu, X.-Q. *OL* **9**, 2277 (2007).
[5]Han, H., Park, S.B., Kim, S.K., Chang, S. *JOC* **73**, 2862 (2008).
[6]Bonge, H.T., Hansen, T. *SL* 55 (2007).

Phenyliodine(III) diacetate – iodine.

Ring cleavage. Several oxidants can cleave lactols at the Cα—Cβ bond, and the PhI(OAc)$_2$–I$_2$ pair is useful for attaching an iodine atom to Cβ in the process. From bicyclic lactols such as those prepared from alkylation of cycloalkanones with epoxides as substrates iodolactones are obtained.[1]

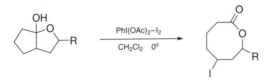

Oxidation of amines. Secondary alkyl amines undergo oxidative cyclization to give pyrrolidines. However, *N*-tosylbenzylamines can only be dehydrogenated.[2]

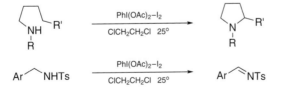

[1]Maio, W.A., Sinishtaj, S., Posner, G.H. *OL* **9**, 2673 (2007).
[2]Fan, R., Pu, D., Wen, F., Wu, J. *JOC* **72**, 8994 (2007).

Phenyliodine(III) dichloride.

Oxidation.   The title reagent is available from chlorination of PhI with $NaClO_n$ in hydrochloric acid. It can be used in conjunction with TEMPO to oxidize alcohols.[1]

[1]Zhao, X.-F., Zhang, C. *S* 551 (2007).

Phenylphosphinic acid.

Ugi reaction.[1]   A new catalyst for the three-component condensation is PhPO(OH)H.

[1]Pan, S.C., List, B. *ACIE* **47**, 3622 (2008).

Phenylselenium triflate.

Glycosylation.[1]   Glycosyl 4-pentenoates are activated by PhSeOTf and smooth glycosylation is achieved (2,4,6-tri-*t*-butylpyrimidine is used as base).

[1]Choi, T.J., Baek, J.Y., Jeon, H.B., Kim, K.S. *TL* **47**, 9191 (2006).

4-Phenyl-1,2,4-triazoline-3,5-dione.

Disulfides.[1]   The title heterocycle rapidly oxidizes thiols at room temperature.

[1]Christoforou, A., Nicolaou, G., Elemes, Y. *TL* **47**, 9211 (2006).

Pinacolatoboryl azide.

C—H insertion.[1]   Hydrocarbons such as cyclohexane are functionalized (converted to amines) in good yields by co-irradiation with the title reagent to afford *B*-aminoboryl pinacolates.

[1]Bettinger, H.F., Fiethaus, M., Bornemann, H., Oppel, I.M. *ACIE* **47**, 4744 (2008).

Piperidine.

Condensation.[1]   A remarkable solvent dependence of the product geometry in the condensation of aldehydes with nitroalkanes has been unraveled.

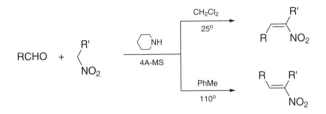

[1]Fioravanti, S., Pellacani, L., Tardella, P.A., Vergari, M. C. *OL* **10**, 1449 (2008).

Platinum and complexes.

Hydrogenation and dehydrogenation. Supported on carbon in a nanofiber form Pt catalyzes selective hydrogenation of nitroarenes to give arylamines in the presence of many other functional groups.[1]

Diphenylamine and 2-aminobiphenyl are dehydrogenated to furnish carbazole on heating with Pt/C and some water at 250°.[2] It is thought that [Pt-H][OH] are formed. Under similar conditions Pd/C is ineffectual.

The Pt-catalyzed hydrogenation of the pyridine nucleus in HOAc is facilitated by microwave.[3]

Addition reactions. Hydrodisilanes generate Pt-coordinated silylenes on exposure to (cod)PtCl$_2$. These silylenoids conjugatively add to enones,[4] except those containing a (Z)-methyl substituent at the β-carbon (in such cases the major products are 1-oxa-2-sila-5-cyclohexenes).[5]

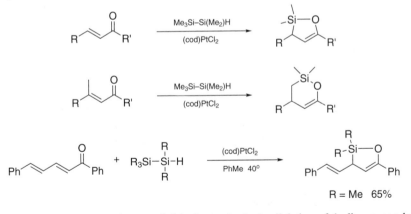

R = Me 65%

A synthesis of (E)-1-triorganosilyl-1-alkenes by hydrosilylation of 1-alkynes can be accomplished in the presence of a carbene-Pt complex in which the metal is also ligated to the two double bonds of diallyl ether.[6]

Allylation. The (cod)PtCl$_2$ complex associated with bis(o-diphenylphosphinophenyl) ether effects monoallylation of amines by allylic alcohols.[7]

[1]Takasaki, M., Motoyama, Y., Higashi, K., Yoon, S.-H., Mochida, I., Nagashima, H. *OL* **10**, 1601 (2008).
[2]Yamamoto, M., Matsubara, S. *CL* **36**, 172 (2007).
[3]Piras, L., Genesio, E., Ghiron, C., Taddei, M. *SL* 1125 (2008).
[4]Okamoto, K., Hayashi, T. *CL* **37**, 108 (2008).
[5]Okamoto, K., Hayashi, T. *OL* **9**, 5067 (2007).
[6]Berthon-Gelloz, G., Schumers, J.-M., De Bo, G., Marko, I.E. *JOC* **73**, 4190 (2008).
[7]Utsunomiya, M., Miyamoto, Y., Ipposhi, J., Ohshima, T., Mashima, K. *OL* **9**, 3371 (2007).

Platinum(II) acetylacetonate.

Addition reactions.[1]   Addition of borylsilanes to the triple bond of a conjugated enyne, catalyzed by Pt(acac)$_2$, is subject to substrate control in that the steric bulk of the terminal substituent exerts an important effect.

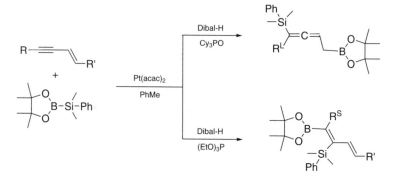

[1]Lüken, C., Moberg, C. *OL* **10**, 2505 (2008).

Platinum(II) chloride.

Addition reactions.   Hydrosilanes add to ethynylarenes regioselectively and stereoselectively, when catalyzed by PtCl$_2$ in the presence of X-Phos.[1]

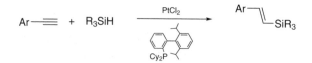

Indoles undergo tetrahydrofuranylation (at C-3) when mixed with 3-alkynols and PtCl$_2$.[2]

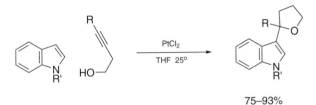

75–93%

Cyclization. Activation of an alkyne by Pt(II) often initiates nucleophilic attack, and with a well-juxtaposed double bond cyclization ensues. The concluding act may then involve addition of an alkoxy unit,[3] or neutralization of the positive charge with a hydride source such as a hydrosilane.[4]

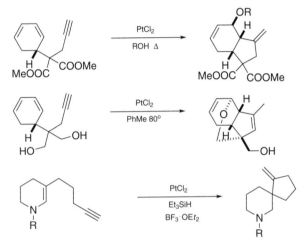

R, electron-withdrawing

The deprotonation option is also available.[5,6]

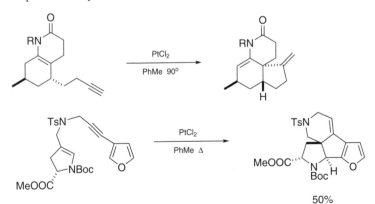

50%

On treatment with PtCl$_2$ under CO in toluene *o*-alkynylaryl alkoxymethyl ethers cyclize to afford benzofurans, by transfer of the alkocymethyl group to C-3.[7]

Cyclization that eventualy places the resulting double bond endocyclic also occurs, as in the annulation of indoles[8] and the formation of spirocyclic furanones and pyrrolones which involves a rearrangement process.[9]

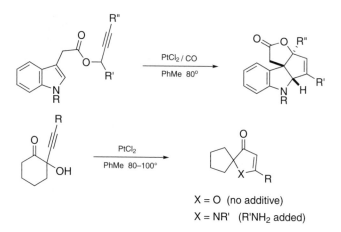

X = O (no additive)
X = NR' (R'NH$_2$ added)

Rearrangement precedes cyclization of α-arylpropargyl acetates to provide 2-acetoxy-indenes.[10]

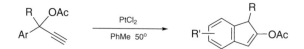

α-(*o*-Allylaminoaryl)propargylic ethers and acetates are converted to 2-(3-butenyl)-3-oxyindoles. A [3,3]sigmatropic rearrangement takes place after nucleophilic attack on the Pt-activated triple bond by the nitrogen atom.[11]

R = Me, Ac

α-Allylpropargyl ethers undergo cyclization to give bicyclo[3.1.0]hex-2-enyl ethers,[12] *N*-(*o*-alkynyl)lactams are transformed into indoles.[13] The net results are equivalent to insertion of alkylidenecarbenes into double and single bonds.

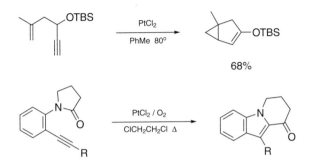

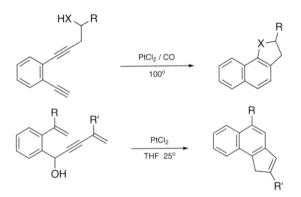

o-Diynylarenes aromatized with participation of the nucleophile appended on the chain extending from one of the triple bonds, naphthannulated heterocycles are thereby created.[14] A more convoluted cyclization is that represented by α-enynyl *o*-alkenylbenzyl alcohols.[15]

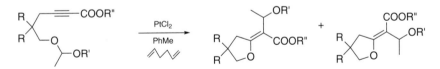

Transformation of 1,2,4-alkatrienes into cyclopentadienes is catalyzed by PtCl₂ at room temperature.[16] Cycloisomerization to break up an acetal unit and re-add the O/C bonding partners to a conjugated triple bond has been observed.[17]

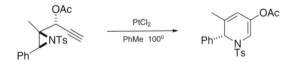

α-(2-Aziridinyl)propargyl acetates are subject to isomerization, which is instigated by the shift of the ester unit. Dihydropyridines are formed.[18]

Cycloaddition.   Many types of cycloaddition reactions catalyzed by PtCl₂ have been discovered. Platinized carbonyl ylides formed in situ from 4-alkynones are intercepted by electron-rich alkenes.[19]

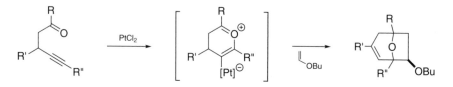

Intramolecular cycloaddition involving conjugated diene and allene units leads to bicyclic products that contain a 1,4-cycloheptadiene nucleus.[20] [6 + 2]Cycloaddition of some cycloheptatrienes appended with a sidechain containing a triple bond occurs at room temperature, with PtCl₂ to promote it. At higher temperature, the formal [6 + 1]-cycloadducts are formed.[21]

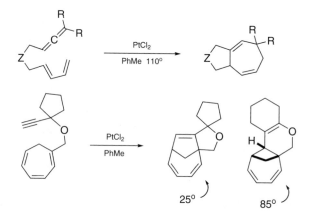

An intramolecular alkenylation is the key step for a synthesis of the zwitterionic dihetero analogue of pyrene in which the two internal carbon atoms are replaced by N and B atoms.[22]

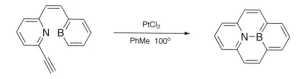

Isomerization.   3-Acyloxyalkynyl ethers undergo rearrangement to afford 2-alkylidene-3-oxoalkanoates. Metal salts that induce the reaction may affect the stereoselectivity differently.[23]

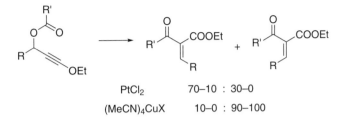

PtCl₂   70–10 : 30–0

(MeCN)₄CuX   10–0 : 90–100

[1]Hamze, A., Provot, O., Brion, J.-D., Alami, M. *TL* **49**, 2429 (2008).
[2]Bhuvaneswari, S., Jeganmohan, M., Cheng, C.-H. *CEJ* **13**, 8285 (2007).
[3]Yeh, M.-C.P., Tsao, W.-C., Cheng, S.-T. *JOC* **73**, 2902 (2008).
[4]Harrison, T.J., Patrick, B.O., Dake, G.R. *OL* **9**, 367 (2007).
[5]Kozak, J.A., Dake, G.R. *ACIE* **47**, 4221 (2008).
[6]Deng, H., Yang, X., Tong, Z., Li, Z., Zhai, H. *OL* **10**, 1791 (2008).
[7]Fürstner, A., Heilmann, E. K., Davies, P.W. *ACIE* **46**, 4760 (2007).
[8]Zhang, G., Catalano, V.J., Zhang, L. *JACS* **129**, 11358 (2007).
[9]Binder, J.T., Crone, B., Kirsch, S.F., Liebert, C., Menz, H. *EJOC* 1636 (2007).
[10]Nakanishi, Y., Miki, K., Ohe, K. *T* **63**, 12138 (2007).
[11]Cariou, K., Ronan, B., Mignani, S., Fensterbank, L., Malacria, M. *ACIE* **46**, 1881 (2007).
[12]Blaszykowski, C., Harrak, Y., Brancour, C., Nakama, K., Dhimane, A.L., Fensterbank, L., Malacria, M. *S* 2037 (2007).
[13]Li, G., Huang, X., Zhang, L. *ACIE* **47**, 346 (2008).
[14]Taduri, B.P., Odedra, A., Lung, C.-Y., Liu, R.-S. *S* 2050 (2007).
[15]Abu Sohel, S.M., Lin, S.-H., Liu, R.-S. *SL* 745 (2008).
[16]Funami, H., Kusama, H., Iwasawa, N. *ACIE* **46**, 909 (2007).
[17]Nakamura, I., Chan, C.S., Araki, T., Terada, M., Yamamoto, Y. *OL* **10**, 309 (2008).
[18]Motamed, M., Bunnelle, E.M., Singaram, S.W., Sarpong, R. *OL* **9**, 2167 (2007).
[19]Kusama, H., Ishida, K., Funami, H., Iwasawa, N. *ACIE* **47**, 4903 (2008).
[20]Trillo, B., Lopez, F., Gulias, M., Castedo, L., Mascarenas, J.L. *ACIE* **47**, 951 (2008).
[21]Tenaglia, A., Gaillard, S. *ACIE* **47**, 2454 (2008).
[22]Bosdet, M.J.D., Piers, W.E., Sorensen, T.S., Parvez, M. *ACIE* **46**, 4940 (2007).
[23]Barluenga, J., Riesgo, L., Vicente, R., Lopez, L.A., Tomas, M. *JACS* **129**, 7772 (2007).

Platinum(II) chloride – silver salts.

Addition reaction.   Hydroarylation of 2-alkynoic acids and esters occur on their treatment with ArH and PtCl₂–AgOTf in TFA at room temperature.[1]

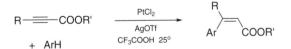

Tetrahydropyrans.   Reductive incorporation of aldehydes into a tetrahydropyran ring by combining with certain unsaturated alcohols is mediated by the PtCl₂–AgOTf couple.[2]

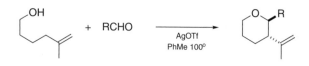

[1] Oyamada, J., Kitamura, T. *T* **63**, 12754 (2007).
[2] Miura, K., Horiike, M., Inoue, G., Ichikawa, J., Hosomi, A. *CL* **37**, 270 (2008).

Platinum(II) iodide.

Indolizines. Cycloisomerization of γ-(2-pyridyl)propargylic esters catalyzed by PtI_2–Ph_3P is found to be affected by substituents at the α-position and the acyl group.[1] Perhaps formation of two different types of products is determined by the degree of loosening of the propargyloxy bond.[1]

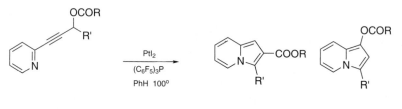

R: e-withdrawing   R: e-donating
R': e-donating   R': e-withdrawing

[1] Hardin, A.R., Sarpong, R. *OL* **9**, 4547 (2007).

Poly(methylhydrosiloxane), PMHS.

Reductive transformations. Reduction of aldehydes to primary alcohols is accomplished by reduction with PMHS which is catalyzed by $Fe(OAc)_2$–Cy_3P.[1]

PMHS also finds use in the reductive amination of β-hydroxy ketones to afford *syn*-1,3-amino alcohols using $(i\text{-PrO})_4Ti$ as a catalyst.[2] With $Pd(OH)_2/C$ as catalyst a mixture of RCN and $ArNX_2$ (X = H or O) is converted by PMHS in ethanol to RCH_2NHAr.[3]

The stabilized copper hydride species **1** is obtained by mixing PMHS with $Cu(OAc)_2 \cdot 2H_2O$ in *t*-BuOH and toluene in the presence of 1,2-bis(diphenylphosphino)-benzene at room temperature.[4] It can be used in lieu of the Stryker reagent.

(1)

Alternatively, carbene-complexed copper hydride species is involved in the transformation of alkynyl epoxides to allenyl carbinols using CuI, *t*-BuONa, PMHS and a imidazolium salt.[5]

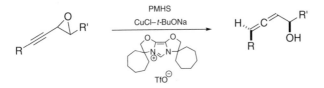

Hydroiodination.   A combination of PMHS and iodine is useful for hydroiodination of alkenes and alkynes, in the Markovnikov sense, at room temperature in $CHCl_3$.[6]

[1]Shaikh, N.S., Junge, K., Beller, M. *OL* **9**, 5429 (2007).
[2]Menche, D., Arikan, F., Li, J., Rudolph, S. *OL* **9**, 267 (2007).
[3]Reddy, C.R., Vijeender, K., Bhusan, P.B., Madhavi, P.P., Chandrasekhar, S. *TL* **48**, 2765 (2007).
[4]Baker, B.A., Boskovic, Z.V., Lipshutz, B.H. *OL* **10**, 289 (2008).
[5]Deutsch, D., Lipshutz, B.H., Krause, N. *ACIE* **46**, 1650 (2007).
[6]Das, B., Srinivas, Y., Holla, H., Narender, R. *CL* **36**, 800 (2007).

Potassium *t*-butoxide.

Vinylation.   Condensation of aldehydes with chloromethyl methyl sulfones results in the formation of 1-alken-3-ols. This late stage of this interesting reaction resembles the Payne rearrangement and it is terminated by extrusion of SO_2.[1]

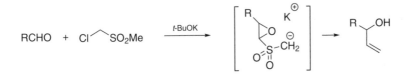

Elimination.   Enamine *N*-oxides are synthesized from β-chloroalkylamines in two steps. *N*-Oxidation by MCPBA and dehydrochlorination with *t*-BuOK.[2]

Aldol reaction.   Cyclization of a keto ester effected by *t*-BuOK in reasonably good yield and stereoselectivity constitutes a key step toward completion of a synthesis of (-)-salinosporamide-A.[3]

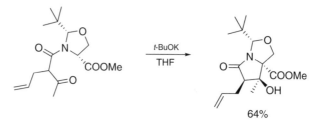

64%

[1]Makosza, M., Urbanska, N., Chesnokov, A.A. *TL* **44**, 1473 (2003).
[2]Bernier, D., Blake, A.J., Woodward, S. *JOC* **73**, 4229 (2008).
[3]Ling, T., Macherla, V.R., Manam, R.R., McArthur, K.A., Potts, B.C.M. *OL* **9**, 2289 (2007).

Potassium fluoride.

N-Allylation.   To achieve monoallylation of arylamines in MeCN a useful base system is KF on Celite.[1]

Dialkyl fluorophosphates.   Oxidative fluorination of dialkyl phosphinites occurs on heating with KF and Cl_3CCN.[2]

Carbamates.   On Hofmann rearrangement of an amide $RCONH_2$ the treatment with KF/Al_2O_3 in MeOH leads to $RNHCOOMe$.[3]

Deformylation.   Arylamines are liberated from formanilides by mixing with KF/Al_2O_3 and microwave irradiation.[4]

[1]Pace, V., Martinez, F., Fernandez, M., Sinisterra, J.V., Alcantara, A.R. *OL* **9**, 2661 (2007).
[2]Gupta, A.K., Acharya, J., Pardasani, D., Dubey, D.K. *TL* **49**, 2232 (2008).
[3]Gogoi, P., Konwar, D. *TL* **48**, 531 (2007).
[4]Ge, Y., Hu, L. *TL* **48**, 4585 (2007).

Potassium monoperoxysulfate, Oxone®.

Amides.   A new protocol for making ArCONHAr′ from ArCHO and Ar′NH₂ is by ball-milling the mixture with Oxone.[1]

Epoxidation.   N-(Bis[diisopropylamino]methoxyphosphonio)oxaziridines **1** are valuable oxygen donors. They are accessible by oxgen transfer from Oxone to the iminium salts. Actually the imine can be used in catalytic quantities for the oxygen transfer reaction.[2]

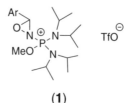

(1)

Quinones.   Oxidation of *p*-methoxyphenols by Oxone is catalyzed by *p*-iodophenoxy-acetic acid.[3]

[1]Gao, J., Wang, G.-W. *JOC* **73**, 2955 (2008).
[2]Prieur, D., El Kazzi, A., Kato, T., Gornitzka, H., Baceiredo, A. *OL* **10**, 2291 (2008).
[3]Yakura, T., Konishi, T. *SL* 765 (2007).

Potassium osmate.

Oxidative cyclization.[1]   The scope of oxidative cyclization by $K_2OsO_2(OH)_4$ in forming 2-hydroxymethylpyrrolidines from aminoalkene derivatives is expanded by adding pyridine-*N*-oxide and citric acid to the reaction medium that contains TFA, acetone and water.

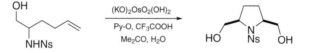

Aminohydroxylation.[2]   Functionalization of the double bond of an allylic alcohol by intramolecular *cis*-aminohydroxylation is attended by dramatic improvement by derivatizing the alcohols into *N*-hydroxycarbamates and thence the *N*-pentafluorobenzoyloxy carbamates.

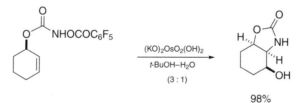

98%

[1]Donohoe, T.J., Wheelhouse, K.M.P., Lindsay-Scott, P.J., Glossop, P.A., Nash, I.A., Parker, J.S. *ACIE* **47**, 2872 (2008).
[2]Donohoe, T.J., Bataille, C.J.R., Gattrelle, W., Kloesges, J., Rossignol, E. *OL* **9**, 1725 (2007).

Potassium permanganate.

α-Acyloxylation.   Heating 2-cycloalkenones (including indanone and α-tetralone) with a carboxylic acid and $KMnO_4$ in benzene furnishes the α′-acyloxy derivatives.[1]

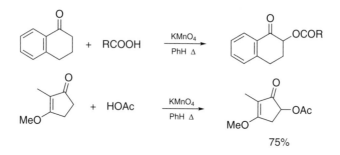

[1]Demir, A.S., Findik, H. *T* **64**, 6196 (2008).

Potassium tetrachloroaurate.

Allenes.   1,3-Chirality transfer in the hydrodeamination of *N*-propargyl(*S*)-prolinols which occurs on treatment with KAuCl$_4$ in MeCN.[1]

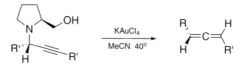

[1]Lo, V.K.-Y., Wong, M.-K., Che, C.-M. *OL* **10**, 517 (2008).

o-(Prenyloxymethyl)benzoic acid.

Hydroxyl protection.   By using the title reagent in the Mitsunobu reaction alcohols are protected. The prenyl group of the derived esters is removable by DDQ and subsequent addition of Yb(OTf)$_3$ promotes lactonization to liberate the alcohols.[1]

[1]Vatèle, J.-M. *T* **63**, 10921 (2007).

(*S*)-Proline.

Aldol reaction.   Aldol reaction catalyzed by proline and derivatives has been reviewed.[1] A ball-mill operation on cycloalkanones, ArCHO with (*S*)-proline leads to predominantly *anti*-aldol products.[2] The aldol reaction between 4-tetrahydrothiapyrone with the racemic 3-aldehyde based on the same heterocycle shows excellent enantiotopic group-selectivity and thence manifesting dynamic kinetic resolution.[3]

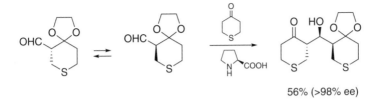

56% (>98% ee)

(S)-Proline is effective in mediating an asymmetric transannular aldol reaction.[4]

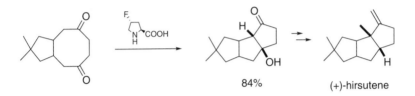

84%   (+)-hirsutene

The *anti*-selective aldol reaction between cyclohexanone and ArCHO reaches >99% ee if it is conducted in the presence of *trans*-4-(4-*t*-butylphenoxy)-L-proline and sulfated β-cyclodextrin in water at room temperature.[5] Another catalyst is *cis*-4-(1-adamantane-carboxamido)-(S)-proline (**1**) in conjunction with β-cyclodextrin.[6]

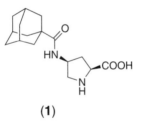

(1)

It appears best to carry out aldol reaction between two aldehydes in dry conditions, and between a ketone and an aldehyde under wet conditions.[7]

The use of proline to accomplish the Friedländer reaction to produce 2-substituted 4-trifluoromethylquinolines takes advantage of its efficiency rather than chiroptical results (for there is none).[8]

Miscellaneous reactions.   The starting point of a practical route to β²-amino acids is the proline-catalyzed Mannich reaction of aldehydes with a chiral iminium salt derived from *N*-benzyl-*N*-(α-phenethyl)amine.[9] Condensation of aldehydes with *N*-Boc imines furnishes mainly *syn*-adducts.[10,11]

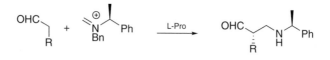

Asymmetric Michael reactions have been conducted with assistance of C_2-symmetric malonamides derived from (S)-proline esters.[12] 2-Methyl-3,4,5,6-tetrahydropyridine and 2-cyclopentenone are condensed to afford a tricyclic alcohol. The reaction starts from Michael reaction of the endocyclic enamine isomer and as the double bond shifts to the exo-cyclic position an intramolecular aldol reaction follows. If the imine is lithiated, the initial Michael reaction (CuBr-catalyzed) then involves the exocyclic carbon.[13]

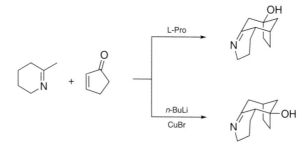

Perhaps through enamine formation from ketones proline catalyzes a Diels–Alder reaction with 1,2,4,5-tetrazines.[14]

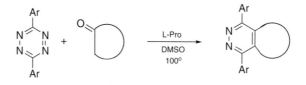

A new use of proline is in the aromatic substitution for converting ArX to aryl cyanides (CuCN, DMF, $80° - 120°$).[15]

[1]Guillena, G., Najera, C., Ramon, D.J. *TA* **18**, 2249 (2007).
[2]Rodriguez, B., Bruckmann, A., Bolm, C. *CEJ* **13**, 4710 (2007).
[3]Ward, D.E., Jheengut, V., Akinnusi, O.T. *OL* **7**, 1181 (2005).
[4]Chandler, C.L., List, B. *JACS* **130**, 6737 (2008).
[5]Huang, J., Zhang, X., Armstrong, D.W. *ACIE* **46**, 9073 (2007).
[6]Liu, K., Häussinger, D., Woggon, W.-D. *SL* 2298 (2007).
[7]Hayashi, Y., Aratake, S., Itoh, T., Okano, T., Sumiya, T., Shoji, M. *CC* 957 (2007).
[8]Jiang, B. Dong, J., Jin, Y., Du, X., Xu, M. *EJOC* 2693 (2008).

[9]Chi, Y., English, E.P., Pomerantz, W.C., Horne, W.S., Joyce, L.A., Alexander, L.R., Fleming, W.S., Hopkins, E.A., Gellman, S.H. *JACS* **129**, 6050 (2007).
[10]Yang, J.W., Stadler, M., List, B. *ACIE* **46**, 609 (2007).
[11]Vesely, J., Rios, R., Ibrahem, I., Cordova, A. *TL* **48**, 421 (2007).
[12]Kim, S.-J., Lee, K., Jew, S., Park, H., Jeong, B.-S. *CL* **37**, 432 (2008).
[13]Movassaghi, M., Chen, B. *ACIE* **46**, 565 (2007).
[14]Xie, H., Zu, L., Queis, H.R., Li, H., Wang, J., Wang, W. *OL* **10**, 1923 (2008).
[15]Wang, D., Kuang, L., Li, Z., Ding, K. *SL* 69 (2008).

(S)-Proline amides.

Aldol reaction.   New amides of (S)-proline and 4-substituted prolines continue to be tested for their effectiveness in promoting enantioselective aldol reactions, mainly based on the model reaction of cyclohexanone with an ArCHO. The long list of compounds includes those of N-aryl amides **1**,[1] **2**,[2] and those bearing additional chiral elements such as **3**[3] and **4**.[4]

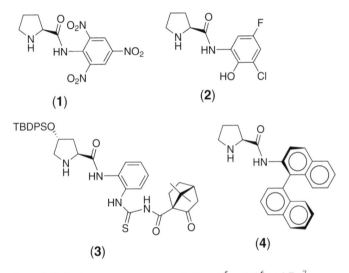

Derivatives of aliphatic amines are represented by **5**,[5] and **6**,[6] and **7a**.[7]

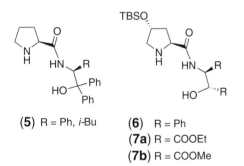

(5) R = Ph, *i*-Bu

(6) R = Ph
(7a) R = COOEt
(7b) R = COOMe

The prolinamide **8** is synthesized from methyl cholate.[8] The diamide **9** containing two prolyl residues is said to fulfill the demand for cross-aldol condensation of aldehydes.[9]

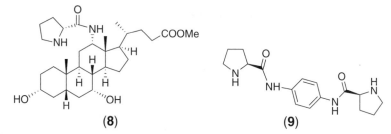

(8)   (9)

Prolinamides featuring additional chiral elements and functional groups may find special utilities, for example to deal with more complicated substrates. There is a report of aldol reaction between α-methylthio acetone and aldehydes which relies on **7b**.[10] For promoting reactions of ethyl glyoxylate, the dipeptide **10** has been employed.[11]

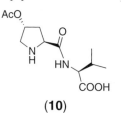

(10)

(S)-Prolyl derivative of BINAMINOL (amide) shows catalytic activities for *anti*-selective aldol reaction.[12]

When 1,1,1-trifluoro-3-alken-2-ones serve as aldol acceptors, the simple N-benzenesulfonylprolinamide plays an adequate catalytic role to guide the reaction.[13]

Other condensations.   The tripeptide H-(R)-Pro-Pro-Asp-NH$_2$ proves highly efficient in catalyzing the Michael reaction of aldehydes with conjugated nitroalkenes.[14]

Biginelli reaction involves a Mannich reaction of iminium species derived from aldehydes and ureas. It is activated by N-(1-adamantyl)-4-hydroxyprolinamide (**11**) and 2-chloro-4-nitrobenzoic acid.[15]

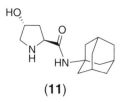

(11)

The Ullmann ether synthesis also benefits from the presence of N-methylprolinamide.[16]

Epoxidation.   For asymmetric epoxidation of alkenes capable of H-bonding by aqueous H_2O_2 and N,N'-diisopropylcarbodiimide the proline-based tripeptide **12** plays a catalytic role by transforming the aspartyl residue into a chiral peracid.[17]

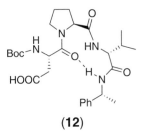

(12)

[1]Sato, K., Kuriyama, M., Shimazawa, R., Morimoto, T., Kakiuchi, K., Shirai, R. *TL* **49**, 2402 (2008).
[2]Sathapornvajana, S., Vilaivan, T. *T* **63**, 10253 (2007).
[3]Tzeng, Z.-H., Chen, H.Y., Huang, C.-T., Chen, K. *TL* **49**, 4134 (2008).
[4]Russo, A., Botta, G., Lattanzi, A. *T* **63**, 11886 (2007).
[5]Maya, V., Raj, M., Singh, V.K. *OL* **9**, 2593 (2007).
[6]He, L., Jiang, J., Tang, Z., Cui, X., Mi, A.-Q., Jiang, Y.-Z., Gong, L.-Z. *TA* **18**, 265 (2007).
[7]Zhao, J.-F., He, L., Jiang, J., Tang, Z., Cun, L.-F., Gong, L.-Z. *TL* **49**, 3372 (2008).
[8]Puleo, G.L., Iuliano, A. *TA* **18**, 2894 (2007).
[9]Xiong, Y., Wong, F., Dong, S., Liu, X., Feng, X. *SL* 73 (2008).
[10]Xu, X.-Y., Wang, Y.-Z., Cun, L.-F., Gong, L.-Z. *TA* **18**, 237 (2007).
[11]Dodda, R., Zhao, C.-G. *SL* 1605 (2007).
[12]Wang, C., Jiang, Y., Zhang, X., Huang, Y., Li, B., Zhang, G. *TL* **48**, 4281 (2007).
[13]Wang, X.-J., Zhao, Y., Liu, J.-T. *OL* **9**, 1343 (2007).
[14]Wiesner, M., Revell, J.D., Wennemers, H. *ACIE* **47**, 1871 (2008).
[15]Xin, J., Chang, L., Hou, Z., Shang, D., Liu, X., Feng, X. *CEJ* **14**, 3177 (2008).
[16]Liu, X., Fu, H., Jiang, Y., Zhao, Y. *SL* 221 (2008).
[17]Berkessel, A. *ACIE* **47**, 3677 (2008).

(S)-Prolinol derivatives.

Aldol reaction.   Over the years, relatively scanty attention has been paid to the application of prolinols to catalyze the aldol reaction, compared to the massive efforts devoted to proline and prolinamides. However, the activity of **1** has been scrutinized.[1]

(1A) Ar = Ph
(1B) Ar = 3,5-$(F_3C)_2C_6H_3$

The triethylsilyl ether of α,α-diphenylprolinol **2B** induces the Michael-aldol reaction tandem that combines *N*-protected *o*-aminobenzaldehydes and 2-alkenals to form chiral 2-substituted 3-formyl-1,2-dihydroquinolines.[2]

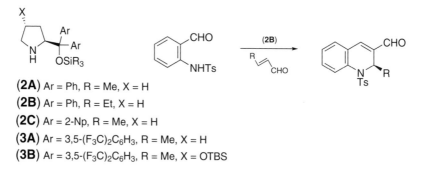

(**2A**) Ar = Ph, R = Me, X = H
(**2B**) Ar = Ph, R = Et, X = H
(**2C**) Ar = 2-Np, R = Me, X = H
(**3A**) Ar = 3,5-(F$_3$C)$_2$C$_6$H$_3$, R = Me, X = H
(**3B**) Ar = 3,5-(F$_3$C)$_2$C$_6$H$_3$, R = Me, X = OTBS

More significant results are the multi-component condensation by way of Michael-aldol[3] and Michael–Michael-aldol reaction sequences[4,5] and the Michael–Baylis-Hillman tandem,[6] each leading to cyclohexenes bearing multiple functional groups and stereogenic centers.

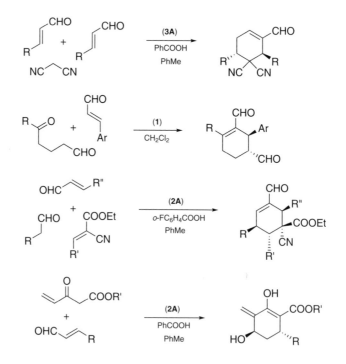

Michael reaction. Silyl ether **2A** and the enantiomeric *ent-***2A** have been employed in inducing the Michael reaction of *N*-Boc hydroxylamine[7] and malonate esters,[8] respectively, to enals, whereas **3A** has been affirmed to promote the addition of *S*-(2,2,2-trifluoroethyl) alkanethioates to enals.[9] Furthermore, **3A** and **3B** find use in asymmetric conjugate addition of benzaldoxime[10] (products are for conversion into chiral 1,3-alkanediols) and RSH,[11] also to enals.

The aldehyde–quinone pair is just another combination.[12]

Homologous series of silyl ethers of α,α-dialkylprolinols **4** have been screened for optimal performance in catalyzing asymmetric Michael reaction of aldehydes and alkenals. There appears some variance in the matching of the chain length and the substituents on the silicon atom.[13]

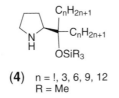

(4)   n = !, 3, 6, 9, 12
R = Me

An intramolecular Michael reaction (catalyzed by **2C**) following the oxidative dearomatization of 4-substituted phenols provides valuable octalones.[14] A method involving two consecutive Michael reactions to form optically active polysubstituted cyclopentanes[15] should be highly rated.

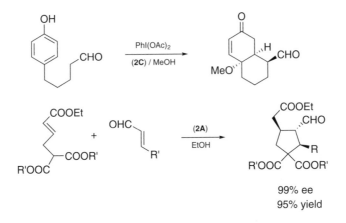

99% ee
95% yield

Much work has also been devoted to Michael reaction of nitroalkanes to conjugated carbonyl compounds using **2A** for catalysis. However, different additives (e.g., PhCOOH,[16]

LiOAc[17]) and/or reaction conditions are involved. 1,3-Dinitropropane and especially 2-substituted homologues, react with enals to provide 2,4-dinitrocyclohexanols with up to five contiguous stereocenters and ee up to 94%.[18]

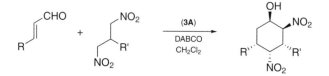

The combinations of aldehydes and nitroalkenes also form chiral adducts, which are valuable precursors of γ-amino acids, in the presence of **2A**[19,20,21] or *ent*-**2A**.[22] Prolinol TBDPS ether (without α-substituents) also mediates Michael reaction of ketones with β-nitrostyrenes.[23]

Mannich and related reactions.   A two-step procedure gives access to amino alcohols bearing a chirality center at the α-carbon atom, the crucial step being a Mannich reaction.[24]

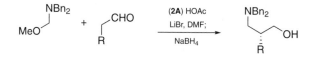

In hydroxylamination of aldehydes with PhN=O under the influence of **2A**, external hydrogen-bond donors are not required.[25]

Cycloaddition reactions.   Asymmetric induction in cycloaddition of enals is based on the formation of conjugated iminium salts with bulky prolinol derivatives. Reaction partners include enamides[26] and cyclopentadiene.[27] The Diels–Alder reaction (catalyst: **3B**) is *exo*-selective.

Asymmetric aziridination of enals by *N*-acetoxycarbamates in the presence of **2A** has been demonstrated.[28]

Addition to ArCHO.   Synthesis of chiral secondary benzylic alcohols by addition of organometallic reagents (including Me_2Zn and Ar_3Bi) can be asymmetrically directed by (S)-α,α-diphenylprolinol.[29]

Redox reactions.   Two prolinol derivatives **5**[30] and **6**[31] with polyfluorinated α-substituents have been developed for use in the CBS-type reduction. Immobilization of **6** in hydrofluoroether perhaps contributes to the attainment of high ee of the reduction and it facilitates recovery of the adjuvant (for reuse).

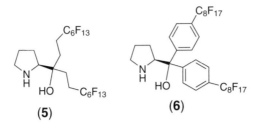

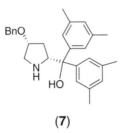

(5)   **(6)**

The prolinol **7** assists epoxidation of enones (reagent: *t*-BuOOH), but good ee are obtained only with chalcones.[32]

(7)

A series of bicyclic P-chiral ligands for the Rh metal has been prepared from (*S*)-α,α-diphenylprolinol and RPCl$_2$. Various degrees of success are seen with the derived catalysts for asymmetric hydrogenation.[33]

[1]Hayashi, Y., Itoh, T., Aratake, S., Ishikawa, H. *ACIE* **47**, 2082 (2008).

[2]Li, H., Wang, J., Xie, H., Zu, L., Jiang, W., Duesler, E., Wang, W. *OL* **9**, 965 (2007).

[3]Hong, B.-C., Nimje, R.Y., Sadani, A.A., Liao, J.-H. *OL* **10**, 2345 (2008).

[4]Carlone, A., Cabrera, S., Marigo, M., Jorgensen, K.A. *ACIE* **46**, 1101 (2007).

[5]Penon, O., Carlone, A., Mazzanti, A., Locatelli, M., Sambri, L., Bartoli, G., Melchiorre, P. *CEJ* **14**, 4788 (2008).

[6]Cabrera, S., Aleman, J., Bolze, P., Bertelsen, S., Jorgensen, K.A. *ACIE* **47**, 121 (2008).

[7]Ibrahem, I., Rios, R., Vesely, J., Zhao, G.-L., Cordova, A. *CC* 849 (2007).

[8]Ma, A., Zhu, S., Ma, D. *TL* **49**, 3075 (2008).

[9]Alonso, D.A., Kitagaki, S., Utsumi, N., Barbas III, C.F. *ACIE* **47**, 4588 (2008).

[10]Bertelsen, S., Diner, P., Johansen, R.L., Jørgensen, K.A. *JACS* **129**, 1536 (2007).

[11]Ishino, T., Oriyama, T. *CL* **36**, 550 (2007).

[12]Aleman, J., Cabrera, S., Maerten, E., Overgaard, J., Jørgensen, K.A. *ACIE* **46**, 5520 (2007).

[13]Palomo, C., Landa, A., Mielgo, A., Oiarbide, M., Puente, A., Vera, S. *ACIE* **46**, 8431 (2007).

[14]Vo, N.T., Pace, R.D.M., O'Hara, F., Gaunt, M.J. *JACS* **130**, 404 (2008).

[15]Zu, L., Li, H., Xie, H., Wang, J., Jiang, W., Tang, Y., Wang, W. *ACIE* **46**, 3732 (2007).

[16]Gotoh, H., Ishikawa, H., Hayashi, Y. *OL* **9**, 5307 (2007).

[17]Wang, Y., Li, P., Liang, X., Zhang, T.Y., Ye, J. *CC* 1232 (2008).

[18]Reyes, E., Jiang, H., Milelli, A., Elsner, P., Hazell, R.G., Jorgensen, K.A. *ACIE* **46**, 9202 (2007).

[19]Garcia-Garcia, P., Ladepeche, A., Halder, R., List, B. *ACIE* **47**, 4719 (2008).

[20]Hayashi, Y., Itoh, T., Ohkubo, M., Ishikawa, H. *ACIE* **47**, 4722 (2008).

[21]Chi, Y., Guo, L., Kopf, N.A., Gellman, S.H. *JACS* **130**, 5608 (2008).

[22]Zhu, S., Yu, S., Ma, D. *ACIE* **47**, 545 (2008).

[23]Liu, F., Wang, S., Wang, N., Peng, Y. *SL* 2415 (2007).
[24]Ibrahem, I., Zhao, G.-L., Cordova, A. *CEJ* **14**, 683 (2008).
[25]Palomo, C., Vera, S., Velilla, I., Mielgo, A., Gomez-Bengoa, E. *ACIE* **46**, 8054 (2007).
[26]Hayashi, Y., Gotoh, H., Masui, R., Ishikawa, H. *ACIE* **47**, 4012 (2008).
[27]Gotoh, H., Hayashi, Y. *OL* **9**, 2859 (2007).
[28]Vesely, J., Ibrahem, I., Zhao, G.-L., Cordova, A. *ACIE* **46**, 778 (2007).
[29]Sato, I., Toyota, Y., Asakura, N. *EJOC* 2608 (2007).
[30]Goushi, S., Funabiki, K., Ohta, M., Hatano, K., Matsui, M. *T* **63**, 4061 (2007).
[31]Chu, Q., Yu, M.S., Curran, D.P. *OL* **10**, 749 (2008).
[32]Li, Y., Liu, X., Yang, Y., Zhao, G. *JOC* **72**, 288 (2007).
[33]Bondarev, O., Goddard, R. *TL* **47**, 9013 (2006).

3-Pyridinecarboxylic anhydride.

Ester and amide synthesis.   Nicotinic anhydride activates carboxylic acids (by forming mixed anhydrides) to be transformed into esters[1] and amides,[2] with alcohols and amines, respectively (catalytic DMAP).

[1]Mukaiyama, T., Funasaka, S. *CL* **36**, 326 (2007).
[2]Funasaka, S., Mukaiyama, T. *CL* **36**, 658 (2007).

3-Pyridinesulfonyl chloride.

Amide synthesis.[1]   The title reagent is newly developed for facilitating amide formation from carboxylic acid and amines. In the condensation DMAP serves as catalyst.

[1]Funasaka, S., Kato, K., Mukaiyama, T. *CL* **37**, 506 (2008).

4-Pyrrolidinopyridine.

Transesterification.[1]   Zwitterionic adduct of 4-pyrrolidinopyridine and an electron-deficient aryl isothiocyanate (e.g., *p*-nitrophenyl and 3,5-bis(trifluoromethyl)phenyl isothiocyanates) catalyzes transesterification of methyl esters, requiring only stoichiometric quantity of the alcohol. The reaction is best performed by azeotropic refluxing, with assistance of 5A-molecular sieves to absorb the liberated MeOH.

β-Lactones.[2]   A synthesis of salinosporamide-A highlights the creation of the β-lactone unit from a keto acid.

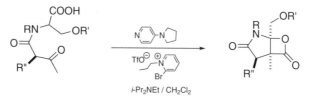

[1]Ishihara, K., Niwa, M., Kosugi, Y. *OL* **10**, 2187 (2008).
[2]Ma, G., Nguyen, H., Romo, D. *OL* **9**, 2143 (2007).

(S)-(2-Pyrrolidinyl)methylamines.

Aldol reaction.   Triflyl and nonaflyl derivatives of (S)-(2-pyrrolidinyl)methylamine mediate asymmetric aldol reactions in aqueous media.[1,2] The latter is a typically recyclable fluorous catalyst.[2]

Also reusable is the salts of N-(2-pyrrolidinylmethyl) derivatives **1** of cyclic amines.[3]

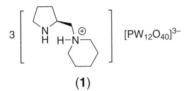

(1)

Michael reaction.   The greatest number of research reports pertaining to methodology development for the asymmetric Michael reaction are based on the addition of ketones to β-nitrostyrene. Among catalysts **2**,[4] **3**,[5] and **4**,[6] the first one is the simplest.

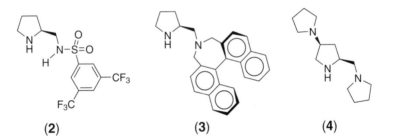

(2)   **(3)**   **(4)**

By means of the "click reaction" compounds containing a 1,2,3-triazole unit (**5**,[7] **6**[8]) have been prepared.

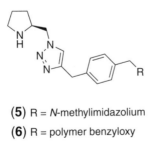

(5) R = N-methylimidazolium

(6) R = polymer benzyloxy

N-Heterocycles quaternized by the (*S*)-(2-pyrrolidinyl)methyl group constitute another series of useful catalysts. These include **7**,[9] **8**,[10] and **9**.[11] Reactions involving **8** and **9** are carried out in ionic liquids.

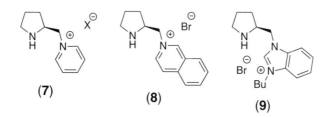

(7)   (8)   (9)

The thiourea derivative **10** is active in promoting conjugate addition of ketones to β-nitrostyrenes.[12]

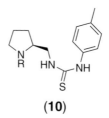

(10)

N-Arylation.   (*S*)-2-(Imidazolylmethyl)pyrrolidine **11** is yet another derivative of the series, and it functions as a catalyst for *N*-arylation of heterocyclic amines.[13]

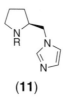

(11)

[1]Mei, K., Zhang, S., He, S., Li, P., Jin, M., Xue, F., Luo, G., Zhang, H., Song, L., Duan, W., Wang, W. *TL* **49**, 2681 (2008).
[2]Zu, L., Xie, H., Li, H., Wang, W. *OL* **10**, 1211 (2008).
[3]Luo, S., Li, J., Xu, H., Zhang, L., Cheng, J.-P. *OL* **9**, 3675 (2007).
[4]Ni, B., Zhang, Q., Headley, A.D. *TA* **18**, 1443 (2007).
[5]Vishnumaya, Singh, V.K. *OL* **9**, 1117 (2007).

[6]Chen, H., Wang, Y., Wei, S., Sun, J. *TA* **18**, 1308 (2007).
[7]Wu, L.-Y., Yan, Z.-Y., Xie, Y.-X., Niu, Y.-N., Liang, Y.-M. *TA* **18**, 2086 (2007).
[8]Alza, E., Cambeiro, X.C., Jimeno, C., Pericas, M.A. *OL* **9**, 3717 (2007).
[9]Ni, B., Zhang, Q., Headley, A.D. *TL* **49**, 1249 (2008).
[10]Xu, D.Q., Wang, B.-T., Luo, S.-P., Yue, H.-D., Wang, L.-P., Xu, Z.-Y. *TA* **18**, 1788 (2007).
[11]Luo, S., Zhang, L., Mi, X., Qiao, Y., Cheng, J.-P. *JOC* **72**, 9350 (2007).
[12]Cao, Y.-J., Lai, Y.-Y., Wang, X., Li, Y.-J., Xiao, W.-J. *TL* **48**, 21 (2007).
[13]Zhu, L., Cheng, L., Zhang, Y., Xie, R., You, J. *JOC* **72**, 2737 (2007).

(*S*)-(2-Pyrrolidinyl)azoles.

Michael reaction.[1] Enantioselective addition of ketones to β-nitrostyrene is performed with the disubstituted triazole **1**. Surprisingly, **1** is not a good catalyst for the aldol reaction.

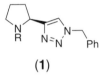

(1)

Mannich reaction. With the 2-(5-tetrazolyl)-4-siloxypyrrolidine **2**, enantioselective and diastereoselective Mannich reaction proceeds in water.[2]

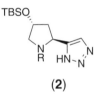

(2)

[1]Chandrasekhar, S., Tiwari, B., Parida, B.B., Reddy, C.R. *TA* **19**, 495 (2008).
[2]Hayashi, Y., Urushima, T., Aratake, S., Okano, T., Obi, K. *OL* **10**, 21 (2008).

R

Rhenium carbonyl clusters.

N-Alkenylattion.[1]   Heating a lactam and a 1-alkyne with $Re_2(CO)_{10}$ in toluene gives the *N*-alkenyl derivative.

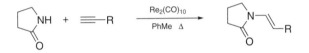

[1]Yudha S, S., Kuninobu, Y., Takai, K. *OL* **9**, 5609 (2007).

Rhodium/alumina.

Hydrogenation.   The Rh/Al_2O_3 catalyst is well suited for hydrogenation of pyrrole derivatives to furnish pyrrolidines.

[1]Jiang, C., Frontier, A.J. *OL* **9**, 4939 (2007).

Rhodium(II) carboxamidates.

Cycloadditions.   The carboxamidate **1** has been used to catalyze the 1,3-dipolar cyclo-addition of nitrones with enals.[1]

(1)

Oxidation.   Caprolactamate of Rh catalyzes oxidation of alkynes to furnish conjugated alkynones with *t*-BuOOH in water.[2]

[1]Wang, Y., Wolf, J., Zavalij, P., Doyle, M.P. *ACIE* **47**, 1439 (2008).
[2]McLaughlin, E.C., Doyle, M.P. *JOC* **73**, 4317 (2008).

Fiesers' Reagents for Organic Synthesis, Volume 25. By Tse-Lok Ho

Rhodium(II) carboxylates.

Transylidation.   Aryliodonium ylides of the ArI=CHTf$_2$ type can be prepared from heating ArI (in large excess) with PhI=CHTf$_2$ and Rh$_2$(OAc)$_4$ at 90°.[1]

Transylidation is involved in the *O*-benzylation of 2-aryliodonio-1,3-cycloalkanedionates with retention of configuration.[2]

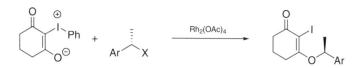

Bond insertion.   A gainful application of the intramolecular Rh-carbenoid insertion into a C—H bond is described in a synthesis of *clasto*-lactacystin β-lactone.[3]

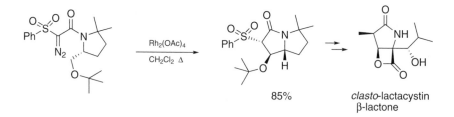

85%   *clasto*-lactacystin
                 β-lactone

Divergent pathways are pursued by Rh-carbenoids generated from α-diazoarylacetic esters in the reaction with 1-arylpropenes. Steric crowding favors C—H bond insertion.[4]

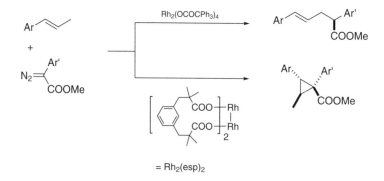

= Rh$_2$(esp)$_2$

Diastereoselective formation of 2-alkoxy-3-aminoalkanoic esters is observed from a reaction of α-diazo esters, alcohols, and imines. Oxonium ylides are formed and trapped.[5]

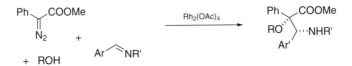

Formation of dioxolanes as the major products from α-diazoalkanoic esters and ArCHO is the result of a 1,3-dipolar cycloaddition.[6]

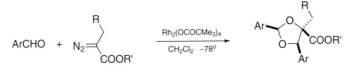

A useful Rh-nitrenoid is generated from TsONHCOOCH$_2$CCl$_3$ and it can be used to functionalize tetrahydrofuran.[7]

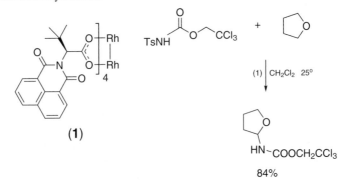

Bicyclization via nitrenoid-to-carbenoid transition followed by O—C bond insertion[8] serves to demonstrate the power of the reactions catalyzed by Rh carboxylates and the benefit of substrate design to accommodate functional participations.

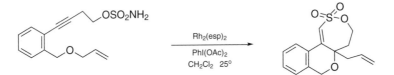

(Z)-α-Azidocinnamic esters cyclize to afford 2-indolecarboxylic esters on thermolysis (~145°). The reaction temperature is lowered to 30°–60° by catalysis of Rh$_2$(OCOC$_3$F$_7$)$_4$ [but not by Rh$_2$(OAc)$_4$].[9] Similarly, 2-arylindoles are synthesized from o-azidostilbenes by warming with Rh$_2$(OCOC$_3$F$_7$)$_4$ and 4A-molecular sieves in toluene at 60°.[10]

Under oxidative conditions sultones are formed from alkoxysulfonylacetic esters.[11] The active methylene group is transformed into a Rh-carbenoid to accomplish the bond insertion.

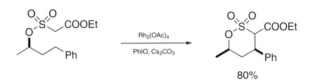

80%

Ligand control of the reaction pathway is manifested in the formation of either a cyclo-propene or a dihydroazulene.[12]

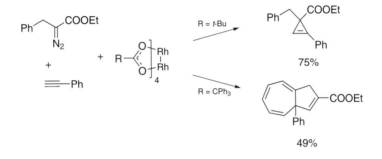

75%

49%

Bicyclization. Carbonyl ylides generated via decomposition of diazoketones and internal trapping can be put to good use. Accessibility of oxabridged tricyclic by an intra-molecular [3+2]cycloaddition has profound significance to the elaboration of the core struc-ture of platensimycin, and the possibility has been studied. Initial experimentation showed the preponderant formation of an isomeric skeleton but by halogen substitution (change of HOMO coefficient) on the dipolarophilic alkene the desired intermediate can be prepared as the major product.[13]

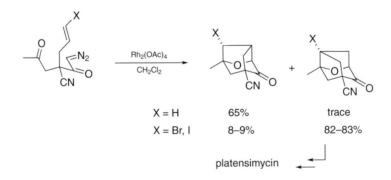

| X = H | 65% | trace |
| X = Br, I | 8–9% | 82–83% |

platensimycin

Benzyl alcohols. After transmetallation arylboronic acids are converted into aryl-rhodium species that are nucleophilic toward various aldehydes. A very active catalyst (**1**) is that derived from $Rh_2(OAc)_4$, 1,3-bis(2,6-diisopropylphenyl)imidazolium chloride and *t*-BuOK.[14]

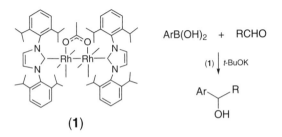

(1)

N-Sulfonylcarboxamides. A mixture of $Rh_2(esp)_2$ and $PhI(piv)_2$ can be used to bring about oxidative sulfamidation of aldehydes.[15]

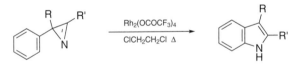

Isomerization. 3-Aryl-3*H*-azirines undergo isomerization to furnish indoles by heating with $Rh_2(OCOCF_3)_4$ in 1,2-dichloroethane.[16]

[1]Ochiai, M., Okada, T., Tada, N., Yoshimura, A. *OL* **10**, 1425 (2008).
[2]Moriarty, R.M., Tyagi, S., Ivanov, D., Constantinescu, M. *JACS* **130**, 7564 (2008).
[3]Yoon, C.H., Flanigan, D.L., Yoo, K.S., Jung, K.W. *EJOC* 37 (2007).
[4]Davies, H.M.L., Coleman, M.G., Ventura, D.L. *OL* **9**, 4971 (2007).
[5]Huang, H., Guo, X., Hu, W. *ACIE* **46**, 1337 (2007).
[6]DeAngelis, A., Panne, P., Yap, G.P.A., Fox, J.M. *JOC* **73**, 1435 (2008).
[7]Lebel, H., Huard, K. *OL* **9**, 639 (2007).
[8]Thornton, A.R., Blakey, S.B. *JACS* **130**, 5020 (2008).
[9]Stokes, B.J., Dong, H., Leslie, B.E., Pumphrey, A.L. *JACS* **129**, 7500 (2007).
[10]Shen, M., Leslie, B.E., Driver, T.G. *ACIE* **47**, 5056 (2008).
[11]Wolckenhauer, S.A., Devlin, A.S., Du Bois, J. *OL* **9**, 4363 (2007).
[12]Panne, P., Fox, J.M. *JACS* **129**, 7500 (2007).
[13]Kim, C.H., Jang, K.P., Choi, S.Y., Chung, Y.K., Lee, E. *ACIE* **47**, 4073 (2008).

[14]Trindale, A.F., Gois, P.M.P., Veiros, L.F., Andre, V., Duarte, M.T., Afonso, C.A.M., Caddick, S., Cloke, F.G.N. *JOC* **73**, 4076 (2008).
[15]Chan, J., Baucom, K.D., Murry, J.A. *JACS* **129**, 14106 (2007).
[16]Chiba, S., Hattori, G., Narasaka, K. *CL* **36**, 52 (2007).

Rhodium(III) chloride.

De-N-allylation.[1]   The *N*-allyl group of an amide is selectively removed by heating with RhCl$_3$ in isopropanol.

[2+2+2]Cycloaddition.   In the presence of RhCl$_3 \cdot$3H$_2$O and *i*-Pr$_2$NEt alkynes are trimerized to provide substituted benzenes,[2] a mixture of a diyne and an alkyne forms a 1 : 1-adduct.[3]

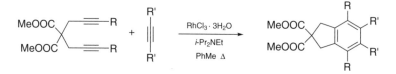

Oxidative coupling.[4]   2,3-Bis(β-indolyl)maleimides are converted into hexacyclic products on heating with RhCl$_3 \cdot$ 3H$_2$O and Cu(OAc)$_2$ in DMF.

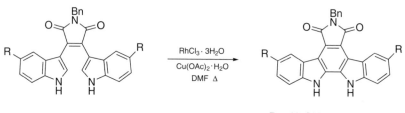

R =, H, OMe, F   73–78%

[1]Zacuto, M.J., Xu, F. *JOC* **72**, 6298 (2007).
[2]Yoshida, K., Morimoto, I., Mitsudo, K., Tanaka, H. *CL* **36**, 998 (2007).
[3]Yoshita, K., Morimoto, I., Mitsudo, K., Tanaka, H. *T* **64**, 5800 (2008).
[4]Witulski, B., Schweikert, T. *S* 1959 (2005).

Rhodium hydroxide/alumina.

Amide formation.[1]   Redox reaction between aldehydes and hydroxylamine leads to carboxamides. The reaction, catalyzed by Rh(OH)$_n$ in water at 160°, perhaps proceeds via oxime formation, dehydration and rehydration.

[1]Fujiwara, H., Ogasawara, Y., Yamaguchi, K., Mizuno, N. *ACIE* **46**, 5202 (2007).

Rhodium(III) iodide.

Strecker reaction.[1]   Aldehydes and amines are condensed with Me₃SiCl under the influence of $RhI_3 \cdot H_2O$ in MeCN at room temperature.

[1]Majhi, A., Kim, S.S., Kadam, S.T. *T* **64**, 5509 (2008).

Rhodium(I) tetrafluoroborate.

Isomerizations.   The Rh(I) salt causes intramolecular redox transformation of propargylic alcohols to provide conjugated carbonyl compounds. Depending on the nature of the carbon unit attached to the far end of the triple bond different ligands are required for optimal reaction.[1]

R = Ar,   ligand: BINAP, solvent: $ClCH_2CH_2Cl$
R = Alkenyl,   ligand: $Ph_2PCH_2CH_2PPh_2$, solvent: $ClCH_2CH_2Cl$
R = Alkyl,   ligand: $Cy_2PCH_2CH_2PCy_2$, solvent: CH_2Cl_2

[1]Tanaka, K., Shoji, T., Hirano, M. *EJOC* 2687 (2007).

Ruthenium-carbene complexes.

General aspects and new metathesis catalysts.   For alkene metathesis Grubbs I (**1**) and Grubbs II (**2, 3**) complexes, and the Grubbs-Hoveyda catalyst (**4A**) and Grela catalyst (**4B**) remain the workhorses.

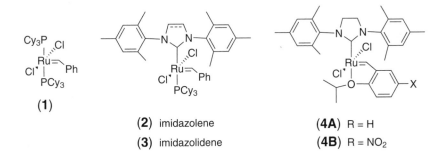

(**1**)

(**2**) imidazolene
(**3**) imidazolidene

(**4A**) R = H
(**4B**) R = NO₂

Insights for asymmetric metathesis of alkenes[1] should guide future developments. Chiral catalysts **5**[2] and **6**[3] have been synthesized.

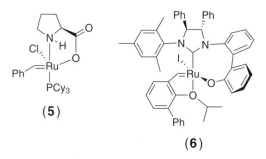

(5)   **(6)**

A review has summarized the positive effects of microwaves.[4] Acetic acid is found to be a good solvent for RCM using Grubbs II catalyst.[5] The annoying problem of double bond migration in substrates that contain hydrogen-bonding groups is solved by adding phenyl-phosphoric acid.[6] Cross metathesis in water at room temperature based on Grubbs II catalyst is facilitated by a nonionic amphiphile, which is a PEG linked to vitamin-E through sebacoyl chloride.[7]

At the end of a metathesis reaction treatment with $NCCH_2COOK$ in MeOH renders the Grubbs II catalyst inactive by transforming it into **7**.[8]

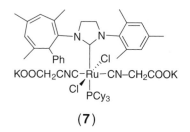

(7)

Several modifications of the commonly used catalysts have been reported during the 2007–2008 period. For example, one with a PEG chain bonded to the imidazolidine ring of the conventional Grubbs II catalyst has the advantage of easy removal by aqueous extraction.[9] The presence of an amide residue in the Grubbs–Hoveyda catalyst (as in **8**) also makes its separation from products easier.[10] As for the development of catalysts for use in aqueous environment, the strategy of attaching a chain containing ammonium group(s) has borne mixed results. Complexes like **9** can be employed in ROMP (ring opening metathetic polymerization) and RCM, but they are not stable enough to effect cross-metathesis.[11]

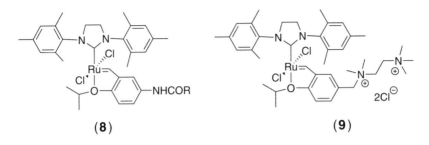

(8)   (9)

Ruthenium complex **10** is a dormant RCM catalyst, activatable by various acids.[12] After the metathesis reaction, it is re-formed by treatment with SiO_2. Catalysts with lessening steric congestion adjacent to the carbene center (e.g., **11**[13] and **12**[14]) are favorable to RCM reactions in order to form tetrasubstituted alkenes.

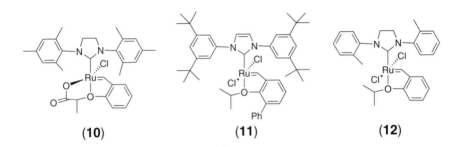

(10)   (11)   (12)

For cross metathesis of alkenes, it is found that in the preparation of disubstituted alkenes with one or more allylic substituents Grubbs–Hoveyda catalysts possessing *N*-(*o*-tolyl) groups in the azolecarbene unit are more efficient than those with the *N*-mesityl groups. But for the formation of trisubstituted alkenes the *N*-mesityl catalysts are superior due to discrimination between productive and nonproductive reaction pathways.[15]

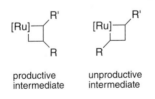

productive
intermediate

unproductive
intermediate

There is a catalyst (**13**) in which the phosphine ligand is ionophilic.[16] The Hoveyda-type complexes **14** possessing a hindered pyrrolidine carbene ligand show comparable activities to the standard catalysts.[17]

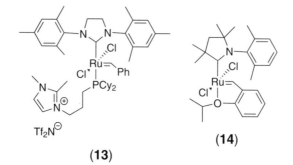

(13)   **(14)**

3-Arylthiazol-2-ylidene ligands have also been investigated as replacements for the imidazole/imidazolidene portion of the established catalysts for alkene metathesis. Promising results have been obtained from **15**.[18] Another novel catalyst is exemplified by **16**.[19]

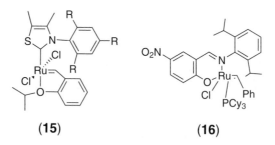

(15)   **(16)**

Metathesis catalysts with one chiral *N*-substituent (e.g., **17**) are useful for ROMP.[20] In alternating copolymerization of norbornene with other cycloalkenes the steric interactions of the growing polymer chain dictate selectivity.

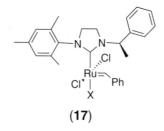

(17)

Novel and highly active RCM catalysts that contain anionic or neutral carborane tags have been developed. Their uses in promoting standard RCM in CH_2Cl_2 at 30° have been demonstrated.[21]

The modified Grubbs I catalyst **18** is effective for RCM of functionalized dienes and enynes.[22]

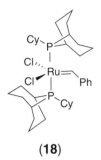

(18)

Cross-metathesis reactions.   The valuable cinnamylation reagent and other homologous allylation reagents are readily accessible from cross-metathesis using the Grubbs II catalyst.[23]

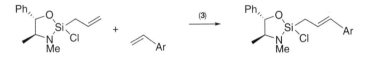

Cross-metathesis of alkenes with 1,2-dichloroethene under the influence of **4B** provides 1-chloroalkenes.[24] The cross-metathesis approach to α-alkylidene γ-butyrolactones is fraught with danger of double bond migration, and it is circumvented by addition of catecholchloroborane[25] or 2,6-dichloro-1,4-benzoquinone.[26] However, the method cannot be used to synthesize α-alkylidene δ-valerolactones.[26]

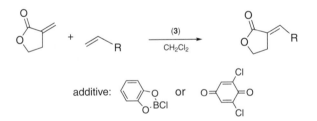

Grubbs II catalyst does not lose activity during cross-metathesis of ω-alkenols with 1-alkynes.[27]

Bridged ring systems with a strained double bond are highly susceptible to cross metathesis accompanied by ring opening. A synthesis of unsymmetrical *cis*-2,6-dialkenyl-piperidines is an application based on this nature.[28]

An ester group located at a bridgehead (C-1) of 7-azanorbornenes has strong directing effect on the ring-opening cross-metathesis.[29]

When the Diels–Alder adduct of cyclopentadiene and allyl nitrosyl ketone undergoes ring-opening cross-metathesis it is converted into a fused ring system.[30]

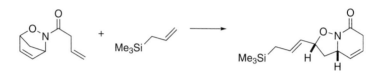

3,3-Disubstituted cyclopropenes undergo ring-opening cross-metathesis with certain alkenes under the influence of **6** to produce (3S)-1,4-alkadienes.[3]

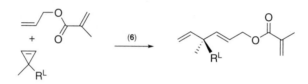

1,3-Transposition of Ru-carbenoid prior to cross-metathesis is an important feature for the attachment of an allylic ester unit to the terminus of a diyne.[31]

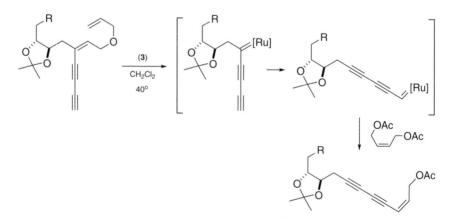

In case one of the cross-metathesis partners is attached to a polymer, it is better to place a longer linker in between in order to maximize the reaction efficiency.[32] Using the Grubbs catalysts to mediate cross-metathesis of more intransigent substrates, the application of microwaves often shows improvements.[33]

The cause of the failure of the Grubbs II catalyst to achieve cross-metathesis of vinyl chloride is ascribed to the rapid elimination of HCl from the chlorovinyldene carbenoid or the exchange of the chlorine atom into a Cy_3P group.[34]

Stepwise cross-metathesis of divinyl sulfones readily provides unsymmetrical dialkenyl sulfones.[35]

Metathetic ring closure. Based on Grubbs I catalyst RCM involving a terminal alkyne unit and an allylic alcohol is actually accelerated by the presence of the OH group.[36]

Closure of macrocycles under kinetic control and thermodynamic control might lead to different ratios of (E/Z)-isomers. It is interesting that reactions under the influence of Grubbs I catalyst and Grubbs II catalyst can be so differentiated.[37]

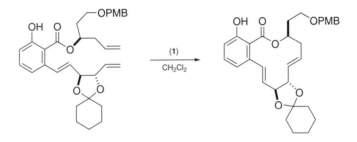

Stereoselective access to 1,6-diol derivatives from siloxanes bearing two unsaturated carbon chains is shown to be amenable to asymmetric induction.[38] This achievement has profound implications to the synthesis of certain insect pheromones.

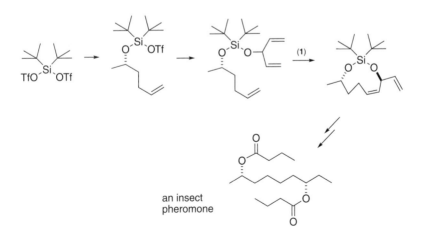

an insect
pheromone

The nature of the *N*-substituent is of critical importance for macrocyclic RCM of a *trans*-2-vinyl-1-amidocyclopropane because it affects the choice of the initial site of Ru-carbenoid formation. An acyl group favors RCM even at high substrate concentration.[39]

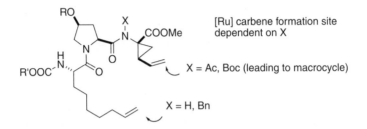

RCM is shown to be applicable to synthesis of benzannulated quinolizines.[40]

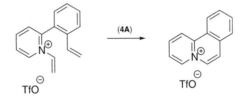

Rapid access to salient heteroannular dienes by RCM of dienynes is represented by synthetic approaches to colchicine[41] and the bicyclo[5.3.1]undeca-1,9-diene system.[42]

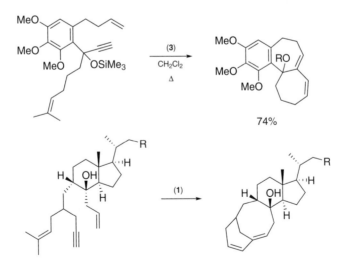

The RCM of polypeptides containing terminal alkene units performed under microwaves gives better yields of the desired products.[43] It is also significant that 1-alkenylcyclobutenes are formed by the RCM method.[44]

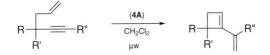

Tandem reactions. The multiple activities of the Ru catalysts enable development of valuable tandem reactions. Further extension of the carbon chain of a conjugated enal obtained from a cross-metathesis reaction is readily achieved on treatment with diazoacetic esters to give alkadienoic esters.[45]

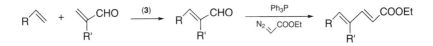

A use of **4A** is in the synthesis of 4-substituted tetrahydrocarbazoles and tetrahydro-β-carbolines,[46] involving cross-metathesis and intramolecular hydroarylation.

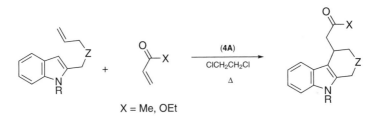

The stability of **4A** in the presence of $BF_3 \cdot OEt_2$ is apparently critical to the synthesis of azacycles containing a 2-oxoalkyl chain at C-2 by cross-metathesis of 1-alken-3-ones with ω-alkenamines. The Lewis acid is responsible for the intramolecular Michael reaction.[47]

Sequential Heck reaction and hydrosilylation of carbonyl have been carried out with the Grubbs I catalyst.[48] Allylic alcohols can be synthesized via conjugated carbonyl compounds prepared from cross-metathesis in situ by reduction with *i*-Bu$_2$AlH.[49] An access to enantiomeric 2,3-dihydroxyalkanoic esters is based on cross-metathesis, dihydroxylation and methanolysis.[50]

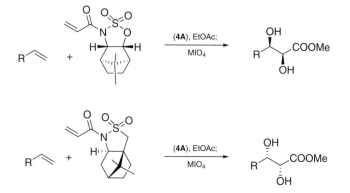

Isomerization. A terminal double bond migrates by one carbon inward along the carbon chain (but no further) on heating with Grubbs II catalyst in MeOH.[51] Thus, 2-allylcyclohexanone is converted to 2-propenylcyclohexanone.

Maleic esters. Diazoacetic esters furnish maleic esters on treatment with Grubbs II catalyst. Such reaction is applicable to the synthesis macrodilactides.[52]

[1]Berlin, J.M., Goldberg, S.D., Grubbs, R.H. *ACIE* **45**, 7591 (2006).

[2]Samec, J.S.M., Grubbs, R.H. *CC* 2826 (2007).

[3]Giudici, R.E., Hoveyda, A.H. *JACS* **129**, 3824 (2007).

[4]Coquerel, Y., Rodriguez, J. *EJOC* 1125 (2008).

[5]Adjiman, C.S., Clarke, A.J., Cooper, G., Taylor, P.C. *CC* 2806 (2008).

[6]Gimeno, N., Formentin, P., Steinke, J.H.G., Vilar, R. *EJOC* 918 (2007).

[7]Lipshutz, B.H., Aguinaldo, G.T., Ghorai, S., Voigtritter, K. *OL* **10**, 1325 (2008).

[8]Galan, B.R., Kalbarczyk, K.P., Szczepankiewicz, S., Keister, J.B., Diver, S.T. *OL* **9**, 1203 (2007).

[9]Hong, S.H., Grubbs, R.H. *OL* **9**, 1955 (2007).

[10]Rix, D., Caijo, F., Laurent, I., Boeda, F., Clavier, H., Nolan, S.P., Mauduit, M. *JOC* **73**, 4225 (2008).

[11]Jordan, J.P., Grubbs, R.H. *ACIE* **48**, 5152 (2007).

[12]Gawin, R., Makal, A., Wozniak, K., Mauduit, M., Grela, K. *ACIE* **46**, 7206 (2007).

[13]Berlin, J.M., Campbell, K., Ritter, T., Funk, T.W., Chlenov, A., Grubbs, R.H. *OL* **9**, 1339 (2007).

[14]Stewart, I.C., Ung, T., Pletnev, A.A., Berlin, J.M., Grubbs, R.H., Schrodi, Y. *OL* **9**, 1589 (2007).

[15]Stewart, I.C., Douglas, C.J., Grubbs, R.H. *OL* **10**, 441 (2008).

[16]Consorti, C.S., Aydos, G.L., Ebeling, G., Dupont, J. *OL* **10**, 237 (2008).

[17]Anderson, D.R., Lavallo, V., O'Leary, D.J., Bertrand, G., Grubbs, R.H. *ACIE* **46**, 7262 (2007).

[18]Vougioukalakis, G.C., Grubbs, R.H. *JACS* **130**, 2234 (2008).

[19]Occhipinti, G., Jensen, V.R., Bjorsvik, H.-R. *JOC* **72**, 3561 (2007).

[20]Vehlow, K., Wang, D., Buchmeiser, M.R., Blechert, S. *ACIE* **47**, 2615 (2008).

[21]Liu, G., Zhang, J., Wu, B., Wang, J. *OL* **9**, 4263 (2007).

[22]Boeda, F., Clavier, H., Jordaan, M., Meyer, W.H., Nolan, S.P. *JOC* **73**, 259 (2008).

[23]Huber, J.D., Perl, N.R., Leighton, J.L. *ACIE* **47**, 3037 (2008).

[24]Sashuk, V., Samojlowicz, C., Szadkowska, A., Grela, K. *CC* 2468 (2008).

[25]Moise, J., Arseniyadis, S., Cossy, J. *OL* **9**, 1695 (2007).

[26]Raju, R., Allen, L.J., Le, T., Taylor, C.D., Howell, A.R. *OL* **9**, 1699 (2007).

[27]Clark, D.A., Clark, J.R., Diver, S.T. *OL* **10**, 2055 (2008).

[28]Cortez, G.A., Baxter, C.A., Schrock, R.R., Hoveyda, A.H. *OL* **9**, 2871 (2007).

[29]Carreras, J., Avenoza, A., Busto, J.H., Peregrina, J.M. *OL* **9**, 1235 (2007).

[30]Calvert, G., Blanchard, N., Kouklovsky, C. *OL* **9**, 1485 (2007).

[31]Cho, E.J., Lee, D. *OL* **10**, 257 (2008).

[32]Garner, L., Koide, K. *OL* **9**, 5235 (2007).

[33]Michaut, A., Boddaert, T., Coquerel, Y., Rodriguez, J. *S* 2867 (2007).

[34]Macnaughton, M.L., Johnson, M.J.A., Kampf, J.W. *JACS* **129**, 7708 (2007).

[35]Bieniek, M., Koloda, D., Grela, K. *OL* **8**, 5689 (2006).

[36]Imahori, T., Ojima, H., Tateyama, H., Mihara, Y., Takahara, H. *TL* **49**, 265 (2008).

[37]Matsuya, Y., Takayanagi, S., Nemoto, H. *CEJ* **14**, 5275 (2008).

[38]Hooper, A.M., Dufour, S., Willaert, S., Pouvreau, S., Pickett, J.A. *TL* **48**, 5991 (2007).

[39]Shu, C., Zeng, X., Hao, M.-H., Wei, X., Yee, N.K., Busacca, C.A., Han, Z., Farina, V., Senanayake, C.H. *OL* **10**, 1303 (2008).

[40]Nunez, A., Cuadro, A.M., Alvarez-Builla, J., Vaquero, J.J. *OL* **9**, 2977 (2007).

[41]Boyer, F.-D., Hanna, I. *OL* **9**, 2293 (2007).

[42]Aldegunde, M.J., Garcia-Fandino, R., Castedo, L., Granja, J.R. *CEJ* **13**, 5135 (2007).

[43]Chapman, R.N., Arora, P.S. *OL* **8**, 5825 (2006).

[44]Debleds, O., Campagne, J.-M. *JACS* **130**, 1562 (2008).

[45]Murelli, R.P., Snapper, M.L. *OL* **9**, 1749 (2007).

[46]Chen, J.-R., Li, C.-F., An, X.-L., Zhang, J.-J., Zhu, X.-Y., Xiao, W.-J. *ACIE* **47**, 2489 (2008).

[47]Fustero, S., Jimenez, D., Sanchez-Rosello, M., del Pozo, C. *JACS* **129**, 6700 (2007).

[48]Ackermann, L., Born, R., Alvarez-Bercedo, P. *ACIE* **46**, 6364 (2007).

[49]Paul, T., Sirasani, G., Andrade, R.B. *TL* **49**, 3363 (2008).

[50]Neisius, N.M., Plietker, B. *JOC* **73**, 3218 (2008).

[51]Hanessian, S., Giroux, S., Larsson, A. *OL* **8**, 5481 (2006).

[52]Hodgson, D.M., Angrish, D. *CEJ* **13**, 3470 (2007).

Ruthenium(III) chloride.

Alkylation.   Indoles are alkylated by epoxides at C-3 under solvent-free conditions in the presence of $RuCl_3 \cdot xH_2O$.[1]

Hydrogenation.   β-Hydroxy ketones are hydrogenated in MeOH to give *anti*-1,3-diols using $RuCl_3–Ph_3P$ as catalyst.[2] Hydrogenation of monosubstituted and 1,2-disubstituted alkenes is achieved by $NaBH_4$ when $RuCl_3 \cdot xH_2O$ is also added.[3]

Coupling reactions.   *N*-Atom directed *o*-arylation of the aromatic nucleus of a 2-aryl-4,5-dihydrooxazole proceeds readily on heating with ArBr, $RuCl_3 \cdot xH_2O$, and K_2CO_3 in NMP.[4] 2-Phenylpyridine and congeners also undergo analogous arylation. Arylation at both unsubstituted *o*-positions is observed when $(PhCOO)_2$ is present.[5]

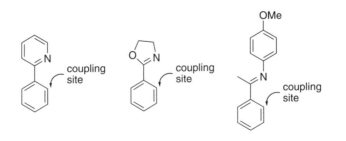

[1]Tabatabaeian, K., Mamaghani, M., Mahmoodi, N.O., Khorshidi, A. *TL* **49**, 1450 (2008).
[2]Labeeuw, O., Roche, C., Phansavath, P., Genet, J.-P. *OL* **9**, 105 (2007).
[3]Sharma, P.K., Kumar, S., Kumar, P., Nielsen, P. *TL* **48**, 8704 (2007).
[4]Ackermann, L., Althammer, A., Born, R. *T* **64**, 6115 (2008).
[5]Cheng, K., Zhang, Y., Zhao, J., Xie, C. *SL* 1325 (2008).

Ruthenium carbonyl clusters.

α-Pyrones.[1]   A synthesis of α-pyrones is based on assemblage of CO, alkynes and conjugated carbonyl compounds in the presence of $Ru_3(CO)_{12}$.

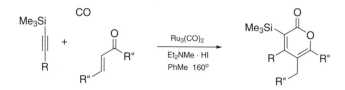

[1]Fukuyama, T., Higashibeppu, Y., Yamaura, R., Ryu, I. *OL* **9**, 587 (2007).

Ruthenium oxide–sodium periodate.

Oxidation.   *N*-Heterocycles such as isoxazolines are oxidized to give the corresponding lactams.[1] *N*-Boc derivatives of cyclic amines undergo ring opening to afford *N*-Boc ω-amino acids.[2]

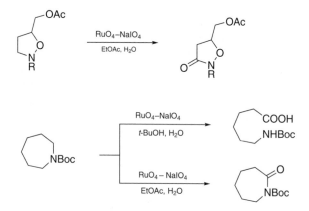

[1]Piperno, A., Chiacchio, U., Iannazzo, D., Giofre, S.V., Romeo, G., Romeo, R. *JOC* **72**, 3958 (2007).
[2]Kaname, M., Yoshifugi, S., Sashida, H. *TL* **49**, 2786 (2008).

S

Samarium.

    Coupling reaction. α-Bromo-α-halo ketones are converted into α-allyl-α-halo ketones by treatment with Sm and allyl bromide.[1]

    Cyclopropanation. A Sm-carbenoid is formed from Sm and CHI$_3$, which adds to conjugated carboxylic acids stereoselectively. The products also suffer hydrodeiodination.[2]

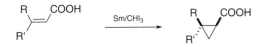

    Reduction. Reduction of RCOCl to RCHO is achieved by samarium and a stoichiometric quantity of Bu$_3$P in MeCN at $-20°$.[3] It does not seem to be a practical method.

[1]Di, J., Zhang, S. *SL* 1491 (2008).
[2]Concellon, J.M., Rodriguez-Solla, H., Simal, C. *OL* **9**, 2685 (2007).
[3]Jia, X., Liu, X., Li, J., Zhao, P., Zhang, Y. *TL* **48**, 971 (2007).

Samarium(III) chloride.

    C-Acylation.[1] 1,3-Dicarbonyl compounds undergo *C*-acylation at room temperature when mediated by SmCl$_3$–Et$_3$N.

[1]Shen, Q., Huang, W., Wang, J., Zhou, X. *OL* **9**, 4491 (2007).

Samarium(II) iodide.

    Alcoholysis. *N*-Acyloxazolidinones are cleaved and turned into esters by reaction with ROH in the presence of SmI$_2$.[1]

    Addition to C=O. Organosamarium reagents, formed by treatment of *t*-butyl α-haloalkyl ketones with SmI$_2$ in THF at $-78°$, are active toward carbonyl compounds.[2]

    Elimination. 2-Bromo-3-hydroxy-1-nitroalkanes, readily prepared from RCHO and bromonitromethane, are converted into 1-nitroalkenes with SmI$_2$.[3] (*Z*)-Allyltrimethylsilanes are obtained by treatment of 2-acetoxy-3-chloro-1-trimethylsilylalkanes with SmI$_2$.[4]

Fiesers' Reagents for Organic Synthesis, Volume 25. By Tse-Lok Ho
Copyright © 2010 John Wiley & Sons, Inc.

Elimination of two vicinal *p*-trifluoromethylbenzoyloxy group to place an exocyclic double bond in a polyfunctional cyclononadiyne is a synthetically challenging task. Its accomplishment by using SmI_2 is pleasing.[5]

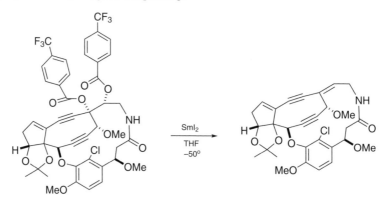

Reduction–addition. On reductive opening of epoxides in the presence of pyridines the radicals are intercepted, therefore 2-(β-hydroxyalkyl)pyridines such as chiral ligands **1** are readily accessible on the basis of this reaction.[6]

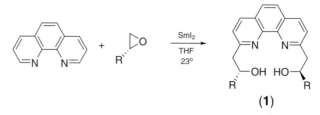

(1)

An important service of SmI_2 is in the diastereoselective reductive closure to form a spirooxindole.[7]

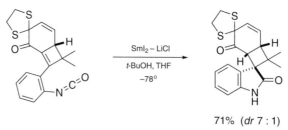

71% (*dr* 7 : 1)

The reduction of alkenyl sulfones accompanied by intramolecular trapping by an aldehyde to form 3-hydroxytetrahydropyrans is synthetically useful because the products are adorned with desirable functional groups.[8]

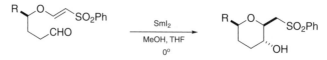

Reduction.   The selective reduction of δ-lactones to diols by SmI_2–H_2O in THF has been observed.[9] It is attributed to stabilization of the radical anions by interaction with the lone pair electrons on both endocyclic and exocyclic oxygen atoms. γ-Lactones survive the reaction conditions.

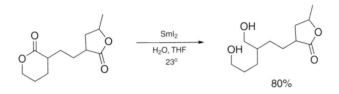

80%

2,3-Diaryl-1,3-butadienes are obtained from ArCOMe directly by the reaction with SmI_2 and Ac_2O in refluxing THF.[10] Transannular addition of ketyl species generated from meso-cyclic ketones to a double bond is quite efficient.[11]

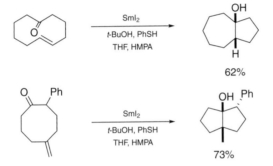

62%

73%

In the tandem addition involving ketyl radicals and two unsaturated CC bonds, the HMPA additive plays an important role in completing the spiro[4.5]decanes. If Sm instead of HMPA is present the reaction stops at monocyclic products.[12]

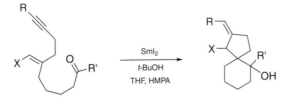

It is highly significant that the naphthalene system is susceptible to an intramolecular attack by a ketyl radical generated from a ketone.[13]

Z = CH₂, NR, C(OCH₂)₂

The observation that reduction of the cyclopentanone in a precursor of (+)-brefeldin-A by SmI₂ with *i*-PrOH as proton source to give predominantly the α-alcohol is a desirable result. Many other conditions favor the production of the β-alcohol (e.g., *t*-BuLi, Dibal-H: 100% β-isomer).[14]

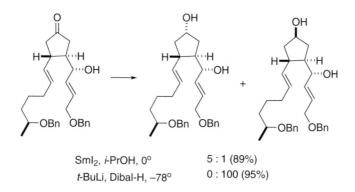

SmI₂, *i*-PrOH, 0°   5 : 1 (89%)

t-BuLi, Dibal-H, −78°   0 : 100 (95%)

Both nitroalkanes and conjugated nitroalkenes are reduced to saturated primary amines by SmI₂ in THF with *i*-PrNH₂ and H₂O as additives.[15]

Intramolecular reductive cyclization involving a conjugated ester and a δ-keto group leads to the formation of cyclopropanolactone.[16]

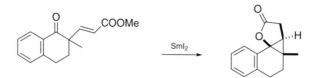

[1]Magnier-Bouvier, C., Reboule, I., Gil, R., Collin, J. *SL* 1211 (2008).

[2]Sparling, B.A., Moslin, R.M., Jamison, T.F. *OL* **10**, 1291 (2008).

[3]Concellon, J.M., Bernad, P.L., Rodriguez-Solla, H., Concellon, C. *JOC* **72**, 5421 (2007).

[4]Concellon, J.M., Rodriguez-Solla, H., Simal, C., Gomez, C. *SL* 75 (2007).

[5]Komano, K., Shimamura, S., Inoue, M., Hirama, M. *JACS* **129**, 14184 (2007).

[6]Plummer, J.M., Weitgenant, J.A., Noll, B.C., Lauher, J.W., Wiest, O., Helquist, P. *JOC* **73**, 3911 (2008).

[7]Reisman, S.E., Ready, J.M., Weiss, M.M., Hasuoka, A., Hirata, M., Tamaki, K., Ovaska, T.V., Smith, C.J., Wood, J.L. *JACS* **130**, 2087 (2008).
[8]Kimura, T., Nakata, T. *TL* **48**, 43 (2007).
[9]Duffy, L.A., Matsubara, H., Proctor, D.J. *JACS* **130**, 1136 (2008).
[10]Li, J., Li, S., Jia, X. *SL* 1529 (2008).
[11]Molander, G.A., Czako, B., Rheam, M. *JOC* **72**, 1755 (2007).
[12]Inui, M., Nakazaki, A., Kobayashi, S. *OL* **9**, 469 (2007).
[13]Aulenta, F., Berndt, M., Brüdgam, I., Hartl, H., Sörgel, S., Reissig, H.-U. *CEJ* **13**, 6047 (2007).
[14]Wu, Y., Gao, J. *OL* **10**, 1533 (2008).
[15]Ankner, T., Hilmersson, G. *TL* **48**, 5707 (2007).
[16]Zriba, R., Bezzenine-Lafollee, S., Guibe, F., Magnier-Bouvier, C. *TL* **48**, 8234 (2007).

Samarium(III) triflate.

Friedel–Crafts reaction.[1]   Activation by $Sm(OTf)_3$ an electrophilic halogenating agent attacks alkenes and thereby gravitating the alkylation of arenes.

[1]Hajra, S., Maji, B., Bar, S. *OL* **9**, 2783 (2007).

Scandium(III) fluoride.

Hydroxymethylation.[1]   A new catalyst for hydroxymethylation (with HCHO) of dimethylsiloxyalkenes is ScF_3. The reaction is performed at room temperature in THF–H_2O (9 : 1).[1]

[1]Kokubo, M., Kobayashi, S. *SL* 1562 (2008).

Scandium(III) triflate.

Ring cleavage.   Catalyzed by $Sc(OTf)_3$, alcoholysis of epoxides[1] and aziridines[2] proceeds at room temperature. The ring opening of *meso*-epoxides is rendered asymmetric if a chiral ligand such as **1** is added to the reaction medium.[1] Lactones give polymers via alcoholysis.[3]

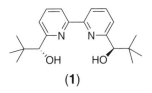

(1)

(Z)-Chloroalkenes.   Ketones are stereoselectively converted into (Z)-chloroalkenes by PhCOCl using $Sc(OTf)_3$ as catalyst. Cyanuric chloride is the chloride source and dehydrating agent (for reverting PhCOOH into PhCOCl).[4]

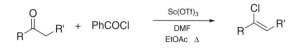

Substitution.   Introduction of a C—S bond at an α-position of a diketopiperazine by replacement of an acetoxy group is achieved in a reaction catalyzed by Sc(OTf)$_3$ using MeCN as solvent, which is critical.[5]

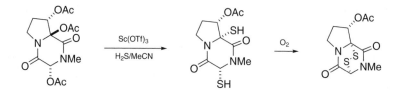

Mannich reaction.   When mixed with Sc(OTf)$_3$ an ArNH$_2$, ethyl glyoxylate, and an enol ester assemble to give a γ-keto-*N*-aryl-α-amino acid ester.[6]

The Mannich reaction of ketene *O,S*-acetals can give rise to β-lactams.[7] Reaction with enol esters proceeds reasonably well, and that involving ArNH$_2$, ethyl glyoxylate and vinyl acetate gives quinoline-2-carboxylic esters.[8]

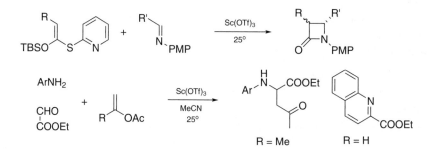

Michael reaction.   (*Z*)-Enones are synthesized from siloxyallenes by reaction with alkylidenemalonic esters under mild conditions.[9] The broad scope of the reaction makes these synthetically valuable polyfunctional products readily available.

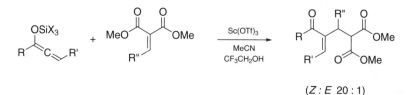

(Z : E 20 : 1)

A highly significant functionalization of cycloalkenes, including cyclobutene, indene, dihydrofurans, at the sp^2-carbon atom(s) via dinitrosation and coordination to cobalt, and Michael reaction is terminated by decomposition of the adducts by an exchange reaction. Activation of the Michael acceptors by Sc(OTf)$_3$ gives the best result, in the presence of a strong base such as (Me$_3$Si)$_2$NLi.[10]

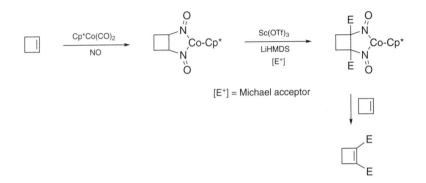

[E⁺] = Michael acceptor

α-Hydroxyalkyl-γ-lactams. Cyclopropanecarboxamides undergo ring opening in the presence of Sc(OTf)$_3$, NaI and Me$_3$SiCl. The resulting ketene *O,N*-acetal complexes bearing an iodoethyl chain are reactive toward RCHO as aldol donors. Adducts are readily formed and in turn they cyclize to afford γ-lactams with high diastereoselectivity.[11]

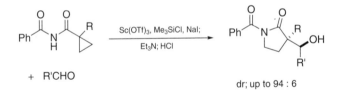

+ R'CHO

dr; up to 94 : 6

Cycloaddition. Aldimines and those prepared in situ combine with ethyl propynoate readily. On applying the reaction to arylamines several quinoline-3-carboxylic esters have been prepared.[12] The formation of 1,4-dihydropyridine-3,5-dicarboxylic esters[13] involves electrocyclic opening of the 1 : 1-cycloadducts and a Diels–Alder reaction.

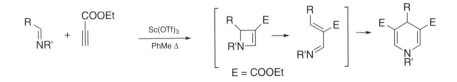

E = COOEt

Cycloisomerization.   2-(4-Pentynyl)furan undergo ring closure to afford a bicyclic isomer in 80% yield.[14] The catalyst system contains Hg(OAc)$_2$ and 0.1 mol% of Sc(OTf)$_3$. Because using Hg(OTf)$_2$ alone gives the product only in a low yield, it is interesting to find out whether Hg(OTf)OAc is a particularly active catalyst.

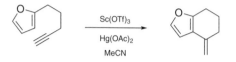

Nazarov cyclization onto a heteroaromatic nucleus employs Sc(OTf)$_3$ (5 mol%) and LiClO$_4$ (1 equiv.) in hot dichloroethane.[15]

Transesterification.   Because of its Lewis acidity, Sc(OTf)$_3$ catalyzes transesterification.[16]

[1]Tschöp, A., Marx, A., Sreekanth, A.R., Schneider, C. *EJOC* 2318 (2007).
[2]Peruncheralathan, S., Henze, M., Schneider, C. *SL* 2289 (2007).
[3]Nomura, N., Taira, A., Nakase, A., Tomioka, T., Okada, M. *T* **63**, 8478 (2007).
[4]Su, W., Jin, C. *OL* **9**, 993 (2007).
[5]Overman, L.E., Sato, T. *OL* **9**, 5267 (2007).
[6]Isambert, N., Cruz, M., Arevalo, M.J., Gomez, E., Lavilla, R. *OL* **9**, 4199 (2007).
[7]Benaglia, M., Cozzi, F., Puglisi, A. *EJOC* 2865 (2007).
[8]Isambert, N., Cruz, M., Arevalo, M.J., Gomez, E., Lavilla, R. *OL* **9**, 4199 (2007).
[9]Reynolds, T.E., Binkley, M.S., Scheidt, K.A. *OL* **10**, 2449 (2008).
[10]Schomaker, J.M., Boyd, W.C., Stewart, I.C., Toste, F.D., Bergman, R.G. *JACS* **130**, 3777 (2008).
[11]Wiedemann, S.H., Noda, H., Harada, S., Matsunaga, S., Shibasaki, M. *OL* **10**, 1661 (2008).
[12]Kikuchi, S., Iwai, M., Fukuzawa, S. *SL* 2639 (2007).
[13]Kikuchi, S., Iwai, M., Murayama, H., Fukuzawa, S. *TL* **49**, 114 (2008).
[14]Yamamoto, H., Sasaki, I., Imagawa, H., Nishizawa, M. *OL* **9**, 1399 (2007).
[15]Malona, J.A., Colbourne, J.M., Frontier, A.J. *OL* **8**, 5661 (2006).
[16]Remme, N., Koschek, K., Schneider, C. *SL* 491 (2007).

Silica gel.

Isomerization.   On exposure to silica gel 1-alkenylcyclopropyl sulfonates are converted into extended allyl sulfonates.[1] Such products can be used to synthesize amines, among numerous other functional molecules.

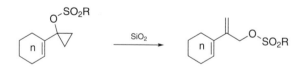

Catalyst support.   A review of the chemistry that employs silica-supported catalysts for the Heck reaction[2] and a report on development of a diphosphine ligand linked to mesoporous silica MCM-41 for PdCl$_2$ (with CuI to promote the Sonogashira coupling[3]) have appeared.

Perchloric acid supported on silica is a useful catalyst for carbamoylation of alcohols and phenols under solvent-free conditions (ROH + NaOCN),[4] and sulfuric acid on silica catalyzes formation of imides from nitriles and acid anhydrides.[5]

Molybdenum imido alkylidene catalysts supported by silica gel remain stable and highly active for alkene metathesis.[6]

Friedel–Crafts reaction.   An intramolecular alkylation with an *N*-tosylaziridine unit as electrophile is apparently favored by entropic and stereochemical factors. Thus heating with SiO_2 at 120° accomplishes this feat.[7]

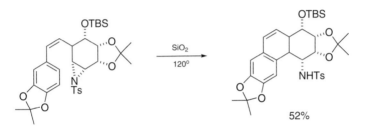

[1]Quan, L.G., Lee, H.G., Cha, J.K. *OL* **9**, 4439 (2007).
[2]Polshettiwar, V., Molnár, A. *T* **63**, 6949 (2007).
[3]Cai, M., Sha, J., Xu, Q. *T* **63**, 4642 (2007).
[4]Modarresi-Alam, A.R., Khamooshi, F., Nasrollahzadeh, M. *T* **63**, 8723 (2007).
[5]Habibi, Z., Salehi, P., Zolfigol, M.A., Yousefi, M. *SL* 812 (2007).
[6]Blanc, F., Thivolle-Cazat, J., Basset, J.-M., Coperet, C., Hock, A.S., Tonzetich, Z.J., Schrock, R.R. *JACS* **129**, 1044 (2007).
[7]Collins, J., Drouin, M., Sun, X., Rinner, U., Hudlicky, T. *OL* **10**, 361 (2008).

Silver acetate.

Cyclocarboxylation.[1]   Propargylic alcohols add to CO_2 in the presence of DBU and the salts of the carbonic acid monoesters rapidly cyclize when AgOAc is present to activate the triple bond.

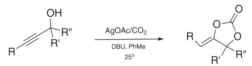

[1]Yamada, W., Sugawara, Y., Cheng, H.M., Ikeno, T., Yamada, T. *EJOC* 2604 (2007).

Silver fluoride.

Alkynylation.   The reaction of 1-alkynes with trifluoromethyl ketones in water is promoted by AgF (with ligand Cy_3P).[1]

[1]Deng, G.-J., Li, C.-J. *SL* 1571 (2008).

Silver hexafluoroantimonate.

Cyclopropanation. Cycloaddition involving carbenoids bearing both donor and acceptor substituents is effectively catalyzed by AgSbF$_6$. Alkenes that are reluctant to react with catalysis by Rh$_2$(OAc)$_4$ (e.g., those prefer to undergo C—H insertion or with a sterically hindered double bond) are more susceptible to participate.[1]

By way of cycloisomerization the formation of a tetracycle containing two cyclopropane units from a 1,5-dien-10-yne is an intriguing transformation, whereas a truncated 1,6-enyne affords a simple alkenylcyclopentene.[2]

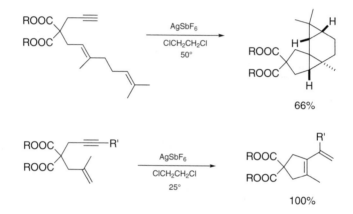

66%

100%

Pyrazoles. *N*-Propargyl-*N*-tosyl hydrazones cyclize with migration of the Ts group upon coordination with AgSbF$_6$ to furnish pyrazole derivatives.[3]

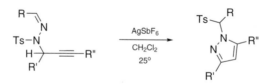

[1]Thompson, J.L., Davies, H.M.L. *JACS* **129**, 6090 (2007).
[2]Porcel, S., Echavarren, A.M. *ACIE* **46**, 2672 (2007).
[3]Lee, Y.T., Chung, Y.K. *JOC* **73**, 4698 (2008).

Silver mesylate.

Isomerization. Propargylic alcohols are converted into 1,3-transposed conjugated carbonyl compounds with mediation of CO$_2$ and DBU, and catalyzed by AgOMs in formamide at room temperature.[1] The process involves C—O bond cleavage, in contrast to the report of a similar reaction catalyzed by AgOAc (*loc. cit.*).

[1]Sugawara, Y., Yamada, W., Yoshida, S., Ikeno, T., Yamada, T. *JACS* **129**, 12902 (2007).

Silver nitrate.

Hydroamination.   Formation of the *N*-Boc derivative of 3-vinyl-5-phenylisoxazolidine from an *O*-alkylhydroylamine that contain a β-allenyl group paves the way toward elaboration of sedamine.[1] The intramolecular hydroamination is a high-yielding reaction.

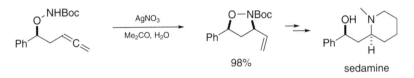

Propargylic amines.   To assemble 1-alkynes, aldehydes, and amines into propargylic amines the heating with a AgNO₃/zeolite catalyst is a convenient method.[2]

Oxidative decarboxylation.   When α-amido acids suffer degradation to lose CO_2, imides are produced. The transformation is mediated by $AgNO_3$, $(NH_4)_2S_2O_8$ and $CuSO_4 \cdot 5H_2O$.[3]

[1]Bates, R.W., Nemeth, J.A., Snell, R.H. *S* 1033 (2008).
[2]Maggi, R., Bello, A., Oro, C., Sartori, G., Soldi, L. *T* **64**, 1435 (2008).
[3]Huang, W., Wang, M., Yue, H. *S* 1342 (2008).

Silver(I) oxide.

Oxidation.[1]   A combination of Ag_2O and pyridine-*N*-oxide in MeCN can be used to convert benzylic halides into aroic esters and primary allylic halides into enals.

Cyclization.[2]   The modes of Ag(I)-catalyzed cyclization of 2-alkynyl-3-formyl-quinolines in MeOH have been correlated with the pKa values of the silver salts used. Both 5-exo-dig and 6-endo-dig pathways are equally favored in the presence of AgOAc, products arising from the former pathway predominate in using more basic silver salts (e.g., Ag_2O, AgO).

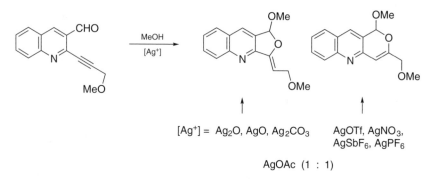

[Ag⁺] = Ag₂O, AgO, Ag₂CO₃   AgOTf, AgNO₃,
                                  AgSbF₆, AgPF₆

AgOAc (1 : 1)

[1]Chen, D.X., Ho, C.M., Wu, Q.Y.R., Wu, P.R., Wong, F.M., Wu, W. *TL* **49**, 4147 (2008).
[2]Godet, T., Vaxelaire, C., Michel, C., Milet, A., Belmont, P. *CEJ* **13**, 5632 (2007).

Silver perchlorate.

O—H bond insertion.[1]   It is significant that AgClO₄ and Rh₂(OAc)₄ show a regio-chemical difference in the trapping of the carbenoid generated from a conjugated α-diazo-ester by BnOH.

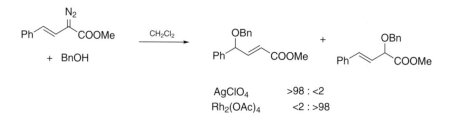

AgClO₄   >98 : <2
Rh₂(OAc)₄   <2 : >98

[1]Yue, Y., Wang, Y., Hu, W. *TL* **48**, 3975 (2007).

Silver tetrafluoroborate.

Nazarov–Friedel-Crafts reaction tandem.[1]   1-Alkenyl-2,2-dichlorocyclopropyl trialkylsilyl ethers are subject to Nazarov cyclization. If the alkenyl chain is terminated with an aromatic ring at a proper distance, bicyclization becomes a possibility.

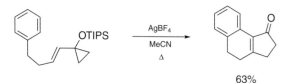

63%

[1]Grant, T.N., West, F.G. *OL* **9**, 3789 (2007).

Silver triflate.

Dihydroisoquinolines.   Aldimines derived from *o*-alkynylaraldehydes and amines are amenable to cyclize under the influence of an Ag(I) species (e.g., AgOTf), barring stereo-chemical inhibition. Accordingly, reduction of the imines facilitates the formation of the heterocyclic compounds.[1] Imines bearing an *N*-methoxycarbonylmethyl group readily generate 1,3-dipolar species and their trapping has been observed.[2]

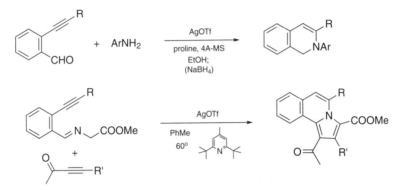

Silyl transfer.   1,1-Di(*t*-butyl)silacyclopropanes readily submit the di-*t*-butylsilyl residue to α-keto esters to form 4-alkoxy-1,3,2-dioxasiloles. In the case of an allyloxy ester the situation is set up for the Ireland–Claisen rearrangement.[3]

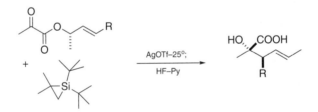

The silyl transfer to imines delivers silaaziridines that can enter cross-coupling with alkynes to afford alylic amines.[4]

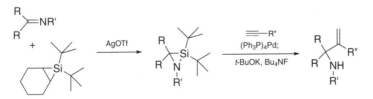

[1]Ding, Q., Yu, X., Wu, J. *TL* **49**, 2752 (2008).
[2]Su, S., Porco J.A. Jr *JACS* **129**, 7744 (2007).
[3]Howard, B.E., Woerpel, K.A. *OL* **9**, 4651 (2007).
[4]Nevarez, Z., Woerpel, K.A. *OL* **9**, 3773 (2007).

Sodium borohydride.

Reduction. The reduction of α-(benzoxazol-2-ylthio) ketones with NaBH$_4$ gives thiiranes via spirocyclic intermediates.[1]

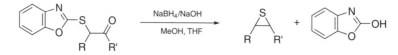

With NaBH$_4$–Br$_2$ malonic esters are reduced to 1,3-diols.[2] *N*-Alkylation of hydroxy-alkylamines is achieved by adding NaBH$_4$ to their premixture with ketones and (*i*-PrO)$_4$Ti.[3]

Differences in the stereochemical course for reduction of imines by NaBH$_4$ and Zn(BH$_4$)$_2$, as controlled by a 1,3-related stereogenic center[4] is synthetically significant.

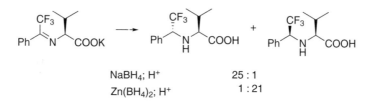

| NaBH$_4$; H$^+$ | 25 : 1 |
| Zn(BH$_4$)$_2$; H$^+$ | 1 : 21 |

Hydrogenation of monosubstituted and 1,2-disubstituted alkenes is accomplishable by NaBH$_4$ and catalytic amounts of RuCl$_3$ · xH$_2$O in aqueous THF.[5]

[1] Yamada, N., Mizuochi, M., Takeda, M., Kawaguchi, H., Morita, H. *TL* **49**, 1166 (2008).
[2] Tudge, M., Mashima, H., Savarin, C., Humphrey, C., Davies, I. *TL* **49**, 1041 (2008).
[3] Salmi, C., Loncle, C., Letourneux, Y., Brunel, J.M. *T* **64**, 4453 (2008).
[4] Hughes, A., Devine, P.N., Naber, J.R., O'shea, P.D., Foster, B.S., McKay, D.J., Volante, R.P. *ACIE* **46**, 1839 (2007).
[5] Sharma, P.K., Kumar, S., Kumar, P., Nielsen, P. *TL* **48**, 8704 (2007).

Sodium borohydride–dimethyl ditelluride.

Elimination. 1,2-Cyclonucleosides are opened on reaction with NaBH$_4$ and catalytic amounts (0.1 equiv.) of MeTeTeMe. If the 3-hydroxyl is acetylated, the tellurides undergo elimination in situ. The corresponding seleno derivatives are stable under similar reaction conditions.[1]

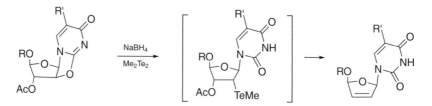

[1] Sheng, J., Hassan, A.E.A., Huang, Z. *JOC* **73**, 3725 (2008).

Sodium dichloroiodate.
Nitrile synthesis.[1]   A direct method for conversion of aldehydes to nitriles entails reaction with aq. ammonia in the presence of $NaICl_4$.

[1]Telvekar, V.N., Patel, K.N., Kundaikar, H.S., Chaudhari, H.K. *TL* **49**, 2213 (2008).

Sodium hydride.
Williamson synthesis.[1]   Ether synthesis from alcohols and RBr (esp. $ArCH_2Br$) is accomplished with NaH in DMF at room temperature.

[1]Jin, C.H., Lee, H.Y., Lee, S.H., Kim, I.S., Jung, Y.H. *SL* 2695 (2007).

Sodium iodide.
Reduction.[1]   Aryl azides are reduced to arylamines by $NaI-BF_3 \cdot OEt_2$ in MeCN at room temperature.

Henry reaction.[2]   In the presence of NaI bromonitromethane condenses with aldehydes in THF. The usefulness of the process is noted in the stereoselective assemblage of the *anti,anti*-isomer of 1-bromo-2-hydroxy-3-dibenzylamino-1-nitrobutane.

[1]Kamal, A., Shankaraiah, N., Markandeya, N., Reddy, C.S. *SL* 1297 (2008).
[2]Concellon, J.M., Rodriguez-Solla, H., Concellon, C., Garcia-Garcia, S., Diaz, M.R. *OL* **8**, 5979 (2006).

Sodium nitrite.
Sandmeyer reaction.[1]   The arylamine to aryl iodide transformation is accomplished by $NaNO_2$, KI, TsOH in MeCN at $10-25°$.

[1]Krasnokutskaya, E.A., Semenischeva, N.I., Filimonov, V.D., Knochel, P. *S* 81 (2007).

Sodium periodate.
Epoxide cleavage.   In aqueous THF or MeCN epoxides are cleaved by $NaIO_4$ as shown by the following example.[1]

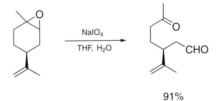

91%

N-Oxy radicals.   Oxy radicals are readily generated from *N*-hydroxycarboximides such as *N*-hydroxyphthalimide by treatment with $NaIO_4$ on wet SiO_2.[2]

[1]Binder, C.M., Dixon, D.D., Almaraz, E., Tius, M.A., Singaram, B. *TL* **49**, 2764 (2008).
[2]Coseri, S. *EJOC* 1725 (2007).

Sodium phenoxide.

Mukaiyama aldol reaction.   The base modified by the phosphine oxide ligand **1** is effective for catalyzing the aldol reaction.[1]

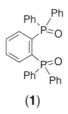

(1)

[1]Hatano, M., Takagi, E., Ishihara, K. *OL* **9**, 4527 (2007).

Sodium tetracarbonylferrate(II).

Carbonylative coupling.[1]   Dibenzyl ketones are obtained in moderate to high yields by the reaction of $ArCH_2Br$ with $Na_2Fe(CO)_4$ in NMP. If unsymmetrical ketones are required the reaction is performed in two stages: first with one bromide at $0°$, then the second bromide at room temperature.

[1]Potter, R.G., Hughes, T.S. *OL* **9**, 1187 (2007).

Sodium tetrachloroaurate.

β-Alkoxy enones.[1]   α-Hydroxypropargyl epoxides undergo interesting transformations on treatment with $NaAuCl_4$ in refluxing benzene. Perhaps an activation of the triple bond to induce attack of the epoxy atom triggers 1,2-rearrangement or C—C bond cleavage.

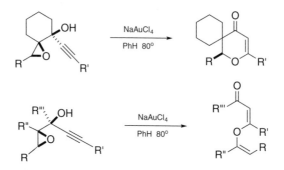

1-Aminoindolizines.   Pyridine-2-carbaldehyde and homologues condense with amines and 1-alkynes in water or under solvent – free conditions to generate 1-aminindolizines.

Apparently, the Au salt is responsible for inducing cycloisomerization of the propargylic amines.[2]

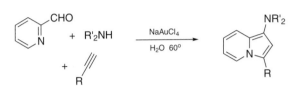

[1]Shu, X.-Z., Liu, X.-Y., Ji, K.-G., Xiao, H.-Q., Liang, Y.-M. *CEJ* **14**, 5282 (2008).
[2]Yan, B., Liu, Y. *OL* **9**, 4323 (2007).

Sodium tetrakis[3,5-bis(trifluoromethyl)phenyl]borate.

Deacetalization.[1]   A colloidal suspension of the title compound in water is capable of hydrolyzing acetals.

[1]Chang, C.-C., Liao, B.-S., Liu, S.-T. *SL* 283 (2007).

Sodium tetramethoxyborate.

Michael reaction.   Conjugate addition involving stabilized *C*-nucleophiles is readily achieved using NaB(OMe)$_4$ [3 mol%] as catalyst in MeCN.[1]

[1]Campana, A.G., Fuentes, N., Gomez-Bengoa, E., Mateo, C., Oltra, J.E., Echavarren, A.M., Cuerva, J.M. *JOC* **72**, 8127 (2007).

Strontium.

Ketones.[1]   Treatment of sodium carboxylates (from RCOOH + NaH) with Sr gives species that react with MeI to afford methyl ketones. Scope and mechanism of this reaction has yet to be established.

[1]Miyoshi, N., Matsuo, T., Asaoka, M., Matsui, A., Wada, M. *CL* **36**, 28 (2007).

Sulfamic acid.

t-Butyl carbamates.   For derivatization of amines with *t*-Boc$_2$O at room temperature without any solvent, ultrasound irradiation enhances the catalysis of sulfamic acid.[1]

[1]Upadhyaya, D.J., Barge, A., Stefania, R., Cravotto, G. *TL* **48**, 8318 (2007).

Sulfur.

Reduction.[1]   Nitroarenes are reduced to the corresponding amines by sulfur – NaHCO$_3$ (3 equiv. each) in DMF at 130°.

[1]McLaughlin, M.A., Barnes, D.M. *TL* **47**, 9095 (2006).

Sulfuric acid.

Porphyrin.[1]   Access to the parent macroheterocycle (in 77% yield) is facilitated by heating the readily available *meso*-tetrakis(hexyloxycarbonyl) derivative with sulfuric acid containing some water at 180°.

[1]Neya, S., Quan, J., Hata, M., Hoshino, T., Funasaki, N. *TL* **47**, 8731 (2006).

T

Tantalum(V) dimethylamide.

Aminoalkylation.[1]   The α-carbon of an *N*-alkylarylamine is activated by Ta(NMe$_2$)$_5$ such that it adds to simple alkenes.

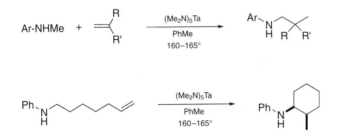

[1]Herzon, S.B., Hartwig, J.F. *JACS* **129**, 6690 (2007).

Tellerium chloride.

Preparation.   The title reagent can be prepared by heating Te(0) in sulfuryl chloride.[1]

[1]Petragnani, N., Mendes, S.R., Silveira, C.C. *TL* **49**, 2371 (2008).

Tetrabenzyl pyrophosphate.

Amide formation.   As a dehydrating agent [(BnO)$_2$PO]$_2$O is active in condensing RCOOH and amines. The reaction is carried out in chloroform and also with catalytic quantities of DMAP.[1]

[1]Reddy, Y.T., Reddy, P.N., Reddy, P.R., Crooks, P.A. *CL* **37**, 528 (2008).

Tetrabutylammonium dichloroiodate.

Iodination.   In providing electrophilic iodine, Bu$_4$NICl$_2$ in sulfuric acid constitutes another reagent system for nuclear iodination arenes.[1]

[1]Filimonov, V.D., Semenischeva, M., Krasnokutskaya, E.A., Hwang, H.Y., Chi, K.-W. *S* 401 (2008).

Fiesers' Reagents for Organic Synthesis, Volume 25. By Tse-Lok Ho
Copyright © 2010 John Wiley & Sons, Inc.

Tetrabutylammonium fluoride, TBAF.

Substitution. Activated aryl fluorides are converted into ethers on reaction with various alcohols released from $(RO)_4Si$ in situ by TBAF. The reaction is conducted in refluxing acetone.[1]

Elimination. 2-Bromo-1-alkenes undergo dehydrobromination on warming with TBAF in DMF at 60°.[2]

1,1-Dicyanoalkyl silyl ethers are converted into acyl cyanides by TBAF. Esters and amides are readily formed therefrom.[3]

Generation of sulfenate anion is convenient from $RS(=O)CH_2CH_2SiMe_3$.[4]

Two interchangeable fluoride ion sources used in aryne generation from *o*-trimethylsilylaryl triflates are CsF and TBAF. Many benzannulated heterocycles are now conveniently prepared via cycloaddition reactions of arynes. 3-Indazolecarboxylic esters are accessible from such a reaction.[5]

Alkylation. A synthetic approach to platensimycin entails a step of intramolecular alkylation, in which a phenolate intermediate is generated from a silyl ether. The process can be accomplished with Bu_4NF.[6]

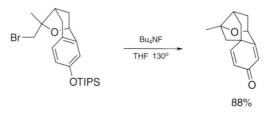

88%

For hydroxyalkylation of 2-fluoro-3-alkynoic esters the amount of TBAF, which is used as base, affects the reaction pathway. Kinetic control and thermodynamic control are observed in the reactions where substoichiometric (e.g., 0.2 equiv.) and excess of TBAF are present, respectively.[7]

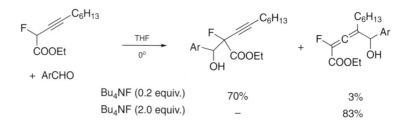

Bu₄NF (0.2 equiv.)	70%	3%
Bu₄NF (2.0 equiv.)	–	83%

To render a γ,δ-unsaturated β-ketolactone in the enolate form by TBAF is critical for realizing an intramolecular Diels–Alder reaction en route to salvinorin-A.[8]

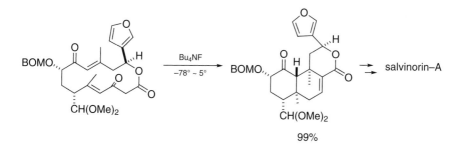

99%

Deprotection.   Aldehydes protected as α-trichloromethylalkyl TBS ethers by reaction with TBS-Cl and Cl$_3$CCOONa are recovered by treatment with TBAF in DMF.[9]

[1]Wang, T., Love, J.A. *S* 2237 (2007).
[2]Okutani, M., Mori, Y. *TL* **48**, 6856 (2007).
[3]Nemoto, H., Moriguchi, H., Ma, R., Kawamura, T., Kamiya, M., Shibuya, M. *TA* **18**, 383 (2007).
[4]Foucoin, F., Caupene, C., Lohier, J.-F., de Oliveira Santos, J.S., Perrio, S., Metzner, P. *S* 1315 (2007).
[5]Liu, Z., Shi, F., Martinez, P.D.G., Raminelli, C., Larock, R.C. *JOC* **73**, 219 (2008).
[6]Lalic, G., Corey, E.J. *OL* **9**, 4921 (2007).
[7]Xu, B., Hammond, G.B. *ACIE* **47**, 689 (2008).
[8]Scheerer, J.R., Lawrence, J.F., Wang, G.C., Evans, D.A. *JACS* **129**, 8968 (2007).
[9]Cafiero, L.R., Snowden, T.S. *TL* **49**, 2844 (2008).

Tetrabutylammonium iodide.

Halogenation.[1]   Results of the stereoselective addition of [I/Cl] to a wider range of alkynes based on Bu$_4$NI in refluxing 1,2-dichloroethane have been obtained. Under the same reaction conditions alkenes give *vic*-dichloroalkanes.

[1]Ho, M.L., Flynn, A.B., Ogilvie, W.W. *JOC* **72**, 977 (2007).

Tetrabutylammonium phenolate.

1,3-Dithian-2-yl anion.   An exceptionally mild base (compared to BuLi) for desilylative generation of the 1,3-dithian-2-yl anion is Bu$_4$NOPh.[1]

[1]Michida, M., Mukaiyama, T. *CL* **37**, 26 (2008).

Tetrabutylammonium tricarbonyl(nitroso)ferrate.

Allylic substitution.   More detailed studies on the reaction of allylic carbonates with active methylene compounds[1] have revealed mechanistic dichotomy that is ligand-dependent.[1]

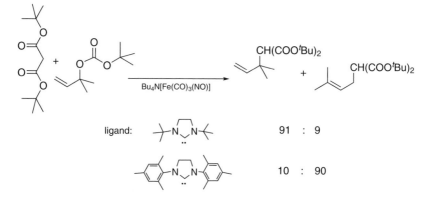

ligand: (structure)   91 : 9

(structure)   10 : 90

Transesterification. Phenyl esters, enol esters, and carbonates undergo transesterification on heating with various alcohols and $Bu_4N[Fe(CO)_3(NO)]$ in hexane.[2]

[1]Plietker, B., Dieskau, A., Möws, K., Jatsch, A. *ACIE* **47**, 198 (2008).
[2]Magens, S., Ertelt, M., Jatsch, A., Plietker, B. *OL* **10**, 53 (2008).

Tetrabutylammonium triphenyldifluorosilicate.

Benzyne generation. As a fluoride source for initiating the generation of benzyne from 2-trimethylsilylphenyl triflate, the effectiveness of $Bu_4N(Ph_3SiF_2)$ is obvious. New possibilities for cycloaddition with enamides leading to indolines and isoquinoline derivatives have been explored.[1]

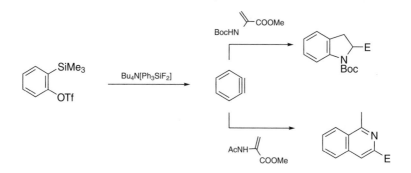

[1]Gilmore, C.D., Allan, K.M., Stoltz, B.M. *JACS* **130**, 1558 (2008).

Tetracarbonylhydridocobalt.

Cyclocarbonylation.[1] Epoxides bearing a β-hydroxyalkyl substituent are shown to form δ-lactones, while catalyzed by $HCo(CO)_4$.

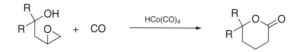

[1]Kramer, J.W., Joh, D.Y., Coates, G.W. *OL* **9**, 5581 (2007).

Tetrachloroauric acid.

Biaryls.[1]   Oxidative dimerization of nonactivated arenes is accomplished with the treatment of PhI(OAc)$_2$ in HOAc, and catalyzed by HAuCl$_4$.

Carbonylation.[2]   A solid catalyst made from HAuCl$_4$ and an ion-exchange resin is used to convert ArNH$_2$ into ArNHCOOMe in MeOH. It can also be used to form ureas from R$_2$NH, CO$_2$ (or CO and O$_2$).

γ-Keto esters.   Acetates of diethynyl carbinols are transformed into γ-keto esters (or lactones) with HAuCl$_4$ · 3H$_2$O in alcoholic solvents.[3]

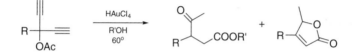

[1]Kar, A., Mangu, N., Kaiser, H.M., Beller, M., Tse, M.K. *CC* 386 (2008).
[2]Shi, F., Zhang, Q., Ma, Y., He, Y., Deng, Y. *JACS* **127**, 4182 (2005).
[3]Kato, K., Teraguchi, R., Kusakabe, T., Motodate, S., Yamamura, S., Mochinda, T., Akita, H. *SL* 63 (2007).

1,1,2,2-Tetrafluoroethanesulfonyl chloride.

Aryl tetraflates.[1]   The title reagent is prepared from tetrafluoroethene. It is used to derivatize phenols to provide products with higher stability than the corresponding triflates. The tetraflates are useful for coupling.

[1]Rostovtsev, V.V., Bryman, L.M., Junk, C.P., Harmer, M.A., Carcani, L.G. *JOC* **73**, 711 (2008).

Tetrakis[chloro(pentamethylcyclopentadienyl)ruthenium(I)].

Cycloaddition.   The title complex is found useful as catalyst for the cycloaddition of organoazides to 1-alkynes to form 1,5-disubstituted 1,2,3-triazoles.[1]

[1]Rasmussen, L.K., Boren, B.C., Fokin, V.V. *OL* **9**, 5337 (2007).

meso-Tetrakis(4-chlorophenylporphyrinato)aluminum tetracarbonylcobaltate.

Carbonylation.[1]   The title reagent (as THF complex) mediates carbonylation of monosubstituted and 2,3-disubstituted epoxides to give succinic anhydrides (22 examples, 90–99% conversion).

[1]Rowley, J.M., Lobkovsky, E.B., Coates, G.W. *JACS* **129**, 4948 (2007).

1,1,3,3-Tetrakis(trifluoromethanesulfonyl)propane.

Michael reaction.[1]   This carbon acid (1) is highly effective for inducing the Michael reaction between 2-siloxyfurans and enones.

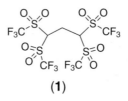

(1)

[1]Takahashi, A., Yanai, H., Taguchi, T. *CC* 2385 (2008).

Tetrakis(triethylphosphite)nickel(0).

Substitution.   The title complex catalyzes retentive substitution of allylic esters with regard to regiochemical and stereochemical senses.[1]

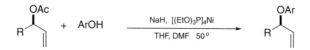

[1]Yatsumonji, Y., Ishida, Y., Tsubouchi, A., Takeda, T. *OL* 9, 4603 (2007).

Tetrakis(triphenylphosphine)nickel(0).

Coupling reactions.   Benzyl ketones and arylacetic acid derivatives of the type ArCH(R)COY are prepared from RCH(X)COY by coupling with ArB(OH)$_2$ using (Ph$_3$P)$_4$Ni as catalyst.[1]

[1]Liu, C., He, C., Shi, W., Chen, M., Lei, A. *OL* 9, 5601 (2007).

Tetrakis(triphenylphosphine)palladium(0).

Coupling reactions.   There is a review on Suzuki coupling based on (Ph$_3$P)$_4$Pd (and other Pd species) with certain concessions such as involving unusual reaction partners, ligandless conditions, catalyst on solid-support, in supercritical CO$_2$, H$_2$O, ionic liquids, fluorous media or micelles, being assisted by microwaves or conducted in microreactors or ballmills.[1] The involvement of a Merrifield resin-linked phosphine ligand is beneficial to Suzuki coupling of ArCl, in terms of catalyst recovery.[2]

Homologative functionalization of alkenes via hydroboration (with 9-BBN) and coupling with carbamoyl chlorides under basic conditions yields carboxamides.[3] Arylcyclopropanes are obtained from a Pd-catalyzed reaction of tricyclopropylbismuth and ArX.[4] α-Bromoalkyl sulfoxides couple with ArB(OH)₂ with inversion of configuration occuring in the oxidative oxidation step (retention of configuration while the Pd-containing intermediates undergo reductive elimination). Of particular significance is that the (RS,RS)-diastereomers fail the coupling.[5]

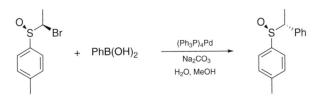

Thiolactams enter Suzuki coupling[6] and Sonogashira coupling[7] with loss of the sulfur atom when copper(I) 2-thienylcarboxylate is also present.

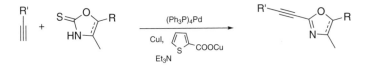

Coupling of ArB(OH)₂ with Ar′₃Sb(OAc)₂ does not require a base.[8]

In copper-free Sonogashira coupling the competition of ligand and amine base determines the reaction mechanism.[9] The oxidative addition of ArI with (Ph₃P)₄Pd is faster when amine is present. With the proposed mechanisms the efficiency of Ph₃P > Ph₃As is explained.

Diaryltetramethyldisiloxanes surrender the aryl groups for coupling with Ar′X in refluxing THF to give biaryls.[10] This reaction also requires stoichiometric Ag₂O and catalytic amounts of TBAF, in addition to (Ph₃P)₄Pd.

A coupling route to 1,3-diaryl-1,2-propadienes is demonstrated by the reaction of α-phenylpropargyl carbonates with ArB(OH)₂. The use of mixed carbonate esters for coupling, instead of acetates and benzoates, are important to give good yields, and in the asymmetric version, high ee.[11]

With an oxidant (benzoquinone) added, α-diazoalkanoic esters and α-diazoketones couple with ArB(OH)₂, providing α-aryl-α,β-unsaturated esters and enones, respectively.[12] On the other hand, diazoacetic esters are arylated and alkenylated.[13]

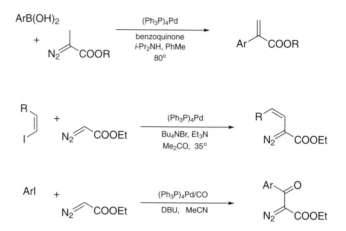

Heterocycle synthesis. In the presence of HCOONa and with microwave irradiation Heck reaction of *N*-alkynoyl-2-(*o*-bromoaryl)ethylamines cyclize to give benzazepine derivatives. Benzannulated eight-membered lactam homologues are also accessible by the method.[14]

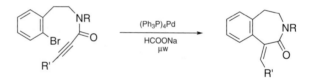

Formation of dihydroquinol-4-ones from 3-(*o*-iodoarylamino)propanoate esters involving arylpalladium intermediates which may exist in the palladacycle form and thereby derive special activity for intramolecular attack on the ester group.[15] Under similar conditions, five-membered ring oxime derivatives are prepared.[16]

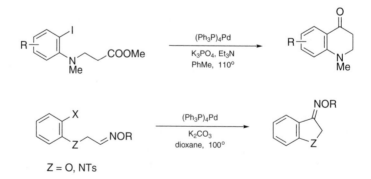

Allenylpalladium compounds are generated from arylpropargyl derivatives, and they behave electrophilically toward nucleophiles. With involvement of such products in inter-molecular coupling the synthetic potential of the catalyst is enhanced.[17]

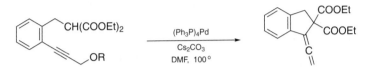

4-Aryltetrahydropyridines are synthesized from homopropargylamines and allenyl-ethylamines via coupling with $ArB(OH)_2$. The products are isomeric at the position of the double bond.[18]

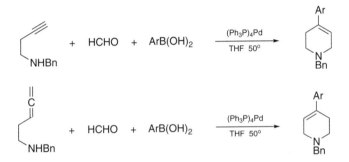

Cleavage of C—O bonds. Allylic ethers are cleaved at room temperature in the presence of $(Ph_3P)_4Pd$. Rates are much higher in protic solvents (MeOH vs. THF) and the cleavage of allyl, methallyl, prenyl groups in succession is possible.[19]

Diallyl malonate esters containing an α-aryl substituent (critical!) undergo decarboxyla-tion to give the allyl 4-pentenoates.[20] 2,3-Alkadienyl 2-alkynoates also decarboxylate, but during the reunion the allenyl and alkynyl fragments are transposed.[21]

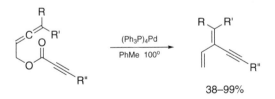

38–99%

In principle, 5-vinyl-2-oxazolidinones are accessible from decomposition of 4-amino-2-butenyl carbonates, with loss of the nonallylic alcohol only. Trapping of the released CO_2 is indeed feasible, although efficiency is not sufficiently high. The heterocyclic products are obtained in much higher yield by conducting the reaction under a CO_2 atmosphere.[22]

Trapping the oxy anion from the Pd-induced ring-opening of a γ,δ-epoxy-α,β-unsaturated ester with triphenyl borate provides allylpalladium species for eventual reaction with nucleophiles. *syn*-3-Alkoxy-4-hydroxy-2-alkenoic esters are readily formed.[23]

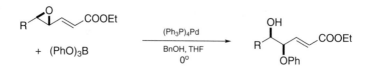

Reductive removal of a trifloxy substituent without affecting an aromatic bromide is important in accessing a precursor for gelsemine synthesis. This can be done by a reaction with Bu₃SnH, catalyzed by (Ph₃P)₄Pd.[24]

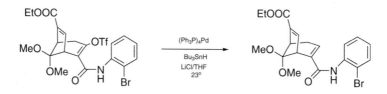

Allyl 2-pyridylacetates is induced to decarboxylate by (Ph₃P)₄Pd. *N*-Allylpyridinium salts and *N*-allyl-2-alkylidene-1,2-dihydropyridines are implicated as reaction intermediates.[25]

Cyclization.   Synthetic exploitation of reactivities of π-allylpalladium complexes that are generated from propargylic[26] and allenic derivatives[27] is shown by the closure of two azacycles in one step.

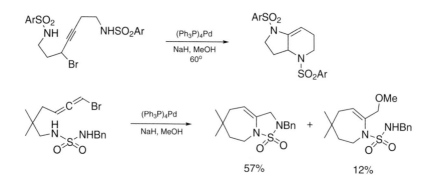

Cyclization attendant by coupling from molecules containing both an amino group and an allene unit has apparent utility in synthesis.[28] Formation of five- or six-membered rings is subject to change of the substituent pattern of the allene moiety. The bond distance between the nitrogen atom and the unsaturation is of necessity important. Diazetidines have been acquired from 4-hydrazinyl-1,2-alkadienes.[29]

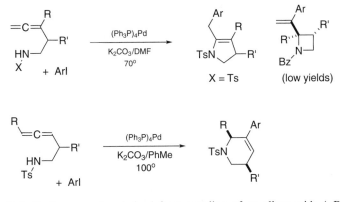

A π-allylpalladium complex derived from coupling of an allene with $ArB(OH)_2$ is reactive toward an aldehyde, therefore formation of cyclic alcohols from 4,5-hexadienal, 5,6-heptadienal and various analogues, in different degrees of efficiency, is expected.[30]

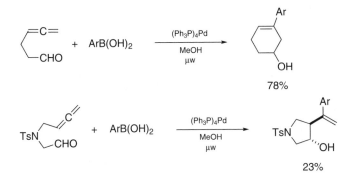

Intramolecular trapping of π-allylpalladium species by enamine to form 2-vinylcyclo-pentanecarbaldehydes or cyclohexanecarbaldehydes is efficient.[31]

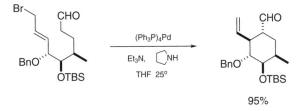

In HCOOH the treatment of molecules containing diyne and enal subunits with $(Ph_3P)_4Pd$ leads to extensive structural reorganization, a furan ring is created in the process.[32]

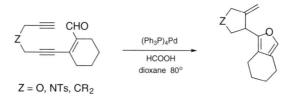

Different reaction patterns are manifested[33] when Meldrum acid that is substituted with a cinnamyl ester is exposed to $(Ph_3P)_4Pd$ and a mixture of $(PhCN)_2PdCl_2 - Yb(OTf)_3$.

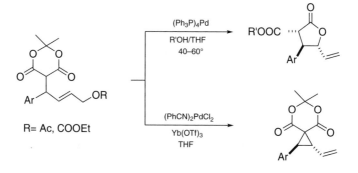

Intramolecular Heck-type reaction initiated by palladation of alkenyl triflates to generate 1,2-dimethylenecyclobutanes has been accomplished. Carbonylation of the cyclic intermediates occurs under CO.[34]

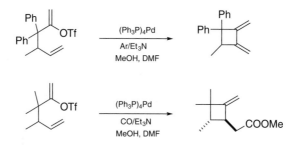

Highly strained ring systems can be generated from a tandem Heck reaction sequence. This achievement attests to the value of the synthetic method.[35]

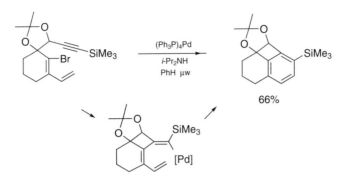

Cycloaddition.   A new trapping partner for the trimethylene-palladium complex is carbon dioxide. 3-Methyl-2-buten-4-olide is formed.[36] *N,N'*-Di-*t*-butyl-1,2-diaziridinone behaves as a 1,3-dipolar species in the presence of (Ph$_3$P)$_4$Pd and such is trapped by alkenes.[37]

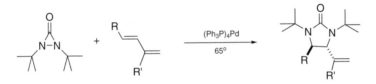

Aminocarbonylation.   A polymer-supported (Ph$_3$P)$_4$Pd catalyzes aminocarbonylation of haloarenecarboxylic acids (with CO and primary or secondary amines) in a flow reactor. Surprisingly, better results than batch process are obtained.[38]

[1]Alonso, F., Beletskaya, I.P., Yus, M. *T* **64**, 3047 (2008).
[2]Schweizer, S., Becht, J.-M., Le Drian, C. *OL* **9**, 3777 (2007).
[3]Yasui, Y., Tsuchida, S., Miyabe, H., Takemoto, Y. *JOC* **72**, 5898 (2007).
[4]Gagnon, A., Duplessis, M., Alsabeh, P., Barabe, F. *JOC* **73**, 3604 (2008).
[5]Rodriguez, N., de Arellano, C.R., Asensio, G., Medio-Simon, M. *CEJ* **13**, 4223 (2007).
[6]Prokopcova, H., Kappe, C.O. *JOC* **72**, 4440 (2007).
[7]Silva, S., Sylla, B., Suzenet, F., Tatibouet, A., Rauter, A.P., Rollin, P. *OL* **10**, 853 (2008).
[8]Yasuike, S., Qin, W., Sugawara, Y., Kurita, J. *TL* **48**, 721 (2007).
[9]Tougerti, A., Negri, S., Jutand, A. *CEJ* **13**, 666 (2007).
[10]Napier, S., Marcuccio, S.M., Tye, H., Whittaker, M. *TL* **49**, 3939 (2008).
[11]Yoshida, M., Okada, T., Shishido, K. *T* **63**, 6996 (2007).
[12]Peng, C., Wang, Y., Wang, J. *JACS* **130**, 1566 (2008).
[13]Peng, C., Cheng, J., Wang, J. *JACS* **129**, 8708 (2007).
[14]Idonets, P.A., Van dr Eycken, E.V. *OL* **9**, 3017 (2007).
[15]Sole, D., Serrano, O. *ACIE* **46**, 7270 (2007).
[16]Ohno, H., Aso, A., Kadoh, Y., Fujii, N., Tanaka, T. *ACIE* **46**, 6325 (2007).
[17]Bi, H.-P., Liu, X.-Y., Gou, F.-R., Guo, L.-N., Duan, X.-H., Shu, X.-Z., Liang, Y.-M. *ACIE* **46**, 7068 (2007).
[18]Tsukamoto, H., Kondo, Y. *ACIE* **47**, 4851 (2008).
[19]Tsukamoto, H., Suzuki, Y., Kondo, Y. *SL* 3131 (2007).

[20]Ohta, T., Ito, Y. *JOC* **72**, 1652 (2007).

[21]Sim, S.H., Park, H.-J., Lee, S.I., Chung, Y.K. *OL* **10**, 433 (2008).

[22]Yoshida, M., Ohsawa, Y., Sugimoto, K., Tokuyama, H., Ihara, M. *TL* **48**, 8678 (2007).

[23]Yu, X.-Q., Yoshimura, F., Ito, F., Sasaki, M., Hirai, A., Tanino, K., Miyashita, M. *ACIE* **47**, 750 (2008).

[24]Grecian, S., Aube, J. *OL* **9**, 3153 (2007).

[25]Waetzig, S.R., Tunge, J.A. *JACS* **129**, 4138 (2007).

[26]Ohno, H., Okano, A., Kosaka, S., Tsukamoto, K., Ohata, M., Ishihara, K., Maeda, H., Tanaka, T., Fujii, N. *OL* **10**, 1171 (2008).

[27]Hamaguchi, H., Kosaka, S., Ohno, H., Fujii, N., Tanaka, T. *CEJ* **13**, 1692 (2007).

[28]Ma, S., Yu, F., Li, J., Gao, W. *CEJ* **13**, 247 (2007).

[29]Cheng, X., Ma, S. *ACIE* **47**, 4581 (2008).

[30]Tsukamoto, H., Matsumoto, T., Kondo, Y. *JACS* **130**, 388 (2008).

[31]Bihelovic, F., Matovic, R., Vulovic, B., Saicic, R.N. *OL* **9**, 5063 (2007).

[32]Oh, C.H., Park, H.M., Park, D.I. *OL* **9**, 1191 (2007).

[33]Fillion, E., Carret, S., Mercier, L.G., Trepanier, V.E. *OL* **10**, 437 (2008).

[34]Innitzer, A., Brecker, L., Mulzer, J. *OL* **9**, 4431 (2007).

[35]Blond, G., Bour, C., Salem, B., Suffert, J. *OL* **10**, 1075 (2008).

[36]Greco, G.E., Gleason, B.L., Lowery, T.A., Kier, M.J., Hollander, L.B., Gibbs, S.A., Worthy, A.D. *OL* **9**, 3817 (2007).

[37]Du, H., Yuan, W., Zhao, B., Shi, Y. *JACS* **129**, 7496 (2007).

[38]Csajagi, C., Borcsek, B., Niesz, K., Kovics, I., Szekelyhidi, Z., Bajko, Z., Urge, L., Darvas, F. *OL* **10**, 1589 (2008).

Tetrakis(triphenylphosphine)platinum(0).

Addition reaction. Stereoselective and regioselective *cis*-addition of aryl and thio groups from ArCOSAr′ to alkynes by catalysis of (Ph₃P)₄Pt has been delineated.[1]

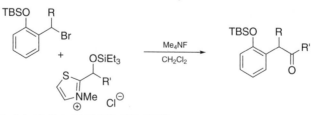

[1]Yamashita, F., Kuniyasu, H., Terao, J., Kambe, N. *OL* **10**, 101 (2008).

Tetramethylammonium fluoride.

o-Hydroxybenzyl ketones.[1] The simultaneous generation of *o*-quinonemethides from *o*-siloxybenzyl bromides initiated by desilylation with Me₄NF on the one hand, and desilylation of 2-(α-siloxyalkyl)thiazolium salts that gives rise to acyl anion equivalents on the other, caters to the formal substitution of the benzyl bromides.

[1]Mattson, A.E., Scheidt, K.A. *JACS* **129**, 4508 (2007).

2,2,6,6-Tetramethyl-1-oxopiperidine salts.

Oxidation.   The oxoammonium salts are mild oxidants for converting 2-cycloalkenols to cyclic enones, with 1,3-transposition of the oxygenated site.[1]

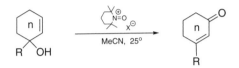

Under anhydrous conditions it is also possible to utilize the salt to oxidize certain unprotected carbohydrates to aldehydes.[2]

[1]Shibuya, M., Tomizawa, M., Iwabuchi, Y. *JOC* **73**, 4750 (2008).
[2]Breton, T., Bashiardes, G., Leger, J.-M., Kokoh, K.B. *EJOC* 1567 (2007).

Thiophosphoryl chloride.

Thionation.   The carbonyl group of amides and ketones are transformed into the thiono group by reaction with PSCl$_3$ under microwave irradiation.[1]

[1]Pathak, U., Pandey, L.K., Tank, R. *JOC* **73**, 2890 (2008).

Thiosuccinic anhydride and homologues.

Diamides.   Cyclic thioanhydrides are a source of unsymmetrical diamides. Aminolysis releases a thiocarboylic acid that can be transformed into another amide function on reaction with an *N*-sulfonyl amine or sulfonyl azide.[1]

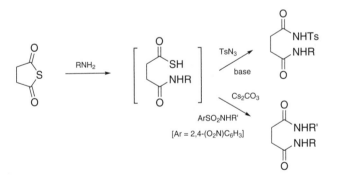

[1]Crich, D., Bowers, A.A. *OL* **9**, 5323 (2007).

Tin(IV) chloride.

Claisen rearrangement.   The aromatic Claisen rearrangement is found to be catalyzed by SnCl$_4$ at room temperature.[1]

Fragmentative defunctionalization.   β,γ-Dioxygenated α-diazo esters undergo fragmentation to give 2-alkynoic esters, in a process that is probably initiated by ionization of the β-hydroxy group.[2]

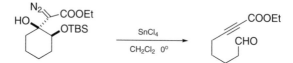

Hetero–Diels–Alder reaction.   The condensation of conjugated α-ketoesters with *N*-alkenyl-2-oxazolidinones shows facial stereodivergence due to the presence of different Lewis acids. For example, stereoisomers are obtained from reactions catalyzed by SnCl₄ and by Eu(fod)₃.[3]

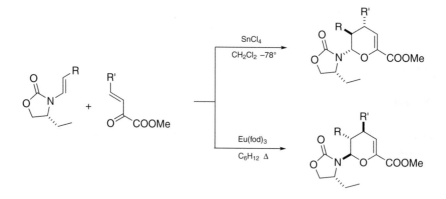

Aromatization.[4]   Enamines of 6-membered cyclic ketones (cyclohexanone, α-tetralone, β-tetralone, . . .) yield arylamines on treatment with SnCl₄ in CH₂Cl₂ at room temperature. [SbCl₅ is less efficient.]

Cyclization.[5]   Efficient cyclization of a polyene initiated from a terminal enol silyl ether shows a remarkable α-selectivity.

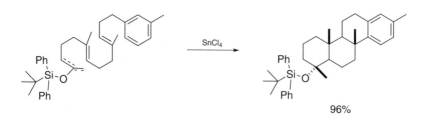

96%

Sn-W mixed hydroxide.   A white precipitate of the composition $Sn_{19}WClO_6 \cdot 9H_2O$, produced by adding $SnCl_4 \cdot 5H_2O$ to $Na_2WO_4 \cdot 2H_2O$, catalyzes the mixture of RCHO and $NH_2OH \cdot HCl$ to convert into RCN in hot *o*-xylene.[6]

[1]Sarkar, D., Venkateswaran, R.V. *SL* 653 (2008).
[2]Draghici, C., Brewer, M. *JACS* **130**, 3766 (2008).
[3]Gohier, F., Bouhadjera, K., Faye, D., Gaulon, C., Maisonneuve, V., Dujardin, G., Dhal, R. *OL* **9**, 211 (2007).
[4]Bigdeli, M.A., Rahmati, A., Abbasi-Ghadim, H., Mahdavinia, G.H. *TL* **48**, 4575 (2007).
[5]Uyanik, M., Ishihara, K., Yamamoto, H. *OL* **8**, 5649 (2006).
[6]Yamaguchi, K., Fujiwara, H., Ogasawara, Y., Kotani, M., Mizuno, N. *ACIE* **46**, 3922 (2007).

Titanium(III) chloride.

Conjugated sulfones.   Low-valent titanium species are generated from $TiCl_3$, Zn(Pb), and ZnI_2 in THF, which promotes the reaction of $PhSO_2CHBr_2$ with carbonyl compounds to afford conjugated sulfones.[1]

[1]Baba, Y., Toshimitsu, A., Matsubara, S. *CL* **36**, 864 (2007).

Titanium(IV) chloride.

New preparation.[1]   From titanium oxide, HCl and a carbon source, microwave heating produces $TiCl_4$. This carbohydrochlorination method also works for generating $SiCl_4$ and BCl_3.

β-Methoxyamino esters.[2]   A Mannich-type reaction between RCH = NOMe and esters is catalyzed by a combination of $TiCl_4$ and *s*-BuNH$_2$.

Rearrangement.[3]   Protic and Lewis acids usually facilitate cleavage of strained cyclic compounds possessing oxygen functionalities and/or unsaturation. 7-Methylenebicyclo[3.2.0]-heptan-2-ones deliver bridged ring products that appear to suggest a synthetic potential to elaboration of the AB-ring segment of taxol.

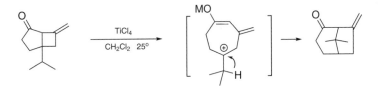

Cyclization.[4]   A pleasing and obviously valuable transformation is the bicyclization involving the reaction of a bis(allylsilane) with oxalyl chloride.

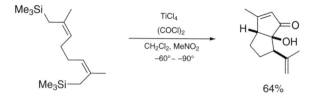

2-Oxalobiaryls give phenanthrenequinones after treatment of their imidazolide derivatives with TiCl$_4$.[5]

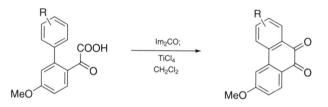

Isoxazolidines.[6] *N*-Nosyloxaziridines engage in 1,3-dipolar cycloaddition in the presence of TiCl$_4$. The process is highly stereoselective.

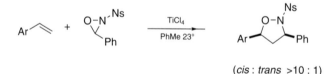

(*cis* : *trans* >10 : 1)

[1]Nordschild, S., Auner, N. *CEJ* **14**, 3694 (2008).
[2]Funatomi, T., Nakazawa, S., Matsumoto, K., Nagase, R., Tanabe, Y. *CC* 771 (2008).
[3]Shimada, Y., Nakamura, M., Suzuka, T., Matsui, J., Tatsumi, R., Tsutsumi, K., Morimoto, T., Kurosawa, H., Kakiuchi, K. *TL* **44**, 1401 (2003).
[4]Auof, C., El Abed, D., Giorgi, M., Santelli, M. *TL* **49**, 3862 (2008).
[5]Yoshikawa, N., Doyle, A., Tan, L., Murry, J.A., Akao, A., Kawasaki, M., Sato, K. *OL* **9**, 4103 (2007).
[6]Partridge, K.M., Anzovino, M.E., Yoon, T.P. *JACS* **130**, 2920 (2008).

Titanium(IV) chloride–magnesium.

Ketones from amides. A methylenating agent is generated from CH$_2$Cl$_2$ in the presence of TiCl$_4$ and magnesium metal and it reacts with amides to give methyl ketones, after hydrolytic workup.[1] Since enamine formation is implicated it can be exploited for the synthesis of δ-keto esters and congeners.[2]

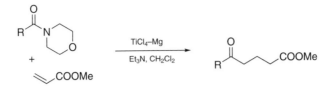

[1]Lin, K.-W., Tsai, C.-H., Hsieh, I.-L., Yan, T.-H. *OL* **10**, 1927 (2008).
[2]Lin, K.-W., Chen, C.-Y., Chen, W.-F., Yan, T.-H. *JOC* **73**, 4759 (2008).

Titanium(IV) chloride–Mischmetal.

De-N-tosylation.[1] Tosylamines are cleaved by heating with the title reagent combination in THF.

[1]Vellemäe, E., Lebedev, O., Mäeorg, U. *TL* **49**, 1373 (2008).

Titanium(IV) chloride–zinc.

Coupling + rearrangement.[1] Pinacol coupling of cyclobutanone with a diaryl ketone, when mediated by TiCl$_4$–Zn, is followed by rapid ring expansion (yields of the products are increased by adding catechol to the reaction media). With TiCl$_4$–Mg(HgCl$_2$) the normal McMurry coupling prevails.

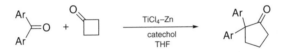

Cyclization and RCM.[2] Low-valent titanium species generated from TiCl$_4$–Zn–PbCl$_2$ and with the presence of MeCHBr$_2$ forms Ti-alkylidenes with terminal alkenes. Intramolecular reaction involving such entities and an ester carbonyl group leads to cyclic enol ethers. Ring-closing metathesis from α,ω-dienes is also effected by such reagents.

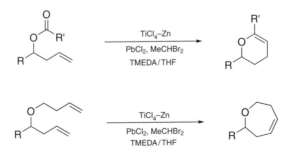

[1]Seo, J.W., Kim, H.J., Lee, B.S., Katzenellenbogen, J.A., Chi, D.Y. *JOC* **73**, 715 (2008).
[2]Iyer, K., Rainier, J.D. *JACS* **129**, 12604 (2007).

Titanium tetraisopropoxide–magnesium.

Ether cleavage.   Propargyl ethers and allyl ethers suffer cleavage on exposure to $(i\text{-PrO})_4\text{Ti–Mg}$ and Me_3SiCl. With addition of EtOAc propargyl ethers are cleaved more rapidly than allyl ethers.[1]

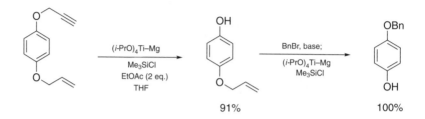

91%   100%

[1]Ohkubo, M., Mochizuki, S., Sano, T., Kawaguchi, Y., Okamoto, S. *OL* **9**, 773 (2007).

Titanium tetrakis(diethylamide).

Hydroamination.[1]   On exchanging two of the Et_2N groups of the title reagent to N-(2,6-diisopropylphenyl)benzamido residues, a precatalyst for anti-Markovnikov hydroamination of 1-alkynes to form aldimines is readily obtained.

[1]Zhang, Z., Leitch, D.C., Lu, M., Patrick, B.O., Schafer, L.L. *CEJ* **13**, 2012 (2007).

Titanocene bis(triethyl phosphite).

Polyene synthesis.   In the presence of titanocene bis(triethyl phosphite), activated alkynes react with (Z)-alkenyl sulfones to give conjugated dienes in a highly regioselective and stereoselective fashion.[1] Mixed unsaturated sulfones react in an analogous manner, and when the titanated coupling adducts are quenched with carbonyl compounds it results in 1,2,4-trienes.[2]

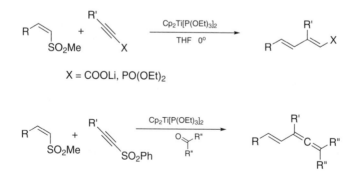

X = COOLi, PO(OEt)$_2$

Homopropargylic alcohols.   A three-component condensation unites an alkynyl sulfone, a carbonyl compound, and a dithioacetal, resulting in a homopropargylic alcohol.[3]

[1]Ogata, A., Nemoto, M., Takano, Y., Tsubouchi, A., Takeda, T. *TL* **49**, 3071 (2008).
[2]Ogata, A., Nemoto, M., Kobayashi, K., Tsubouchi, A., Takeda, T. *CEJ* **13**, 1320 (2007).
[3]Takeda, T., Ando, M., Sugita, T., Tsubouchi, A. *OL* **9**, 2875 (2007).

Titanocene dichloride–manganese.

Aldols from α-haloketones.[1]   The Reformatsky-type reaction is readily effected at room temperature with the assistance of titanocene chloride, which is generated from Cp_2TiCl_2 and manganese in THF. Aromatic aldehydes are not suitable for the reaction as they tend to undergo pinacol coupling under the reaction conditions.

[1]Estevez, R.E., Paradas, M., Millan, A., Jimenez, T., Robles, R., Cuerva, J.M., Oltra, J.E. *JOC* **73**, 1616 (2008).

Titanocene dichloride–zinc.

Allylic substitution.[1]   Alkyltitanocenes are nucleophilic toward allylic acetates such as those derived from Baylis–Hillman adducts. Accordingly, it is applicable to the synthesis of α-alkyl-α,β-unsaturated esters.

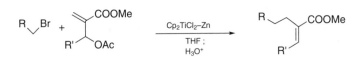

Reductive cyclization.[2]   Epoxides substituted with a carbon chain terminated (at a proper length) in an allenyl group are converted into cyclic products containing vicinal hydroxymethyl and vinyl substituents.

[1]Mandal, S.K., Paira, M., Roy, S.C. *JOC* **73**, 3823 (2008).
[2]Xu, L., Huang, X. *TL* **49**, 500 (2008).

p-Toluenesulfonic anhydride.

2-Aminopyridine.[1]   Tosic anhydride is used in activating pyridine-*N*-oxide and homologues for attack by *t*-butylamine. The resulting 2-*t*-butylaminopyridine is dealkylated by CF_3COOH.

[1]Yin, J., Xiang, B., Huffman, M.A., Raab, C.E., Davies, I.W. *JOC* **72**, 4554 (2007).

p-Toluenesulfonyl chloride.

Tosylation.[1]   A method for selective tosylation of primary alcohols (in the presence of secondary alcohols) consists of grinding with TsCl at room temperature. If KOH is present secondary alcohols are also tosylated.

Isothiocyanates.   Primary amines are converted into isothiocyanates by combining with carbon disulfide and decomposing the dithiocarbamic acid salts with TsCl.[2]

[1]Kazemi, F., Massah, A.R., Javaherian, M. *T* **63**, 5083 (2007).
[2]Wong, R., Dolman, S.J. *JOC* **72**, 3969 (2007).

p-Toluenesulfonyl fluoride.

Tosylates.[1]   A direct transformation of silyl ethers into tosyl esters is accomplished by the treatment with TsF and DBU in MeCN at room temperature.

[1]Gembus, V., Marsais, F., Levacher, V. *SL* 1463 (2008).

1-(*p*-Toluenesulfonyl)imidazole.

Esterification.[1]   Carboxylic acids in the sodium salt form are esterified by various alcohols on heating with the title reagent and Bu_4NI, Et_3N in DMF.

Alkyl azides.   In situ activation of alcohols (to ROTs) for conversion into azides is by heating with the title reagent, reaction with NaN_3 that is present (also Et_3N and Bu_4NBr) completes the transformation.[2]

[1]Rad, M.N.S., Behrouz, S., Faghihi, M.A., Khalafi-Nezhad, A. *TL* **49**, 1115 (2008).
[2]Rad, M.N.S., Behrouz, S., Khalafi-Nezhad, A. *TL* **48**, 3445 (2007).

p-Toluenesulfonyl isocyanate.

N-Tosyl amides.   Carboxylic acids react with TsN=C=O to afford RCONHTs.[1] These amides can be transformed into thioesters in two step: *N*-methylation (MeI, K_2CO_3, DMF) and treatment with RSH.[2]

[1]Manabe, S., Sugioka, T., Ito, Y. *TL* **48**, 787 (2007).
[2]Manabe, S., Sugioka, T., Ito, Y. *TL* **48**, 849 (2007).

O-(*p*-Toluenesulfonyl)-*N*-methylhydroxylamine.

α-Tosyoxylation of ketones.[1]   A tosyloxy group is readily introduced at an α-position of an enolizable ketone by reaction with TsONHMe in MeOH.

[1]John, O.R.S., Killeen, N.M., Knowles, D.A., Yau, S.C., Bagley, M.C., Tomkinson, N.C.O. *OL* **9**, 4009 (2007).

Trialkylboranes.

Carbamoylation.   Tertiary amines give N-phenyl α-aminocarboxamides on reaction with PhNCO in the presence of Et_3B and air.[1]

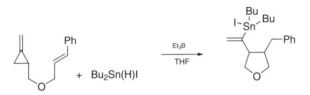

Reduction.   Alkyl iodides are reduced in water by exposure to air and Bu_3B.[2]

N-Heteroarylation.[3]   In the Pd-catalyzed substitution of heteroaryl halides (e.g., bromopyridines) with amides and sulfonamides the yields are greatly increased in the presence of Et_3B. Coordination of the Lewis acid by the nuclear nitrogen atom accelerates the reductive elimination step.

Hydrostannylation.[4]   The free radical mode of addition involving $Bu_2Sn(H)I$ and 2-allyloxymethyl-1-methylenecyclopropanes leads to tetrahydrofurans substituted with an α-stannylvinyl group at C-3. Cleavage of the three-membered ring occurs at the proximal bond.

[1]Yoshimitsu, T., Matsuda, K., Nagaoka, H., Tsukamoto, K., Tanaka, T. *OL* **9**, 5115 (2007).
[2]Medeiros, M.R., Schacherer, L.N., Spiegel, D.A., Wood, J.L. *OL* **9**, 4427 (2007).
[3]Shen, Q., Hartwig, J.F. *JACS* **129**, 7734 (2007).
[4]Hayashi, N., Hirokawa, Y., Shibata, I., Yasuda, M., Baba, A. *JACS* **130**, 2912 (2008).

2,8,9-Trialkyl-1-phospha-2,5,8,9-tetraazabicyclo[3.3.3]undecanes.

Diaryl ethers.[1]   With one of the congeners (**1**, R $= i$-Bu) present a mixture of ArOTBS and Ar'F react under microwave irradiation to give ArOAr'.

[1]Raders, S.M., Verkade, J.G. *TL* **49**, 3507 (2008).

Trialkylphosphines.

Reduction.[1]   Reduction of ArCOCF$_3$ to ArCH(OH)CF$_3$ is observed on mixing with equimolar of Bu$_3$P in toluene at room temperature.

Reaction of oximes.   Ketoximes are transformed into *N*-acetyl enamines by heating with Ac$_2$O in toluene in the presence of a trialkylphosphine.[2] Formation of *N*-phenylthio ketimines from either oximes or nitroalkanes is also catalyzed by Me$_3$P, with the PhS group provided by *N*-phenylthiophthalimide.[3]

Cycloaddition.   Trapping of CO$_2$ (from supercritical carbon dioxide) into propargylic alcohols to form 4-alkylidene-1,3-dioxolan-2-ones is assisted by Bu$_3$P.[4]

The high affinity of allenoic esters to phosphines enables catalytic activation of the γ-carbon to engage in nucleophilic addition. Thus with Me$_3$P the condensation of methyl 2,3-butadienoate with ArCHO to give 6-aryl-5,6-dihydro-2-pyrones, catalytic amounts of the phosphine are required in the presence of ROH or ROLi.[5] (Additional advantage is that such additives/reactants favor the formation of the phosphonium dienolates in the *s-cis*-form.)

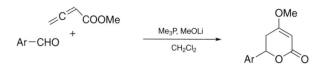

Intramolecular trapping of an adduct derived from R$_3$P and allenoic ester by a conjugated nitroalkene can lead to different results in accordance with the phosphine used.[6]

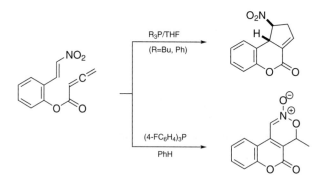

[1]Shi, M., Liu, X.-G., Guo, Y.-W., Zhang, W. *T* **63**, 12731 (2007).
[2]Zhao, H., Vandenbossche, C.P., Koenig, S.G., Singh, S.P., Bakale, R.P. *OL* **10**, 505 (2008).
[3]Bures, J., Isart, C., Vilarrasa, J. *OL* **9**, 4635 (2007).
[4]Kayaki, Y., Yamamoto, M., Ikariya, T. *JOC* **72**, 647 (2007).
[5]Creech, G.S., Kwon, O. *OL* **10**, 429 (2008).
[6]Henry, C.E., Kwon, O. *OL* **9**, 3069 (2007).

Tributyltin hydride.

Hydrostannylation.   Regioselective conversion of 1-alkynes to 2-tributylstannyl-1-alkenes is accomplished with Bu₃SnH in the presence of (*t*-BuNC)₃Mo(CO)₃.[1,2] By changing the catalyst to (*p*-O₂NC₆H₄NC)₃W(CO)₃ propargyl acetate furnishes a distannylated product.[2]

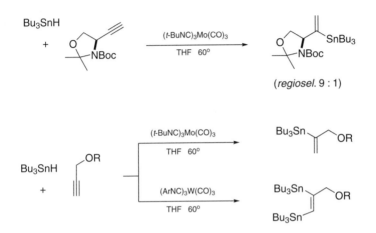

[1] Lin, H., Kazmaier, U. *EJOC* 2839 (2007).
[2] Wesquet, A.O., Kazmaier, U. *ACIE* **47**, 3050 (2008).

Tributyltin hydride–2,2′-azobis(isobutyronitrile).

Cyclizations.   *N*-Diphenylmethylene derivatives of *o*-bromoarylethylamines form aryl radicals that add to the nitrogen atom to give indolines.[1]

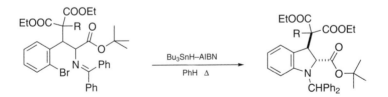

The tricyclic core of stemonamide is formed (55% yield) in one step via sequential Michael reactions initiated by an alkyl radical.[2] The selectivity of the process is due to the much greater tendency, at the start, to form a seven-membered ring than a eight-membered ring. [1,1′-bis(cyclohexanecarbonitrile) instead of AIBN is used as radical initiator in this case.]

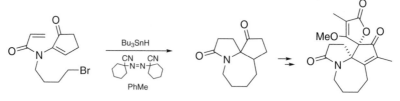

stemonamide

A tandem double cyclization is initiated on forming an aryl radical from an *N*-acrylyl-*N*-cyano-*o*-iodobenzylamine. Addition onto the cyano group is followed by a 1,4-addition of the iminyl radical.[3]

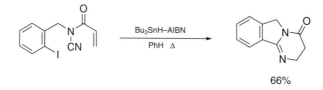

66%

Fragmentation. Hydrogen abstraction is the predominant event for the aryl radicals generated from *o*-iodoarylanilides. When the acyl residue contains an α-azido substituent, the transposed C-radicals rapidly lose N_2 and then fragment.[4]

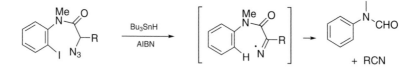

+ RCN

Carbonylation. Dichlorocarbene adducts of alkenes are transformed into cyclopropanecarbaldehydes with Bu_3SnH–AIBN under CO. Radical carbonylation and hydrodechlorination are involved.[5]

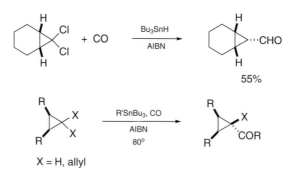

55%

X = H, allyl

[1]Viswanathan, R., Smith, C.R., Prabhakaran, E.N., Johnston, J.N. *JOC* **73**, 3040 (2008).
[2]Taniguchi, T., Tanabe, G., Muraoka, O., Ishibashi, H. *OL* **10**, 197 (2008).
[3]Servais, A., Azzouz, M., Lopes, D., Courillon, C., Malacria, M. *ACIE* **46**, 576 (2007).
[4]Bencivenni, G., Lanza, T., Leardini, R., Minozzi, M., Nanni, D., Spagnolo, P., Zanardi, G. *JOC* **73**, 4721 (2008).
[5]Nishii, Y., Nagano, T., Gotoh, H., Nagase, R., Motoyoshiya, J., Aoyama, H., Tanabe, Y. *OL* **9**, 563 (2007).

Tricarbonyl(cyclopentadienyl)hydridochromium.

Radical cyclization.[1]   The CpCr(CO)$_3$H complex initiates H addition to C=C to form a carbon radical. A 1,6-diene is induced cyclize. The transition metal hydride is used as a catalyst while hydrogen is consumed.

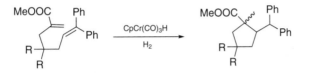

[1]Smith, D.M., Pulling, M.E., Norton, J.R. *JACS* **129**, 770 (2007).

Trichloroacetonitrile.

Dihydrooxazines.[1]   Certain 1,3-diols form the heterocycles on reaction with Cl$_3$CCN in the presence of DBU, by way of an intramolecular displacement.

[1]Rondot, C., Retailleau, P., Zhu, J. *OL* **9**, 247 (2007).

Trichlorosilane.

Enolsilylation.[1]   α-Bromo ketones and esters are transformed into trichlorosilyl enol ethers by HSiCl$_3$ and Et$_3$N (catalytic Ph$_3$PO), which condense with ArCHO. The overall process is superior to the Wittig reaction.

[1]Smith, J.M., Greaney, M.F. *TL* **48**, 8687 (2007).

Triethyl phosphite.

Heterocyclization.[1]   Certain *o*-nitroarylalkenes cyclize on heating with (EtO)$_3$P by involving the nitro group at a reduced state with a double bond of the side chain at an *o*-position. Tetrahydroquinolines, tetrahydroquinoxalines, and dihydrobenzoxazines are obtained.

R = H, Me

[1]Merisor, E., Conrad, J., Klaiber, I., Mika, S., Beifuss, U. *ACIE* **46**, 3353 (2007).

Triethylsilyl chloride.

O-Silylation.[1] This silyl chloride can be used to selectively derivatize an allylic hydroxyl group in the presence of a homoallylic OH.

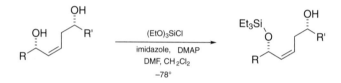

[1]Hicks, J.D., Huh, C.W., Legg, A.D., Roush, W.R. *OL* **9**, 5621 (2007).

Trifluoroacetic acid, TFA.

Dehydration.[1] Dissolution of 2-(α-hydroxyalkyl)cyclopropanols in CF$_3$COOH causes ionization and ring opening, β,γ-unsaturated carbonyl compounds are formed. Since the substrates are usually synthesized from conjugated lower homologues, the overall result is a methylene group insertion between the carbonyl group and the α-carbon atom.

[1]Nomura, K., Matsubara, S. *SL* 1412 (2008).

Trifluoroacetic anhydride, TFAA.

Trifluoromethyl ketones.[1] A simple preparation of RCOCF$_3$ involves heating carboxylic acids with pyridine and TFAA in toluene at 60–100°, then hydrolysis with water at 45°.

[1]Reeves, J.T., Gallou, F., Song, J.J., Tan, Z., Lee, H., Yee, N.K., Senanayake, C.H. *TL* **48**, 189 (2007).

1,1,1-Trifluoroacetone.

Oppenauer oxidation.[1]   Oxidation of secondary alcohols is accomplished with Et_2AlOEt and CF_3COMe at room temperature.

[1]Mello, R., Martinez-Terrer, J., Asensio, G., Gonzalez-Nunez, M.E. *JOC* **72**, 9376 (2007).

2,2,2-Trifluoroethyl isocyanate.

Hofmann–Löffler–Freytag reaction.[1]   Derivatization by the title reagent (prepared from 2,2,2-trifluoroethylamine and phosgene) transforms an alcohol into a N-(2,2,2-trifluoroethyl)carbamate, which on subjecting to a Hofmann–Löffler–Freytag reaction delivers the cyclic carbonate of a 1,3-diol.

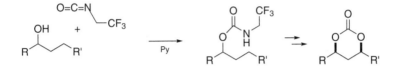

[1]Chen, K., Richter, J.M., Baran, P.S. *JACS* **130**, 7247 (2008).

Trifluoromethanesulfonic acid (triflic acid).

Cyclization.   Cyanoacetic esters undergo O,N-diprotonation in TfOH. If the ester bears a benzyl or phenylethyl group at the α-position, the derived dication would undergo an intramolecular Friedel–Crafts reaction.[1]

n = 1, 2

Smooth aza-Nazarov cyclization to form dihydroisoindolones is observed with TfOH where CF_3COOH is totally ineffectual. Formation of dicationic superelectrophilic species is apparently important for overcoming the energy barrier of the cyclization in such cases.[2]

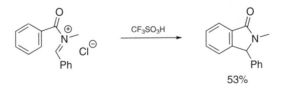

53%

A highly unusual yet simple ring closure of 2-arylnitroethanes to 4*H*-1,2-benzoxazines is accomplished by heating with TfOH.[3]

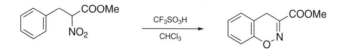

Friedel–Crafts alkylation. Alkenes (styrenes and trisubstituted alkenes) alkylate indoles at the β-position in CH_2Cl_2 containing 5 mol% of TfOH (or an Au(III) species).[4]

Benzyl trifluoromethyl carbinols are obtained from reaction of arenes with trifluoromethyl epoxides. The direction of epoxide ring opening is determined by the electron-withdrawing trifluoromethyl group.[5]

[1]Nakamura, S., Sugimoto, H., Ohwada, T. *JOC* **73**, 4219 (2008).
[2]Klumpp, D.A., Zhang, Y., Oconnor, M.J., Esteves, P.M., de Almeida, L.S. *OL* **9**, 3085 (2007).
[3]Nakamura, S., Sugimoto, H., Ohwada, T. *JACS* **129**, 1724 (2007).
[4]Rozenman, M.M., Kanan, M.W., Liu, D.R. *JACS* **129**, 14933 (2007).
[5]Prakash, G.K.S., Linares-Palomino, P.J., Glinton, K., Chacko, S., Rasul, G., Mathew, T., Olah, G.A. *SL* 1158 (2007).

Trifluoromethanesulfonic anhydride (triflic anhydride).

Conjugate addition. Allyl transfer from allyltributylstannane to enals, enones, and enoates in the presence of Tf_2O and 2,6-di-*t*-butylpyridine affords enol triflates.[1]

Reduction. Carboxamides are reduced to amines with Hantzsch ester, upon conversion (in situ) into *N,O*-ketene triflates with Tf_2O.[2]

Cycloaddition. The combination of enamides with alkynes (or enol ethers)[3] and with nitriles[4] gives substituted pyridines and pyrimidines, respectively, as promoted by Tf_2O.

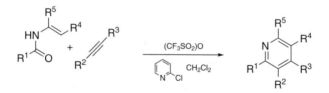

Pummerer rearrangement of 4-benzenesulfinyl-1-naphthols generates naphthoquinone phenylsulfonium ions. These react with enol ethers to afford annulated furanyl ethers.[5] Another Pummerer rearrangement on aryl substituted ketene dithioacetal monoxides provides 2-methylthiobenzothiophenes.[6]

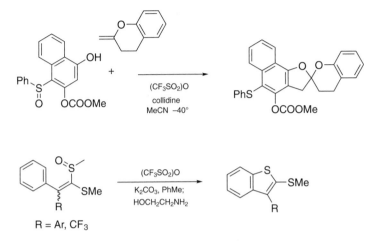

In an approach to stemofoline the elaboration of the tricyclic system involves generation of a 1,3-dipolar species from a vinylogous *N*-trimethylsilylmethyl amide. The design also provides a proper dipolarophile.[7]

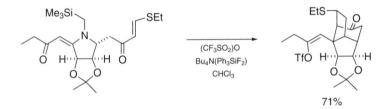

Fragmentative elimination. *N*-Oxides of the tetracyclic segment common to Aspidosperma and Strychnos alkaloids are converted into the tricyclic lactam isomers after treatment with Tf₂O and quenched with water.[8] Cyclic nitrilium ions intervene in this transformation.

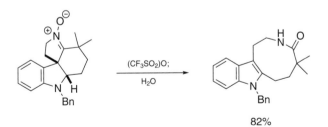

82%

[1]Beaulieu, E.D., Voss, L., Trauner, D. *OL* **10**, 869 (2008).
[2]Barbe, G., Charette, A.B. *JACS* **130**, 18 (2008).

[3]Movassaghi, M., Hill, M.D., Ahmad, O.K. *JACS* **129**, 10096 (2007).
[4]Movassaghi, M., Hill, M.D. *JACS* **128**, 14254 (2006).
[5]Akai, S., Kakiguchi, K., Nakamura, Y., Kuriwaki, I., Dohi, T., Harada, S., Kubo, O., Morita, N. *ACIE* **46**, 7458 (2007).
[6]Yoshida, S., Yorimitsu, H., Oshima, K. *OL* **9**, 5573 (2007).
[7]Carra, R.J., Epperson, M.T., Gin, D.Y. *T* **64**, 3629 (2008).
[8]Murphy, J.A., Mahesh, M., McPheators, G., Anand, R.V., McGuire, T.M., Carling, R., Kennedy, A.R. *OL* **9**, 3233 (2007).

Trifluoromethanesulfonic imide (triflic imide).

Aldol reaction.[1]   Catalyzed by Tf_2NH, the Mukaiyama aldol reaction of tris(trimethyl-silyl) ethers of enolized ketones affords products that react with Grignard reagents diastereoselectively.

Diels–Alder reaction.[2]   The condensation of *N*-arylaldimines and allylsilanes is catalyzed by Tf_2NH. Some of the products also undergo aromatization to give 2-aryl-4-silylmethylquinolines.

[1]Boxer, M.B., Akakura, M., Yamamoto, H. *JACS* **130**, 1580 (2008).
[2]Shindoh, N., Tokuyama, H., Takasu, K. *TL* **48**, 4749 (2007).

Trifluoromethyltrimethylsilane.

Trifluoromethylation.[1]   Reagent **1**, useful for trifluoromethylation of 1,3-dicarbonyl compounds and thiols, is obtained by reaction of the corresponding chloroiodine compound with Me_3SiCF_3.

(1)

[1]Kieltsch, I., Eisenberger, P., Togni, A. *ACIE* **46**, 754 (2007).

Trimethylsilylacetonitrile.

Cyanomethylation.   Carbonyl compounds and imines are subject to cyanomethylation by Me_3SiCH_2CN with tris(2,4,6-trimethoxyphenyl)phosphine as catalyst.[1]

[1]Matsukawa, S., Kitazaki, E. *TL* **49**, 2982 (2008).

Trimethylsilyl azide.

Organic azides.   A method for the preparation of aryl azides from arylamines consists of reaction with *t*-BuONO and Me_3SiN_3 in MeCN.[1]

Arenediazonium tetrafluoroborates can be converted into aryl azides under mild conditions, by reaction with Me_3SiN_3 in an ionic liquid at room temperature.[2] Analogously, aryl bromides and iodides are formed with Me_3SiX (X = Br, I).

Tetrazoles.[3] Nitriles undergo Cu-catalyzed cycloaddition with Me_3SiN_3 and adducts are desilylated in situ to furnish five-substituted tetrazoles.

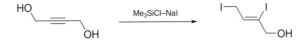

[1] Barral, K., Moorhouse, A.D., Moses, J.E. *OL* **9**, 1809 (2007).
[2] Hubbard, A., Okazaki, T., Laali, K.K. *JOC* **73**, 316 (2008).
[3] Jin, T., Kitahara, F., Kamijo, S., Yamamoto, Y. *TL* **49**, 2824 (2008).

Trimethylsilyl chloride.

Cyanosilylation. Me_3SiCl mediates formation of α-cyanohydrin trimethylsilyl ether using NaCN in DMSO. This combination is more economical than Me_3SiCN.[1]

(Z)-2,4-Diiodo-2-butenol.[2] An apparently valuable allylic alcohol is obtained in one step from 2-butyne-1,4-diol by reaction with Me_3SiCl–NaI.

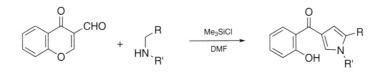

Oxidation.[3] In the presence of Me_3SiCl aromatic sulfur compounds are oxidized by KNO_3, which includes the transformations of ArSH to $ArSO_2Cl$ and $Ar_2S(O)_n$ to $Ar_2S(O)_{n+1}$ where [n = 0, 1].

Condensation.[4] Warming heteroarylmethylamines and 3-formylchromone with Me_3SiCl in DMF at 60° gives 2-heteroaryl-4-(*o*-benzoyl)pyrroles in moderate to excellent yields.

[1] Cabirol, F.L., Lim, A.E.C., Hanefeld, U., Sheldon, R.A., Lyapkalo, I.M. *JOC* **73**, 2446 (2008).
[2] Taber, D.F., Sikkander, M.I., Berry, J.F., Frankowski, K.J. *JOC* **73**, 1605 (2008).
[3] Prakash, G.K.S., Mathew, T., Panja, C., Olah, G.A. *JOC* **72**, 5847 (2007).
[4] Plaskon, A.S., Ryabukhin, S.V., Volochnyuk, D.M., Shivanyuk, A.N., Tolmachev, A.A. *T* **64**, 5933 (2008).

Trimethylsilyl cyanide.

Cyanation reactions. New catalysts for converting carbonyl compounds to *O*-trimethylsilyl cyanohydrins by Me₃SiCN include Cp₂FePF₆[1] and NbF₅.[2] Derivatization of ketones can use the titanium complex of **1**[3] or (Ph₃PBn)Cl.[4]

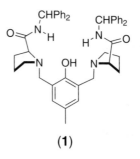

(1)

Strecker reaction can be performed at room temperature without solvent. Products are directly obtained (no workup handling).[5] Still, reports abound with various catalysts (e.g., sulfamic acid[6]).

By ligand exchange to deliver a cyano group from Me₃SiCN to PhI(OCOCF₃)₂ in the presence of BF₃·OEt₂, cyanating agent(s) for heteroaromatic compounds, PhI(CN)X, where X = OCOCF₃ or CN, are formed.[7]

Double trapping. Normally, adducts obtained from trapping of ionized acetonides with Me₃SiCN have little synthetic value. However, the observation that further reaction of the cyano group with an internal azide to form a tetrazole unit[8] is worth attention.

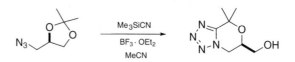

[1]Khan, N.H., Agrawal, S., Kureshy, R.I., Abdi, S.H.R., Singh, S., Suresh, E., Jasra, R.V. *TL* **49**, 640 (2008).

[2]Kim, S.S., Rajagopal, G. *S* 215 (2007).

[3]Shen, K., Liu, X., Li, Q., Feng, X. *T* **64**, 147 (2008).

[4]Wang, X., Tian, S.-K. *TL* **48**, 6010 (2007).

[5]Baeza, A., Najera, C., Sansano, J.M. *S* 1230 (2007).

[6]Li, Z., Sun, Y., Ren, X., Wei, P., Shi, Y., Ouyang, P. *SL* 803 (2007).

[7]Dohi, T., Morimoto, K., Takenaga, N., Goto, A., Maruyama, A., Kiyono, Y., Tohma, H., Kita, Y. *JOC* **72**, 109 (2007).

[8]Hanessian, S., Simard, D., Deschenes-Simard, B., Chenel, C., Haak, E. *OL* **10**, 1381 (2008).

Trimethylsilyldiazomethane.

Methylenation.   Methylenation of carbonyl compounds using Me_3SiCHN_2 in some situations is followed by cyclization of the products, for example, to give indoles from *o*-aminobenzaldehyde and in a synthesis of the indolizines.[1]

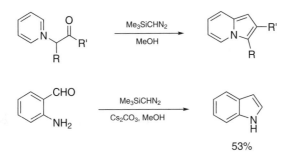

The methylenation in the presence of an imidazolylidene-CuCl complex and Ph_3P is also reported.[2]

1,2,3-Triazoles.   An adduct from Cp_2^*Sm and Me_3SiCHN_2 reacts with RCN to furnish substituted 1,2,3-triazoles.[3] But the usefulness of the adduct for synthesis is unknown.

1-Silyl-3-borylallenes.   Alkynylboranes react with Me_3SiCHN_2 by way of 1,2-insertion and 1,3-borotropic rearrangement.[4] The products are α-silylpropargylating agents.

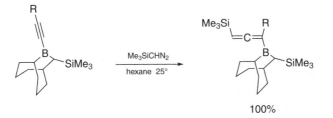

[1]Zhu, L., Vimolratana, M., Brown, S.P., Medina, J.C. *TL* **49**, 1768 (2008).
[2]Lebel, H., Davi, M., Diez-Gonzalez, S., Nolan, S.P. *JOC* **72**, 144 (2007).
[3]Evans, W.J., Montalvo, E., Champagne, T.M., Ziller, J.W., DiPasquale, A.G., Rheingold, A.L. *JACS* **130**, 16 (2008).
[4]Canales, E., Gonzalez, A.Z., Soderquist, J.A. *ACIE* **46**, 397 (2007).

Trimethylsilyl trifluoromethanesulfonate.

Deprotection.   The ring system of an *N,O*-acetonide breaks apart on treatment with Me_3SiOTf.[1]

Cyclocondensation. Two different processes depend on the catalysis of Me₃SiOTf in the construction of a cyclohexanopyrrole intermediate for a synthesis of goniomitine.[2] It assists the cleavage of a push-pull cyclopropane ring and the condensation with a nitrile unit.

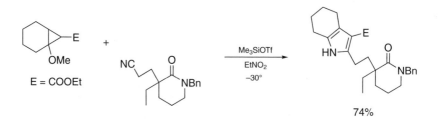

74%

The condensation of benzyl cyclopropyl ketones with ethyl 2,3-butadienoate is interesting. The furocoumarin system is elaborated.[3]

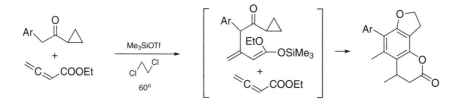

¹Poon, K.W.C., Lovell, K.M., Dresner, K.N., Datta, A. *JOC* **73**, 752 (2008).
²Morales, C.L., Pagenkopf, B.L. *OL* **10**, 157 (2008).
³Shi, M., Tang, X.-Y., Yang, Y.-H. *OL* **9**, 4017 (2007).

Triphenylphosphine.

Substitution. In converting ArOH to *t*-BuOCOOAr by Boc₂O in the neat,[1] and the group exchange in HC(OEt)₃ to form *N,O,O*-ortho esters with lactams,[2] Ph₃P serves as a catalyst. In the latter reaction a cocatalyst Me₃SiCl is present.

Ring strain makes trimethylsilylcyclopropenes susceptible to attack by Ar₃P such that further reaction with aldehyde is realized. A Brook-type rearrangement drives the reaction to completion by expulsion of the Ar₃P from the adducts. In practice, tris(2,4,6-trimethoxy-phenyl)phosphine is used instead of Ph₃P (much better yields).[3]

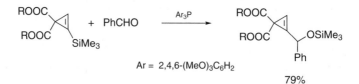

Ar = 2,4,6-(MeO)₃C₆H₂

79%

Cyclization and cycloaddition. Through creation of ylides and related zwitterionic species from activated allenes, Michael reaction is readily initiated. Options for further transformations leading to highly functionalized cyclic systems are as many as designed: reflexive Michael reactions to give bicyclo[n.3.0]cycloalkenes,[4] a regiochemically switchable cycloaddition to alkylidenemalononitriles [using tris(4-fluorophenyl)phosphine as catalyst],[5] and spiroannulation of α-methylenecycloalkanones.[6]

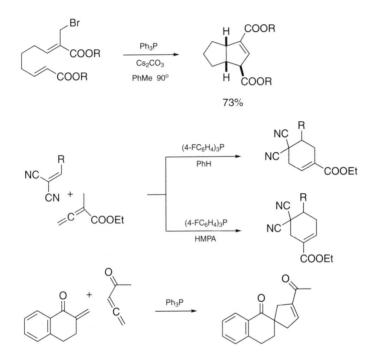

73%

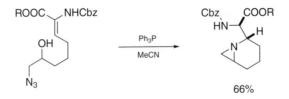

vic-Azido alcohols are converted into aziridines upon treatment with Ph₃P, via a Staudinger reaction, N -> O shift of the phosphino group, and intramolecular Mitsunobu reaction. In one example, the aziridine group acts as a nucleophile to participate in an intramolecular Michael reaction.[7]

66%

Using Ph$_3$P to induce Staudinger reaction of α-azidocinnamic esters as well as Wolff rearrangement of α-diazo ketones, two types of mutually reactive molecules are formed. The natural course for aza-Wittig reaction is pursued to produce cyclization-prone ketene imines.[8]

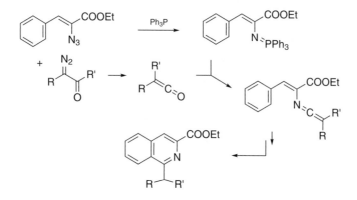

Heating aziridines and diazodicarboxylic esters with Ph$_3$P in toluene gives 5-aminopyrazolines.[9]

Henry reaction.   The 2,4,6-trimethoxy analogue [i.e., tris(2,4,6-trimethoxyphenyl)-phosphine] is found to be a nonbasic catalyst for the synthesis of 2-nitroalkanols from nitroalkanes and aldehydes.[10]

[1]Chebolu, R., Chankeshwara, S.V., Chakraborti, A.K. *S* 1448 (2008).
[2]Motherwell, W.B., Bégis, G., Cladingboel, D.E., Jerome, L., Sheppard, T.D. *T* **63**, 6462 (2007).
[3]Chuprakov, S., Malyshev, D.A., Trofimov, A., Gevorgyan, V. *JACS* **129**, 14868 (2007).
[4]Ye, L.-W., Sun, X.-L., Wang, Q.-G., Tang, Y. *ACIE* **46**, 5951 (2007).
[5]Tran, Y.S., Kwon, O. *JACS* **129**, 12632 (2007).
[6]Wallace, D.J., Sidda, R.L., Reamer, R.A. *JOC* **72**, 1051 (2007).
[7]Wynne, E.L., Clarkson, G.J., Shipman, M. *TL* **49**, 250 (2008).
[8]Yang, Y.-Y., Shou, W.-G., Chen, Z.-B., Hong, D., Wang, Y.-G. *JOC* **73**, 3928 (2008).
[9]Cui, S.-L., Wang, J., Wang, Y.-G. *OL* **10**, 13 (2008).
[10]Weedon, J.A., Chisholm, J.D. *TL* **47**, 9313 (2006).

Triphenylphosphine–dialkyl azodicarboxylate.

Epoxides.[1]   Reaction pathways for the Mitsunobu reaction to convert 1,1-diaryl-2-phenylethanediols to epoxides are phosphine-dependent. Retention of configuration at the secondary carbinolic center is observed in reaction mediated by Ph$_3$P–DIAD, while more electron-rich phosphines (e.g., Bu$_3$P) favor products with inversion of configuration.[1]

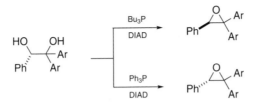

N-Alkylation.[2]   Some *N*-aromatics have been alkylated with ROH, Ph₃P, DIAD in MeCN, products such as *N*-alkylpyridinium and *N*-methyl-*N'*-alkylimidazolium tetrafluoroborates are readily isolated.

Oximes from alcohols.[3]   Primary and secondary alcohols are directly converted into oximes on reaction with *N*-tosylhydroxylamine TBS ether in a conventional Mitsunobu reaction, with succeeding elimination of TsH and desilylation by heating with CsF in MeCN. The reaction conditions are sufficiently mild and many functional groups are left unscathed.

Cycloaddition.[4]   An unusual transformation at the conclusion of a stephacidin-A synthesis involves a cycloaddition that is prosecuted by Bu₃P–DEAD.

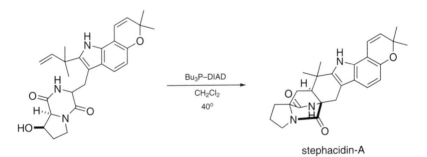

stephacidin-A

Dehydration.[5]   The Mitsunobu reagent (Ph₃P–DEAD) successfully achieves selective dehydration of a secondary alcohol (apparently an equatorial cyclohexanol) in the presence of an angular OH group is achieved at room temperature.

Dialkyl carbonates.[6]   Treatment of alcohols with Ph₃P–DEAD under CO₂ in dry DMSO at 90–100° leads to formation of carbonates.

[1]Garcia-Delgado, N., Riera, A., Verdaguer, X. *OL* **9**, 635 (2007).
[2]Petit, S., Azzouz, R., Fruit, C., Bischoff, L., Marsais, F. *TL* **49**, 3663 (2008).

[3]Kitahara, K., Toma, T., Shimokawa, J., Fukuyama, T. *OL* **10**, 2259 (2008).
[4]Greshock, T.J., Williams, R.M. *OL* **9**, 4255 (2007).
[5]Larionov, O.V., Corey, E.J. *JACS* **130**, 2954 (2008).
[6]Chaturvedi, D., Mishra, N., Mishra, V. *TL* **48**, 5043 (2007).

Triphenylphosphine–bromine.

N-Nitrosation.   On activation by $Ph_3P–Br_2$ the electrophilic NO^+ species is brought out of Bu_4NNO_2. Both secondary and tertiary amines are converted by the reagent mix into nitrosamines, whereas arylhydrazines give aryl azides.[1]

[1]Iranpoor, N., Firouzabadi, H., Nowrouzi, N. *TL* **49**, 4242 (2008).

Triphenylsilyl tetracarbonylcobaltate.

Carbonylation.[1]   Under CO the ring expansion of oxazolines is mediated by $Ph_3SiCo(CO)_4$. Actually, BnOH is also required as an additive to generate $HCo(CO)_4$ for the reaction. Chiral precursors of β-amino acids are readily prepared this way.

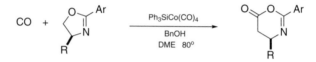

[1]Byrne, C.M., Church, T.L., Kramer, J.W., Coates, G.W. *ACIE* **47**, 3979 (2008).

Triphosphazene.

Beckmann rearrangement.[1]   The title reagent **1** (i.e., 1,3,5-triazo-2,4,6-triphosphorine-2,2,4,4,6,6-hexachloride) is capable of converting oximes into carboxamides, in either MeCN or hexafluoroisopropanol at 70°.

(1)

[1]Hashimoto, M., Obora, Y., Sakaguchi, S., Ishii, Y. *JOC* **73**, 2894 (2008).

Triruthenium dodecacarbonyl.

Allylation.[1]   Linear products are obtained from the Ru-catalyzed allylation of active methylene compounds (deprotonated by LiHMDS) with either primary or secondary allylic acetates. A suitable ligand for the Ru metal is *o*-diphenylphosphinobenzoic acid.

[1]Kawatsura, M., Ata, F., Wada, S., Hayase, S., Uno, H., Itoh, T. *CC* 298 (2007).

Tris(acetonitrile)cyclopentadienylruthenium(I) hexafluorophosphate.

Isomerization.   Uninterrupted long-chain alkenols undergo isomerization to afford saturated carbonyl compounds on treatment with $[CpRu(MeCN)_3]PF_6$ and 1-methyl-2-diisopropylphosphino-4-*t*-butylimidazole (ligand).[1] Migration of the double bond over 30 positions has been noted.

Substitution.   An S_N2' substitution operates in the reaction of cinnamyl chloride with $PhB(OH)_2$, when catalyzed by $[Cp^*Ru(MeCN)_3]PF_6$.[2]

Ene reaction.   Unsymmetrical diynes and polyynes combine with 1-alkenes in the fashion of an ene reaction, leading to enynes/enediynes in which the *sp*-carbon participating in new bonding now bearing an allyl group.[3]

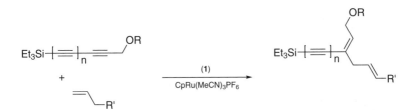

Cyclization.   In HOAc the Ru-catalyzed cyclization of 1,6-diynes proceeds with loss of the terminal *sp*-carbon. If both triple bonds are internal, *vic*-dialkylidenecyclopentanes are obtained.[4] 1,7-Diynes seem to behave differently.

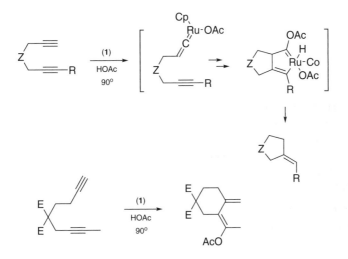

[1]Grotjahn, D.B., Larsen, C.R., Gustafson, J.L., Nair, R., Sharma, A. *JACS* **129**, 9592 (2007).
[2]Bouziane, A., Heliou, M., Carboni, B., Carreaux, F., Demerseman, B., Bruneau, C., Renaud, J.-L. *CEJ* **14**, 5630 (2008).

[3]Cho, E.J., Lee, D. *JACS* **129**, 6692 (2007).
[4]Gonzalez-Rodriguez, C., Varela, J.A., Castedo, L., Saa, C. *JACS* **129**, 12916 (2007).

Tris(dibenzylideneacetone)dipalladium.

Suzuki coupling. β,β-Dibromostyrenes undergo Suzuki coupling and the products are readily converted to diarylethynes.[1] Note that coupling of 2,2-difluoroethenyl tosylate with ArX furnishes β,β-difluorostyrenes.[2]

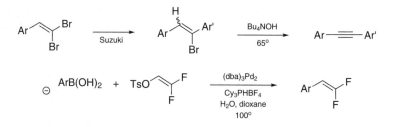

1-Alkenylpyridinium salts such as 1-(3-keto-1-butenyl)pyridinium tetrafluoroborate are active in Suzuki coupling with $ArB(OH)_2$.[3]

The dimorpholinophosphine **1** is an efficient, air-stable ligand for maintaining $(dba)_3Pd_2$ catalytically active in Suzuki coupling,[4] whereas chloro(mesityloxy)-*t*-butylphosphine (**2**) is good for coupling to form biaryls.[5]

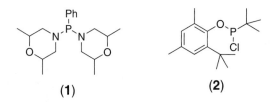

(1)   **(2)**

2-Arylation of pyridine via coupling of 2-PyB(O-*i*Pr)$_3$Li requires KF as additive and a phosphinous acid ligand.[6]

Diaryl ketones are produced from $ArB(OH)_2$ and Ar'CHO in a reaction catalyzed by $(dba)_3Pd_2$, using tris(1-naphthyl)phosphine as ligand and the presence of Cs_2CO_3 in the air.[7]

Availability of $ArB(OH)_2$ is further assured by the development of a general method for coupling ArCl and bis(pinacolato)diboron.[8]

Decarboxylative coupling. Diarylethynes are prepared from two ArX and propynoic acid.[9] The more reactive ArI is engaged in the first stage of the reaction, whereas after decarboxylation of the coupling product the second aryl group (from ArBr) is attached.

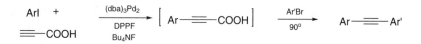

Arylation and alkenylation.   The effectiveness of (dba)$_3$Pd$_2$ for catalyzing Suzuki coupling and *N*-arylation is further demonstrated in an elaboration of carbazoles.[10]

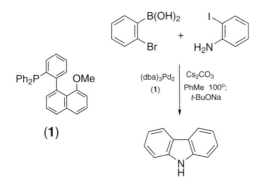

Another carbazole synthesis involves a twofold *N*-arylation, as shown by the preparation of an intermediate for murrayazoline.[11]

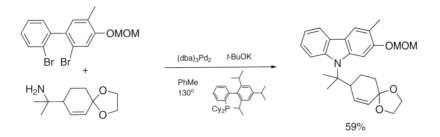

59%

3-Substituted indoles are available from coupling of *o*-bromoiodoarenes with allylic amines.[12] The catalyst system is capable of inducing further *N*-arylation of the resulting indoles.

Double *N*-arylation and *S*-arylation complete the assembly of phenothiazines[13] from an amine and two aryl halides, one of the halides being an *o*-bromoarenethiol.

A study of *N*-arylation of amides has shown the effect of product yields on critical structural feature of the X-Phos-type (2-di-*t*-butylphosphinobiaryl) ligand. It appears that a methyl group C-3 (ortho to the phosphine substituent) changes the outward orientation of the amido-Pd to an inward conformation.[14]

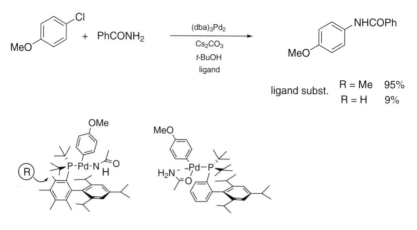

ligand subst.   R = Me   95%
                   R = H   9%

preferred structures of ‡

The coupling of ArCl with benzophenone imine is a fast reaction, therefore it represents an option for preparation of arylamines, as the products are easily hydrolyzed.[15] A simple protocol also avails for the preparation of the N-Boc derivatives of N-alkenylhydrazines from N-haloalkenes.[16]

Miscellaneous coupling reactions. Heck reaction and subsequent intramolecular Michael reaction are probably involved in the aminoarylation of 3-cyclohexenones.[17]

The reaction of tosylhydrazones with ArX under the influence of (dba)$_3$Pd$_2$–X-Phos performs the equivalent of an arylative Shapiro reaction.[18] A mixture of carbonyl compounds and TsNHNH$_2$ can be employed in lieu of the tosyhydrazones.[19]

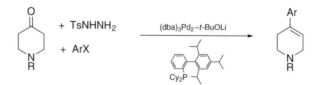

t-Butyl 3-sulfinylpropanoates release the sulfinyl group for coupling with ArI to form aryl sulfoxides (and with a chiral ferrocenylphosphine ligand present asymmetric induction has been observed – to 83% ee).[20] Assisted by KF hexamethyldisilane supplies the Me$_3$Si group to form ArSiMe$_3$.[21]

Two synthetically interesting processes are the S$_N$2' reaction of the esters of allenyl carbinols with organozinc reagents to generate conjugated dienes,[22] and the benzologue version (1,5-transpositional displacement) involving allyltributylstannane.[23]

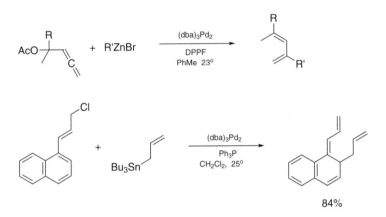

The Heck reaction of oximes derived from allyl ketones is followed by cyclization to give isoxazolines.[24] An intramolecular Heck reaction of substituted 6-chloro-1-hexenes readily affords the methylenecyclopentanes.[25]

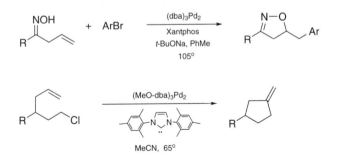

Allylation.   2,3-Disubstituted indoles are allylated at C-3 by the Pd-catalyzed reaction with an alkyl allyl carbonate.[26] The allyl group of allyl *o*-/*p*-nitroarylacetates is recaptured via a *sp*^{3}-*sp*^{3} coupling after decarboxylation.[27]

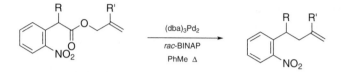

Carbonylation. *N,O*-Acetals in which the amino nitrogen atom is acylated are converted into *O*-chelated palladium species by replacing the oxy group (e.g., PhO group). Incorporation of CO completes the homologation process.[28]

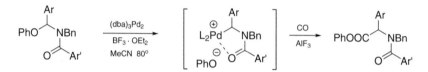

Addition and cycloaddition. In the (dba)$_3$Pd$_2$-catalyzed hydrostannylation of alkynes the regiochemistry is controlled by the phosphine ligands. (*E*)-1-Tributylstannylalkenes are produced preponderantly in the presence of Cy$_3$P or *t*-Bu$_3$P, but much more 2-stannyl-1-alkenes are obtained with Ph$_3$P.[29]

π-Allylplladium zwitterions are generated from 2-vinyl-1,1-cyclopropanedicarboxylic esters on treatment with an analogue of (dba)$_3$Pd$_2$. These species can be trapped by aldehydes.[30]

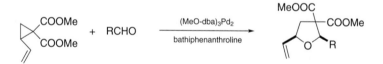

Rearrangement. 1-Allenylcyclobutanols undergo Wagner–Meerwein rearrangement to afford 2-vinylcyclopentanones. The rearrangement can be rendered asymmetric.[31]

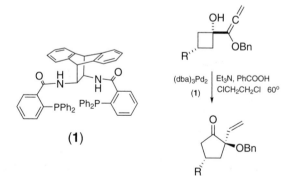

(1)

[1]Chelucci, G., Capitta, F., Baldino, S., Pinna, G.A. *TL* **48**, 6514 (2007).

[2]Gogsig, T.M., Sobjerg, L.S., Lindhardt, A.T., Jensen, K.L., Skrydstrup, T. *JOC* **73**, 3404 (2008).

[3]Buszek, K.R., Brown, N. *OL* **9**, 707 (2007).

[4]Cho, S.-D., Kim, H.-K., Yim, H.-s., Kim, M.-R., Lee, J.-K., Kim, J.-J., Yoon, Y.-J. *T* **63**, 1345 (2007).

[5]Mai, W., Lv, G., Gao, L. *SL* 2247 (2007).

[6]Billingsley, K.L., Buchwald, S.L. *ACIE* **47**, 4695 (2008).

[7]Qin, C., Chen, L., Wu, H., Cheng, J., Zhang, Q., Zuo, B., Su, W., Ding, J. *TL* **49**, 1884 (2008).

[8]Billingsley, K.L., Barder, T.E., Buchwald, S.L. *ACIE* **46**, 5359 (2007).

[9]Moon, J., Jeong, M., Nam, H., Ju, J., Moon, J.H., Jung, H.M., Lee, S. *OL* **10**, 945 (2008).

[10]Kitamura, Y., Yoshikawa, S., Furuta, T., Kan, T. *SL* 377 (2008).

[11]Ueno, A., Kitawaki, T., Chida, N. *OL* **10**, 1999 (2008).

[12]Jensen, T., Pedersen, H., Bang-Andersen, B., Madsen, R., Jorgensen, M. *ACIE* **47**, 888 (2008).

[13]Dahl, T., Tornoe, C.W., Bang-Andersen, B., Jorgensen, M. *ACIE* **47**, 1726 (2008).

[14]Ikawa, T., Barder, T.E., Biscoe, M.R., Buchwald, S.L. *JACS* **129**, 13001 (2007).

[15]Grossman, O., Rueck-Braun, K., Gelman, D. *S* 537 (2008).

[16]Barluenga, L., Moriel, P., Aznar, F., Valdes, C. *OL* **9**, 275 (2007).

[17]Hyde, A.M., Buchwald, S.L. *ACIE* **47**, 177 (2008).

[18]Barluenga, J., Moriel, P., Valdes, C., Aznar, F. *ACIE* **46**, 5587 (2007).

[19]Barluenga, J., Tomas-Gamasa, M., Moriel, P., Aznar, F., Valdes, C. *CEJ* **14**, 4792 (2008).

[20]Maitro, G., Vogel, S., Sadaoui, M., Prestat, G., Madec, D., Poli, G. *OL* **9**, 5493 (2007).

[21]McNeill, E., Barder, T.E., Buchwald, S.L. *OL* **9**, 3785 (2007).

[22]Schneekloth J.S. Jr, Pucheault, M., Cross, C.M. *EJOC* 40 (2007).

[23]Lu, S., Xu, Z., Bao, M., Yamamoto, Y. *ACIE* **47**, 4366 (2008).

[24]Jiang, D., Peng, J., Chen, Y. *OL* **10**, 1695 (2008).

[25]Firmansjah, L., Fu, G.C. *JACS* **129**, 11340 (2007).

[26]Kagawa, N., Malerich, J.P., Rawal, V.H. *OL* **10**, 2381 (2008).

[27]Waetzig, S.R., Tunge, J.A. *JACS* **129**, 14860 (2007).

[28]Lu, Y., Arndtsen, B.A. *OL* **9**, 4395 (2007).

[29]Darwish, A., Lang, A., Kim, T., Chong, J.M. *OL* **10**, 861 (2008).

[30]Parsons, A.T., Campbell, M.J., Johnson, J.S. *OL* **10**, 2541 (2008).

[31]Trost, B.M., Xie, J. *JACS* **130**, 6231 (2008).

Tris(dibenzylideneacetone)dipalladium–chloroform.

Allylic substitution.   Allylic carbonates are found to undergo substitution by benzyl-amine, there is regiochemical dependence on solvent.[1]

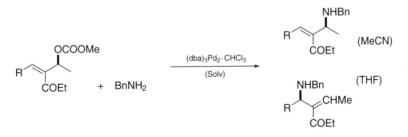

Two complementary methods for allylation of arene derivatives are: Suzuki coupling with allyl acetates,[2] and converting allyl acetates to allylindium reagents in situ for couping with ArX.[3]

Coupling reactions.   Insertion of the trimethylsilylcarbenoid into an Ar—Br bond generates benzylpalladium reagents. Transfer of the Pd unit to a new carbon site starts another homologation reaction.[4]

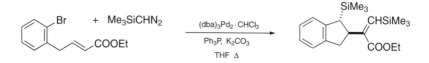

Chain elongation and amination occur on alkenyl iodides, when amines are added as co-reactants.[5]

α-Amino ketones are prepared from $RSnBu_3$, CO, imines (and $ClCOOR'$ for trapping the amino group) at room temperature. The products serve as intermediates of imidazolones.[6]

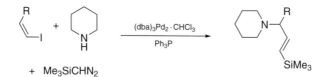

Suzuki coupling of unactivated ArCl is effectively conducted with the Pd complex in the presence of a bowl-shaped phosphine ligand **1**; the deeper the bowl, the higher the effectiveness.[7]

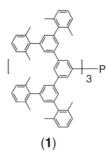

(1)

Rearrangement.   (2-Pyridinioethynyl)triphenylborate undergoes 1,2-phenyl migration while the N—H proton is transferred to the proximal *sp*-carbon.[8]

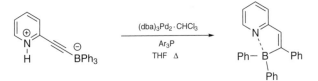

[1]Benfatti, F., Cardillo, G., Gentilucci, L., Mosconi, E., Tolomelli, A. *OL* **10**, 2425 (2008).
[2]Poláčková, V., Toma, Š., Kappe, C.O. *T* **63**, 8742 (2007).
[3]Seomoon, D., Lee, P.H. *JOC* **73**, 1165 (2008).
[4]Kudirka, R., Van Vranken, D.L. *JOC* **73**, 3585 (2008).
[5]Devine, S.K.J., Van Vranken, D.L. *OL* **9**, 2047 (2007).
[6]Siamaki, A.R., Black, D.A., Arndtsen, B.A. *JOC* **73**, 1135 (2008).
[7]Ohta, H., Tokunaga, M., Obora, Y., Iwai, T., Iwasawa, T., Fujihara, T., Tsuji, Y. *OL* **9**, 89 (2007).
[8]Ishida, N., Narumi, M., Murakami, M. *OL* **10**, 1279 (2008).

Tris(pentafluorophenyl)borane.

Tritylation.[1]   Primary and secondary alcohols can be tritylated with TrOH in CH_2Cl_2 at room temperature with $(C_6F_5)_3B$ as catalyst.

[1]Reddy, C.R., Rajesh, G., Balaji, S.V., Chethan, N. *TL* **49**, 970 (2008).

Tris(trimethylsilyl)silane.

Alkenylsilanes.[1]   Addition of $(Me_3Si)_3SiH$ to 1-alkynes in the presence of a Rh(I) complex furnishes the (*E*)-1-silylalkenes, whereas the (*Z*)-isomers are obtained from formal group exchange reaction starting from (*Z*)-1-alkenyl sulfones. These alkenylsilanes can be used in Hiyama coupling after treatment with alkaline H_2O_2.

Oxygenation.[2]   To functionalize a tertiary carbon atom adjacent to an alcohol via derivatization of the OH group to a bromomethyldimethylsilyl ether, radical generation to conduct hydrogen atom abstraction, with $Cu(OAc)_2$ to provide the necessary functionalizing element is involved.

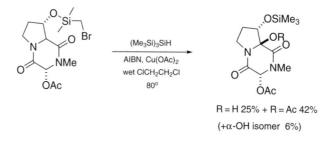

R = H 25% + R = Ac 42%

(+α-OH isomer 6%)

[1]Wang, Z., Pitteloud, J.-P., Montes, L., Rapp, M., Derane, D., Wnuk, S.F. *T* **64**, 5322 (2008).
[2]Overman, L.E., Sato, T. *OL* **9**, 5267 (2007).

Tungsten carbene and carbyne complexes.

Dihydropyran formation. Cyclization of 4-alkynols is induced by the tungsten version of a Fischer carbene complex without the need of UV irradiation, its application to a synthetic approach to altromycin-B disaccharide has been demonstrated.[1]

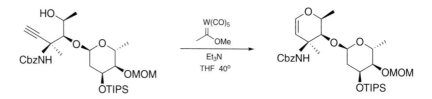

Alkyne metathesis. The complex **1** is a valuable catalyst for alkyne metathesis at room temperature.[2]

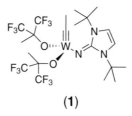

(1)

[1]Ko, B., McDonald, F.E. *OL* **9**, 1737 (2007).
[2]Beer, S., Hrib, C.G., Jones, P.G., Brandhorst, K., Grunenberg, J., Tamm, M. *ACIE* **46**, 8890 (2007).

Tungsten hexacarbonyl.

Cyclization. Activation of alkynes by $W(CO)_6$ under photoirradiation can provoke participation of an amine in the vicinity. A structural transformation of *o*-cycloaminoaryl-alkynes upon the treatment consists of cyclization with reorganization of ring system.[1]

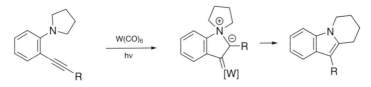

Siloxycycloalkenes substituted by a propargylamino group at the allylic position cyclize by two different modes, depending on the solvent.[2]

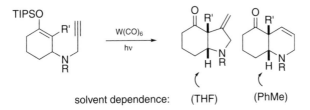

solvent dependence: (THF)   (PhMe)

A formal [4+1]cycloaddition of a diene unit and a terminal alkyne is mediated by $W(CO)_6$.[3] Most interestingly, the addition of Bu_3N to the reaction medium changes the reaction course drmatically.

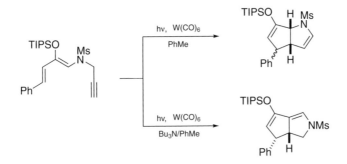

[1]Takaya, J., Udagawa, S., Kusama, H., Iwasawa, N. *ACIE* **47**, 4906 (2008).
[2]Grandmarre, A., Kusama, H., Iwasawa, N. *CL* **36**, 66 (2007).
[3]Onizawa, Y., Kusama, H., Iwasawa, N. *JACS* **130**, 802 (2008).

U

Urea – hydrogen peroxide.
 Oxidation.[1]   Oxidation of imines to nitrones by this reagent is catalyzed by MeReO$_3$.

[1]Soldaini, G., Cardona, F., Goti, A. *OL* **9**, 473 (2007).

Urea nitrate.
 Nitration.[1]   Regioselective mononitration (but no further) of moderately deactivated arenes is accomplished with urea nitrate or nitrourea, despite their use in excess.

[1]Almog, J., Klein, A., Sokol, A., Sasson, Y., Sonenfeld, D., Tamiri, T. *TL* **47**, 8651 (2006).

W

Water.

Deacetalization.[1]   Merely deionized water is the required reagent for removing acetal protection under microwave irradiation.

[1]Procopio, A., Gaspari, M., Nardi, M., Oliverio, M., Tagarelli, A., Sindona, G. *TL* **48**, 8623 (2007).

Wittig reagents.

Methylenation.[1]   The crystalline, salt-free Ph₃P=CH₂, as described 36 years ago,[1] is critical for overcoming the difficulty in methylenating the cyclononenone carbonyl in a synthesis of coraxeniolide-A.[2]

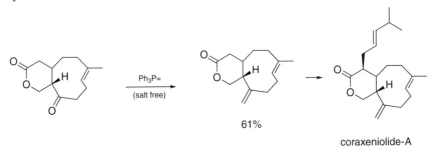

61%

coraxeniolide-A

[1]Schmidbauer, H., Stuhler, H., Vornberger, W. *CB* **105**, 1084 (1972).
[2]Larionov, O.V., Corey, E.J. *JACS* **130**, 2954 (2008).

Y

Ytterbium(III) triflate.

Condensations. Conjugated thioesters are made from aldehydes and monothioesters of malonic acid at room temperature using Yb(OTf)$_3$ as catalyst. Arylacetaldehydes afford thioesters containing a benzylic double bond.[1]

Allylation. N-(δ-Hydroxyalkyl)ureas in which the β-position is branching out by a methylene group are synthesized by Lewis acid-catalyzed reaction of stannylated allylureas. Different diastereomers can be obtained by proper choice of the Lewis acid.[2]

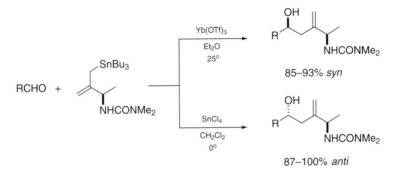

Heterocycles. In the presence of Yb(OTf)$_3$ epoxides are readily transformed into oxazolidines[3] and 2-iminothiazolidines[4] by reaction with imines and thioureas, respectively. 2-(2-Aminoxyethyl)cyclopropane-1,1-dicarboxylic esters condense with aldehydes to give bicyclic isoxazolidines.[5]

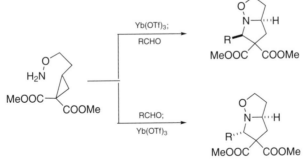

Fiesers' Reagents for Organic Synthesis, Volume 25. By Tse-Lok Ho
Copyright © 2010 John Wiley & Sons, Inc.

An eight-membered lactam is formed on treatment of the amide derived from tryptamine and epoxycinnamic acid.[6]

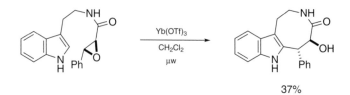

37%

Isobenzofurans and activated cyclopropanes condense to give an oxabridged ring system.[7]

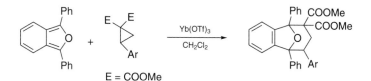

E = COOMe

[1]Berrue, F., Antoniotti, S., Thomas, O.P., Amade, P. *EJOC* 1743 (2007).
[2]Nishigaichi, Y., Tamura, K., Ueda, N., Iwamoto, H., Takuwa, A. *TL* **49**, 2124 (2008).
[3]Yu, C., Dai, X., Su, W. *SL* 646 (2007).
[4]Su, W., Liu, C., Shan, W. *SL* 725 (2008).
[5]Jackson, S.K., Karadeolian, A., Driega, A.B., Kerr, M.A. *JACS* **130**, 4196 (2008).
[6]Johansen, M.B., Leduc, A.B., Kerr, M.A. *SL* 2593 (2007).
[7]Ivanova, O.A., Budynina, E.M., Grishin, Y.K., Trushkov, I.V., Verteletskii, P.V. *ACIE* **47**, 1107 (2008).

Yttrium(III) chloride.

Cyclopentadienylyttrium dihydride. Consecutive reaction of YCl$_3$ with CpNa and Dibal-H leads to CyYH$_2$, which adds to CH$_2$=CHMgCl readily. The dimetallated ethane thus generated is a valuable reagent that can be used to functionalize and homologate alkynes and transforming esters into cyclopropanols (Kulinkovich reaction).[1]

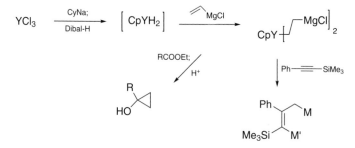

[1]Tanaka, R., Sanjiki, H., Urabe, H. *JACS* **130**, 2904 (2008).

Yttrium(III) triflate.

Condensation.   In the presence of Y(OTf)$_3$ microwave irradiation of a mixture of cinnamic acid and resorcinol leads to a 4-aryl-3,4-dihydrocoumarin.[1]

[1]Rodrigues-Santos, C.E., Echevarria, A. *TL* **48**, 4505 (2007).

Z

Zeolites.
Condensation. Cinnamic acid condenses with benzene in zeolites, but the reaction pattern depends on the type of the zeolite used.[1]

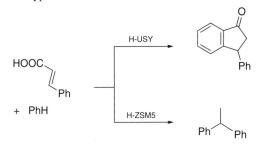

[1]Chassaing, S., Kumarraja, M., Pale, P., Sommer, J. *OL* **9**, 3889 (2007).

Zinc.
Organozinc chlorides. To avoid generation of dibenzyls during preparation of benzylic zinc reagents from ArCH$_2$Cl in THF, a protocol exploits the beneficial effect of LiCl.[1] Allylzinc chlorides are similarly available, and they add to carbonyl compounds with excellent diastereoselectivity.[2]

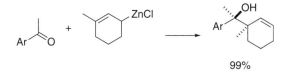

99%

Reduction. *N*-Hydroxymaleimides are reduced to succinimides without affecting the *N*-hydroxyl group by Zn–HOAc. The reduction provides *cis*-isomers from fully substituted maleimides.[3]

Fiesers' Reagents for Organic Synthesis, Volume 25. By Tse-Lok Ho
Copyright © 2010 John Wiley & Sons, Inc.

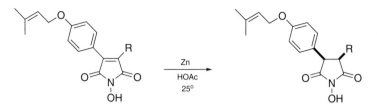

The triple bond of a propargylic sulfone is reduced to give a (Z)-allylic sulfone, on treatment with zinc in a mixture of THF and aqueous NH_4Cl at room temperature.[4]

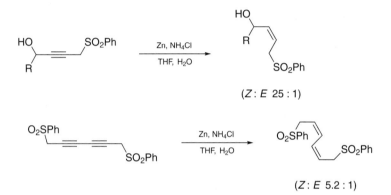

$(Z:E$ 25 : 1$)$

$(Z:E$ 5.2 : 1$)$

Organoazides are reduced by $Zn–NH_4Cl$ in hot aqueous EtOH and react with β-keto esters in situ to give β-amino-α,β-unsaturated esters.[5]

Reductive alkylation.   A previously reported N-methylation method for secondary amines with aqueous HCHO, Zn and HOAc is applicable to amino acids. Mono- or dimethylation can be controlled by adjustment of pH, reagent stoichiometry, and reaction time.[6] α-Branched amines can be prepared from amines, RCHO and alkyl halides by the action of zinc in MeCN.[7]

Amido carbenoids.   Acetals of N-formyllactams (as well as imides and oxazolinones) engage in cyclopropanation with alkenes in the presence of Zn and $CuCl$.[8]

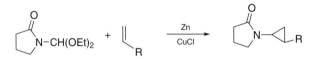

[1]Metzger, A., Schade, M.A., Knochel, P. *OL* **10**, 1107 (2008).
[2]Ren, H., Dunet, G., Mayer, P., Knochel, P. *JACS* **129**, 5376 (2007).
[3]Cheng, C.-F., Lai, Z.-C., Lee, Y.-J. *T* **64**, 4347 (2008).
[4]Sheldrake, H.M., Wallace, T.W. *TL* **48**, 4407 (2007).

[5]Prabhakar, A.S., Sashikanth, S., Reddy, P.P., Cherukupally, P. *TL* **48**, 8709 (2007).
[6]da Silva, R.A., Estevam, I.H.S., Bieber, L.W. *TL* **48**, 7680 (2007).
[7]Sengmany, S., Le Gall, E., Troupel, M. *SL* 1031 (2008).
[8]Motherwell, W.B., Begis, G., Cladingboel, D.E., Jerome, L., Sheppard, T.D. *T* **63**, 6462 (2007).

Zinc bromide.

Propargylmagnesium bromides. [1] In preparation of Grignard reagents from propargylic bromides and Mg the use of $ZnBr_2$ as catalyst is preferable to $HgCl_2$ for apparent reasons (toxicity).

[1]Acharya, H.P., Miyoshi, K., Kobayashi, Y. *OL* **9**, 3535 (2007).

Zinc chloride.

Furans. Propargyl ketones cycloisomerize to provide furans at room temperature, using $ZnCl_2$ as promoter.[1]

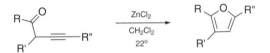

Beckmann rearrangement. Ketoximes rearrange on heating with $ZnCl_2$ and TsOH in MeCN.[2]

Reductive silylation. Carbonyl compounds are converted into silyl ethers by a mixture of R_3SiCl, CaH_2, and $ZnCl_2$ in THF.[3]

β-Amido ketones. [4] When ketones, aldehydes and nitriles are mixed with $ZnCl_2$ and $SiCl_4$ they condense to provide β-amido ketones. A Ritter reaction following aldol reaction accounts for the results.

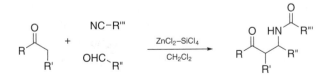

[1]Sniady, A., Durham, A., Morreale, M.S., Wheeler, K.A., Dembinski, R. *OL* **9**, 1175 (2007).
[2]Xiao, L., Xia, C., Chen, J. *TL* **48**, 7218 (2007).
[3]Tsuhako, A., He, J.-Q., Mihara, M., Saino, N., Okamoto, S. *TL* **48**, 9120 (2007).
[4]Salama, T.A., Elmorsy, S.S., Khalil, A.-G.M., Ismail, M.A. *TL* **48**, 6199 (2007).

Zinc iodide.

Cyclization.   Treatment of conjugate dienyl azides with ZnI_2 gives pyrroles.[1]

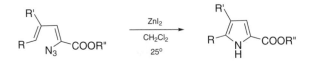

Indolizidinones are created from 6-(3-benzenesulfinylpropyl)-2-piperidone when exposed to ZnI_2 and a ketene silyl ether.[2] This method[3] entails a Pummerer rearrangement to generate a sulfur-stabilized carbocation for inducing the ring closure (N—C bond formation).

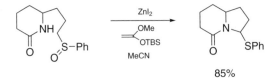

85%

[1]Dong, H., Shen, M., Redford, J.E., Stokes, B.J., Pumphrey, A.L., Driver, T.G. *OL* **9**, 5191 (2007).
[2]Kuhakarn, C., Seehasombat, P., Jaipetch, T., Pohmakotr, M., Reutrakul, V. *T* **64**, 1663 (2008).
[3]Kita, Y., Yasuda, H., Tamura, O., Itoh, F., Tamura, Y. *TL* **25**, 4681 (1984).

Zinc oxide.

Acylation.[1]   Zinc oxide is found to be a good catalyst for acylation of ferrocene.

[1]Wang, R., Hong, X., Shan, Z. *TL* **49**, 636 (2008).

Zinc triflate.

Silylation.[1]   With $Zn(OTf)_2$ as catalyst 1-alkynes are silylated by a mixture of Me_3SiOTf and Et_3N at room temperature.

Michael reaction.[2]   Great activity is exhibited by $Zn(OTf)_2$ for promoting the Mukaiyama version of the Michael addition involving the enol silyl ether of methyl α-diazoacetoacetate and conjugated cycloalkenones.

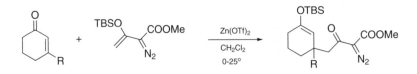

Heterocycle synthesis.   Arylhydrazines and 1-alkynes react to form *N*-heterocycles. 1-Aryl-2-pyrazolene are produced from 3-butynol,[3] whereas 2-methyl-3-alkylindoles are obtained from other 1-alkynes.[4]

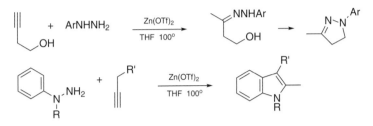

Maleimide formation from 2,3-dienoic esters and isonitriles is a cycloaddition catalyzed by Zn(OTf)$_2$ under O$_2$.[5]

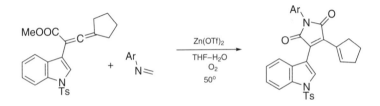

[1]Rahaim R.J., Jr, Shaw, J.T. *JOC* **73**, 2912 (2008).
[2]Liu, Y., Zhang, Y., Jee, N., Doyle, M.P. *OL* **10**, 1605 (2008).
[3]Alex, K., Tillack, A., Schwarz, N., Beller, M. *OL* **10**, 2377 (2008).
[4]Alex, K., Tillack, A., Schwarz, N., Beller, M. *ACIE* **47**, 2304 (2008).
[5]Li, Y., Zou, H., Gong, J., Xiang, J., Luo, T., Quan, J., Wang, G., Yang, Z. *OL* **9**, 4057 (2007).

Zirconia, sulfated.
    Mannich reaction.[1]   Reaction between ketene silyl acetals and aldimines occurs in the presence of sulfated zirconia in MeCN at room temperature. The reusable solid catalyst is easily recovered.

[1]Wang, S., Matsumura, S., Toshima, K. *TL* **48**, 6449 (2007).

Zirconium(IV) bromide.
    Bromination.[1]   Active arenes such as phenols, aryl ethers, and anilines are brominated by ZrBr$_4$ in the presence of diisopropyl azodicarboxylate in CH$_2$Cl$_2$. A free para position is the preferred site for bromination.

[1]Stropnik, T., Bombek, S., Kocevar, M., Polanc, S. *TL* **49**, 1729 (2008).

Zirconium(IV) *t*-butoxide.

Substitution.[1]   A stepwise reaction of carbonate esters with amines leads to unsymmetrical ureas. Using $(t\text{-BuO})_4\text{Zr}$ as catalyst the formation of carbamates is achieved at 80°, and the second step is performed under microwave irradiation. Different additives are indicated for the two steps, they are 2-pyridone and 2-quinolone, respectively.

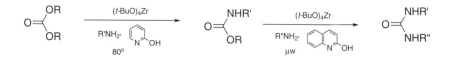

[1]Han, C., Porco Jr, J.A. *OL* **9**, 1517 (2007).

Zirconium (IV) chloride.

Friedel-Crafts alkylation.[1]   Cyclization of 2-(2-arylethyl)cyclohexanols usually gives *cis*-AB tricyclic compounds. The *trans*-fused isomers, which are potential precursors of C-aromatic tricyclic diterpenes, are accessible when ZrCl_4 is used to induce the cyclization.

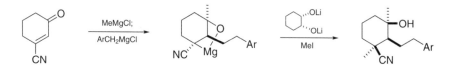

[1]Fleming, F.F., Wei, G., Steward, O.W. *JOC* **73**, 3674 (2008).

Zirconium tetrakis(dimethylamide).

Hydroamination.   A catalyst (**1**) for intramolecular hydroamination is readily made from $(\text{Me}_2\text{N})_4\text{Zr}$ by ligand exchange.[1] Alternatively, a chiral 6,6′-dimethyl-2,2′-amidobiphenyl may be used as ligand.[2]

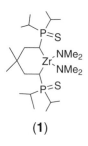

(1)

[1]Kim, H., Livinghouse, T., Lee, P.H. *T* **64**, 2525 (2008).
[2]Wood, M.C., Leitch, D.C., Yeung, C.S., Kozak, J.A., Schafer, L.L. *ACIE* **46**, 354 (2007).

Zirconocene, Zr-alkylated.

Reductive coupling.   After transformation of Cp_2ZrBu_2 into Cp_2ZrEt_2 by reaction with ethylene the sequential treatment with alkynes leads to zirconacyclopentadienes possessing more varied substituents. On oxidative opening of the zironacycles with NCS and CuCl reagents for alkenylcuprations are produced. Dienylcopper species are useful for further synthetic purposes, for example, preparation of linear polyenes.[1]

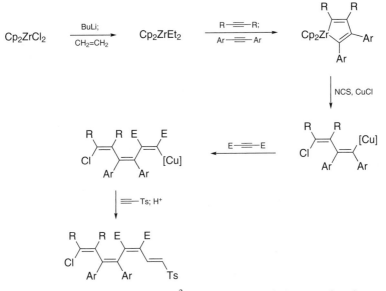

A synthesis of octacyclopropylcubanes[2] from dicyclopropylethyne requires three steps, starting from reaction with Cp_2ZrBu_2.

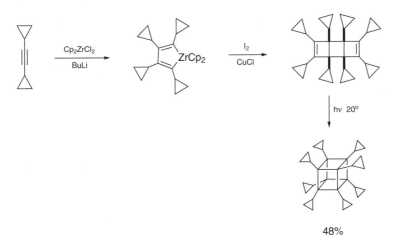

48%

[1]Kanno, K.-I., Igarashi, E., Zhou, L., Nakajima, K., Takahashi, T. *JACS* **130**, 5624 (2008).
[2]De Meijere, A., Redlich, S., Frank, D., Magull, J., Hofmeister, A., Menzel, H., König, B., Svoboda, V. *ACIE* **46**, 4574 (2007).

Zirconocene dichloride.

Homopropargylic alcohols.[1]   Silver(I) alkynides in a CH_2Cl_2 solution or suspension prepared from alkynes and $AgNO_3$ are transformed into alkynylzirconocenes, which can be used to attack epoxides.

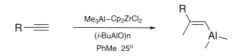

Addition reactions.   The mixed alane generated in situ from Me_3Al and isobutyl-aluminoxane adds to 1-alkynes to afford 2-methyl-1-alkenylaluminum reagents. This process has been applied to a synthesis of coenzyme Q_{10}.[2]

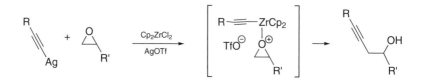

The alkenylalanes can be converted into trisubstituted enolates by peroxyzinc species,[3] thereby endorsing many further synthetic opportunities.

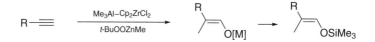

Better regiocontrol for carboalumination is obtained using (ethylenebisindenyl)zirco-nium dichloride, the benzologue of Cp_2ZrCl_2, while adding 5 mol% of MAO to accelerate the reaction.[4]

After forming $Cp_2Zr(CH_2{=}CH_2)$ by Grignard reaction of Cp_2ZrCl_2 to transform dialky-nyldiarylsilanes into silacyclobutenes, conjugated dienes are produced on protodesilylation with TBAF.[5]

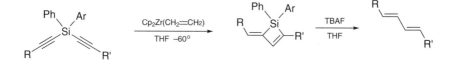

Hydration of fullerene at room temperature is catalyzed by Cp_2MCl_2 (M=Zr, 73%; Hf, 75%; Ti, 66%). The yield is increased to ~90% when the reaction is run at 80° at shorter reaction time (1 hr). Transition metal salts do not have this catalytic activity.[6]

[1]Albert, B.J., Koide, K. *JOC* **73**, 1093 (2008).
[2]Lipshutz, B.H., Butler, T., Lower, A., Servesko, J. *OL* **9**, 3737 (2007).
[3]DeBerg, J.R., Spivey, K.M., Ready, J.M. *JACS* **130**, 7828 (2008).
[4]Lipshutz, B.H., Butler, T., Lower, A. *JACS* **128**, 15396 (2006).
[5]Jin, C.K., Yamada, T., Sano, S., Shiro, M., Nagao, Y. *TL* **48**, 3671 (2007).
[6]Tuktarov, A., Akhmetov, A.R., Pudas, M., Ibragimov, A.G., Dzhemilov, U.M. *TL* **49**, 808 (2008).

Zirconocene hydrochloride.

Hydrozirconation.   Hydrozirconation of nitriles by $Cp_2Zr(H)Cl$ followed by reaction with RCOCl and quenched with nucleophiles gives rise to functionalized amides such as acylaminals and acyl hemiaminals (by adding alcohols and water, respectively).[1]

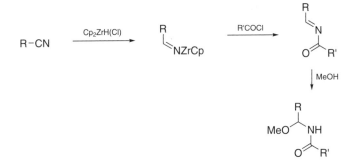

Synthesis of homologated allylic alcohols from propargylic alcohols is easily performed via hydrozirconated intermediates. Chain elongation or functionalization at either end of the original triple bond is feasible because hydrozirconation of the derived alkoxides becomes group-directed.[2]

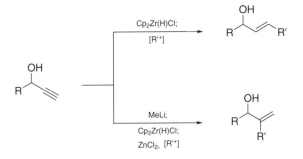

With $Cp_2Zr(H)Cl$ to catalyze the addition of methylpentanediolatoborane to alkynes, air-stable alkenylborates are formed.[3]

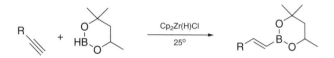

[1]Wan, S., Green, M.E., Park, J.-H., Floreancig, P.E. *OL* **9**, 5385 (2007).
[2]Zhang, D., Ready, J.M. *JACS* **129**, 12088 (2007).
[3]PraveenGanesh, N., d'Hondt, S., Chavant, P.Y. *JOC* **72**, 4510 (2007).

Zirconyl chloride.

Allylation.   With promotion by $ZrOCl_2 \cdot 8H_2O$ allylation of aldehydes with allyltributylstannane is accomplished in water. α-Branched homoallylic amines are obtained with a mixture of RCHO and $ArNH_2$.[1]

[1]Shen, W., Wang, L.-M., Feng, J.-J., Tian, H. *TL* **49**, 4047 (2008).

AUTHOR INDEX

Fiesers' Reagents for Organic Synthesis, Volume 25. By Tse-Lok Ho
Copyright © 2010 John Wiley & Sons, Inc.

SUBJECT INDEX

Reagents for Organic Synthesis, Volume 25. By Tse-Lok Ho
Copyright © 2010 John Wiley & Sons, Inc.